Building VPNs
With IPSec and MPLS

Building VPNs

With IPSec and MPLS

Nam-Kee Tan
CCIE #4307

McGraw-Hill

New York   Chicago   San Francisco   Lisbon
London   Madrid   Mexico City   Milan   New Delhi
San Juan   Seoul   Singapore   Sydney   Toronto

The McGraw·Hill Companies

Library of Congress Cataloging-in-Publication Data

Tan, Nam-Kee.
  Building VPNs: with IPSec and MPLS/by Nam-Kee Tan.
    p. cm.
  Includes bibliographical references and index.
  ISBN 0-07-140931-9
    1. Computer security–Software. 2. Computer networks–Security measures–Software.
  I. Title.

QA76.9.A25T35 2003
005.8–dc21   2003046460

1 2 3 4 5 6 7 8 9 0 DOC/DOC 0 9 8 7 6 5 4 3

ISBN 0-07-140931-9

*The sponsoring editor for this book was Steve Chapman and the production
supervisor was Pamela Pelton. It was set in Century Schoolbook by Best-Set.*

Printed and bound by RR Donnelley.

This book was printed on recycled, acid-free paper containing
a minimum of 50% recycled, de-inked fiber.

McGraw-Hill books are available at special quantity discounts to use as premiums and sales
promotions, or for use in corporate training programs. For more information, please write to the
Director of Special Sales, McGraw-Hill Professional, Two Penn Plaza, New York, NY 10121-2298.
Or contact your local bookstore.

To Chai-Tee, My Number One

Contents

Part 3: Deployment of MPLS VPN for Service Providers

List of Figures

List of Tables

List of Code Listings

Preface

Introduction

Virtual private networks (VPN) are networks deployed on a public network infrastructure that employs the same security, management, and quality of service policies applied in a private network. The benefits of using VPNs include cost savings and extending connectivity to telecommuters, mobile users, and remote offices as well as to new constituencies, such as customers, suppliers and partners.

This is especially the case in IP-based VPNs, which are rapidly becoming the foundation for the delivery of New World services. Two unique and complementary VPN architectures based on IP Security (IPSec) and Multiprotocol Label Switching (MPLS) technologies are emerging to form the predominant foundations for the provision of these New World services. Both IPSec and MPLS VPNs are Layer 3 VPNs that utilize IP as the network layer protocol. However, the deployment of the two VPN architectures is somewhat different. IPSec VPN is based on the overlay VPN model, and MPLS VPN is based on the peer-to-peer VPN model with the combined benefits of traditional overlay and peer-to-peer VPN models.

The difference between IPSec and MPLS VPNs can be narrowed down to one important factor: who owns the VPN. In other words, the implementation of VPN architectures is dependent on whether the VPN is enterprise or service provider managed. IPSec VPNs are enterprise-managed VPNs. They are used are an alternative WAN infrastructure, which replaces or augments existing private networks utilizing leased-line or frame relay/ATM networks.

MPLS VPNs are service provider–managed VPNs. They provide the flexible connectivity and scalability of IP coupled with the privacy and QoS of frame relay and ATM, allowing service providers to cater to a more diverse base of customers and offer a whole range of emerging value-added services, such as e-commerce, application hosting, and multimedia applications to these customers.

This book is devoted to the implementation and deployment of these two VPN architectures.

Purpose of This Book

The VPN technologies that are covered in this book are unique in their own ways. Currently, the technical literature in the marketplace covers IPSec and MPLS VPN in a very superficial and brief manner. No book has yet touched on these two intriguing technologies together in-depth. Neither have these technologies been compared and contrasted.

The purpose of this book is to provide the reader with a practical insight on building a VPN from scratch with IPSec and MPLS. It covers the implementation and deployment of IPSec/MPLS VPNs in enterprise and service provider settings. Practical-oriented case studies are used extensively throughout the text to illustrate the conceptual aspect of these technologies. This "know-how" approach allows readers to relate to and apply what they have learned through the case scenarios to their own network environment.

Intended Audience

This book is most relevant to network engineers and managers who will design implement, or maintain IPSec and MPLS VPNs for enterprises and service providers. This book is also appropriate for sales engineers, systems engineers, account managers, and project managers in the networking industry who are involved with IPSec/MPLS VPN projects or for anyone who wishes to increase knowledge on the deployment of IPSec and MPLS VPNs. In addition, practicing professionals preparing for practical IPSec and MPLS certification tests will find this book particularly useful and informative.

For readers to absorb more information from this book and to understand the IPSec and MPLS VPN concepts better, they should have some prior knowledge of the IPSec protocol framework and MPLS technology and a good understanding of advanced routing protocols such as OSPF, integrated IS-IS, and BGP. Note this prerequisite is preferred but not mandatory since this book will attempt to address these areas progressively through each chapter.

Organization of This Book

This book is divided into three parts:

Part 1 presents a comprehensive background on IPSec and MPLS VPNs:

- Chapter 1 discusses the growing demand for VPN. It provides the reader a brief overview of the different VPN types and models available today. The chapter then moves on to compare and contrast the overlay VPN model versus the peer-to-peer VPN model.

Part 2 is devoted to the implementation of IPSec VPNs in enterprises:

- Chapter 2 provides an introduction to all the nitty-gritty algorithms and technologies associated with IPSec. The chapter then moves on to describe the IPSec protocol framework, followed by a detailed discussion of IPSec operations.

- Chapter 3 presents actual site-to-site IPSec VPN design and implementation together with various IPSec configuration caveats. The text also introduces some of the commands and tools meant for troubleshooting, monitoring, and verifying the operation of an IPSec network. The coverage then moves on to discuss the tunnel endpoint discovery (TED) mechanism.

- Chapter 4 discusses how a site-to-site IPSec VPN can be used in conjunction with the other two IKE peer authentication techniques: RSA encrypted nonces and RSA signatures. The text first presents an overview of RSA encrypted nonces and takes the reader step-by-step through the configuration and monitoring process. The coverage then moves on to identify some of the scalability issues encountered when using pre-shared keys and RSA encrypted nonces peer authentication. Consequently, more sophisticated authentication components such as digital signatures, digital certificates, the certificate authority (CA), and the simple certificate enrollment protocol (SCEP) are brought into the picture to collectively address the scalability issues. The chapter concludes by featuring a site-to-site IPSec VPN using the IKE RSA signatures peer authentication together with some illustrations of the CA server setup, certificate enrollment, configuration, monitoring, and verification processes.

- Chapter 5 focuses on the deployment of multiple IPSec VPN sites in large enterprise networks. The coverage is on the design of scalable IPSec VPNs using the hub-and-spoke topology.

- Chapter 6 discusses how IPSec can interact with IP addressing services such as network address translation (NAT), network resiliency techniques such as hot standby router protocol (HSRP), and tunneling mechanisms such as generic routing encapsulation (GRE). The text also addresses performance optimization parameters such as fragmentation, IKE SA lifetimes, and IKE keepalives.

Part 3 is devoted to the deployment of MPLS VPNs for service providers:

- Chapter 7 provides a brief overview on the MPLS inner workings. The text discusses all the MPLS main concepts and core technologies, which are illustrated with the MPLS unicast IP routing model.

- Chapter 8 focuses on the inner workings of the MPLS VPN architecture by examining the MPLS VPN functional as well as architectural building blocks, the MPLS VPN routing model, and the MPLS VPN forwarding mechanisms.

- Chapter 9 shows the reader how to build an MPLS VPN by first going through the design considerations, followed by the implementation preliminaries and finally the actual implementation. The text uses three different case studies to illustrate the deployment of MPLS VPN. The PE-CE routing protocols used in the case studies include EBGP, static routes, and RIPv2. The chapter also discusses how to prevent potential BGP loops by using the site-of-origin (SOO) extended BGP community attribute when the standard AS-path–based BGP loop prevention has been compromised.

- Chapter 10 covers the MPLS VPN topologies that do not share the same connectivity requirements. The VPN topologies include overlapping topology, central services topology, hybrid topology, and hub-and-spoke topology. The text also describes the scenarios where the BGP Allowas-in feature and BGP AS-override mechanism need to be enabled. The coverage then moves on to the two other important aspects to consider when deploying MPLS VPN: Internet access and performance optimization.

- Chapter 11 discusses how route reflectors can be used to scale MP-BGP route distribution in the MPLS VPN backbone. The coverage includes partitioned route reflectors, which can be deployed to further enhance the MPLS VPN scalability. The case study in this chapter gives the reader some valuable insight on how to partition VPN routes based on standard BGP communities. The concluding section of the chapter describes how to optimize and fine-tune the BGP route reflectors for better performance.

Approach

"What I hear, I forget. What I read, I remember. What I do, I understand."—Confucius

The entire content of the book adopts a practical outlook and approach. It contains a whole series of real-life case studies and utilizes representative topologies as a basis for illustrating the concepts discussed in each chapter. The learning-by-examples approach involves the reader in the IPSec/MPLS VPN deployments and provides a fully working solution down to the basics. As such, this hands-on process helps the reader to remember and understand the IPSec/MPLS VPN concepts and technologies. Readers can then apply what they have learned from these examples and scenarios to their specific situations.

Nam-Kee Tan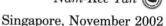

Singapore, November 2002

Acknowledgments

My first acknowledgement goes to Emily Baet from Global Knowledge Network, Singapore, and Jackie Mao as well as James Lin from Global Intelligence Network, Taiwan. They are the good folks who have helped me either directly or indirectly with this book. I probably owe the greatest thanks to those who read the manuscript (or a large part of it) looking for obscurities and errors. They are Boris Choo, Colin Ng, Evan Sim, Jahae Koo, Jerand Seng, Josef Miskulnig, Margaret Neo, Steven Lim, Sam Yeo, and Tina Huang. Much credit also goes to my editor, Marjorie Spencer, who among many other things went through ten chapters of the book and gave me many invaluable comments. These comments have been most useful in helping me spot and include information on some of the areas that I had previously overlooked. I would also like to thank my copyeditor, Peg Markow who meticulously scrutinized every single detail of the book content to ensure that everything is in proper order.

Many people have taught and influenced me in my trade over the last ten years. I have also learned a great deal from hundreds of network engineers, consultants, and managers who are either my students in Cisco training classes or my counterparts in consulting assignments. Special thanks to all these good people, without whom I would not be in the networking industry today.

A special thanks is also in order for my kid sister Sok-Thian and her husband Yuji Suzuki for being such good hosts when I was at their house in Shizuoka working on Chapters 7 and 11 after completing an assignment in Tokyo with Nokia, Japan. Of course, thanks to Emmy, my cuddly and adorable baby niece for lightening up my heavy writing with her stunning smile and playful gurgle. In addition, my elder sister Sok-Ai handled the finance and administrative matters diligently for Couver without complaint while I worked on the book. Thanks, Sis, for all the help and support.

I was on the road so frequently for the first half of 2002 that I really missed and enjoyed the home-cooked meals and local delicacies that Mom never failed to prepare for me when I was back home. But, sadly, Mom passed away in January 2003, just a few months after I had completed this book, I would like to take this opportunity to pay tribute to my mom for everything she has done for me. She will always be in my heart.

Most of all, I thank my partner Chai-Tee, who has relentlessly showered me with all the care and encouragement that I cannot do without. It is an ordeal to put up for over a year with someone who maintains a day job and writes a book in the evenings and over weekends. I know a few line of praise and thanks are not enough to make up for the lost time, but Chai-Tee I really appreciate your support, patience, and love.

About the Author

Nam-Kee Tan (CCIE #4307) has been in the networking industry for 10 years. He is the Managing Principal of Couver Network Consulting, where he provides consulting and training services to corporate and government clients throughout Asia Pacific. His areas of specialization includes advanced IP services, network management, traffic engineering, MPLS, Internet security solutions, VPN implementations, PKI technologies, and Intrusion Detection Systems.

Nam-Kee is also a Certified Cisco Systems Instructor (CCSI #98976) who has delivered a whole range of Cisco certified courses to hundreds of networking professionals from most of the Fortune 500 companies. He writes actively and has two book titles published by McGraw-Hill. In addition, he holds an M.S. in Data Communications from the University of Essex, UK, and an MBA from the University of Adelaide, Australia.

The VPN Overview

VPN-in-Brief

1.1 VPN Overview

This is the information age. We no longer have to commute physically from one place to another to complete a set of tasks or to gather pieces of information. Everything can be done virtually with a mouse click on an online host. In a way, everything we do in our daily lives is related in one way or another to information access. This has made information sharing almost mandatory and indispensable. These days, a customer can retrieve and compare products or services information promptly online, anytime, anywhere. For competitive reasons, organizations that provide this information have to make the information readily available online. In other words, the concept of a shared infrastructure is undisputedly important. A shared infrastructure is none other than a public network. At present, the biggest public network is the Internet, which has over 100,000 routes and is still growing rapidly.

As more and more companies link up their corporate network to the Internet, we are faced with an inevitable issue—information security. Sharing information on a public network also implies giving access and visibility to everyone who wants to retrieve these data. What if the person who has the accessibility and visibility to the information decides to create havoc? Some of the general threat types that are posed by malicious hackers include eavesdropping, denial of service, unauthorized access, data manipulation, masquerade, session replay, and session hijacking.

How do we ensure the safe passage of data across a shared infrastructure? The answer is to deploy a secured virtual private network (VPN). VPNs are networks deployed on a public network infrastructure that utilize the same security, management, and quality of service policies that are applied in a private network. VPNs provide an alternative to building a private network for site-to-site communication over a public network or the Internet. Because they operate across a shared infrastructure rather than a private network, companies can cost effectively extend the corporate WAN to telecommuters, mobile users, and remote offices as well as to new constituencies, such as customers, suppliers, and business partners.

Traditional private WANs connect customer sites via dedicated point-to-point links. This means that multiple independent circuits have to terminate at the corporate network egress, making the deployment nonscalable and difficult to maintain. VPNs extend the classic WAN by replacing the physical point-to-point links with logical point-to-point links sharing a common infrastructure, allowing all the traffic to be aggregated into a single physical connection. This scenario results in potential bandwidth and cost savings at the network egress. Because customers no longer need to maintain a private network, and because a VPN itself is cheaper to own and offers significant cost savings over private WANs, operation costs are reduced.

VPNs provide an alternative WAN infrastructure that can replace or augment commercial private networks that use leased-line or frame relay/ATM networks. There are two ways business customers can implement and manage their VPNs. They can either roll out their own VPNs and manage them internally, or outsource the VPN management to their service providers for a total VPN package that is tailored to their particular business needs.

Last but not least, from the service providers' perspective, VPNs are a fundamental building block in delivering new value-added services that benefit their business customers as well as themselves. In this instance, the service providers deploy the VPNs for their customers, and the customers need only subscribe to the service providers for the VPN services.

1.2   VPN Types and Solutions

In this section we address three types of VPNs: remote access, site-to-site, and firewall-based (a site-to-site variation). The variation between remote access and site-to-site VPNs will become more ambiguous as new devices such as hardware VPN clients, become more prevalent. These appear as a single device accessing the network, albeit there may be a network with several devices behind it. In all cases, the VPN comprises two endpoints that may be represented by routers, firewalls, client workstations, or servers.

1.2.1   Remote access

Remote access VPNs or virtual private dialup network (VPDN) are deployed for individual remote users, commonly referred to as mobile users and telecommuters. In the past, corporations supported these remote users via dialup networks, which required the remote users to make toll calls to access the corporate network directly. This was not a cost-effective solution, especially when an international user made a call back.

With the introduction of remote access VPNs, a mobile user can make a local call to their Internet service provider (ISP) to access the corporate network with their PC via the Internet wherever they may be. Remote access VPNs are an extension of the traditional dialup networks. In this case, the software on the PC provides a secure connection, often known as a tunnel, back to the corporation. Since the users need only make local calls, the operation cost is reduced. Remote access VPNs and their corresponding technologies are beyond the scope of this book.

1.2.2 Site-to-site

Site-to-site VPNs are deployed for interconnecting corporate sites. In other words, the network of one location (site) is connected to the network of another location (site) via a VPN. In the past, a leased line or frame relay connection was required to connect the sites; however, these days, most corporations have Internet access. With Internet access, leased lines and frame relay circuits can be replaced with site-to-site VPNs. Site-to-site VPNs are an extension of legacy WAN networks.

Site-to-site VPNs can be further viewed as intranet VPNs or extranet VPNs. Intranet VPNs refer to connections between sites that all belong to the same organization. User access between these sites is less restraining than for extranet VPNs. Extranet VPNs refer to connections between an organization and its business partners. User access between these sites should be tightly controlled by both entities at their respective sites.

1.2.3 Firewall-based

A firewall-based VPN is intrinsically a site-to-site implementation. Firewall-based VPN solutions are not a technical but a security issue. They are deployed when a corporation requires more advanced perimeter security measures for its VPNs. Corporations can enhance their existing firewalls to support firewall-based VPNs.

1.3 VPN Terminology

This section lists some of the common VPN terminology that is used in the subsequent sections:

Provider network (P-Network): the service provider infrastructure that is used to provide VPN services.

Customer network (C-Network): the part of the network that is still under customer control.

Customer site: a contiguous part of the C-Network that can comprise many physical locations.

Provider (P) device: the device in the P-Network with no customer connectivity and without any "knowledge" of the VPN. This device is usually a router and is commonly referred as the P router.

Provider edge (PE) device: the device in the P-Network to which the CE devices are connected. This device is usually a router and is often referred as the PE router.

Customer edge (CE) device: the device in the C-network that links into the P-network; also known as customer premises equipment (CPE). This device is usually a router and is normally referred as the CE router.

Virtual circuit (VC):   logical point-to-point link that is established across a shared layer-2 infrastructure.

1.4   VPN Models

VPN services can be offered based on two major paradigms:

1. The overlay VPNs, whereby the service provider furnishes virtual point-to-point links between customer sites.
2. The peer-to-peer VPNs, whereby the service provider participates in customer routing.

In the following sections, we discuss these two different models in details.

1.4.1   Overlay model

The overlay VPN is deployed via private trunks across a service provider's shared infrastructure. These VPNs can be implemented at layer-1 using leased/dialup lines, at layer-2 using X.25/frame relay/ATM Virtual Circuits, or at layer-3 using IP (GRE) tunneling.

In the overlay VPN model, the service provider network is a connection of point-to-point links or virtual circuits (VCs). Routing within the customer network is transparent to the service provider network, and routing protocols run directly between customer routers. The service provider has no knowledge of the customer routes and is simply responsible for providing point-to-point transport of data between the customer sites.

Figure 1.1 illustrates the deployment of an overlay VPN. The scenario adopts a hub-and-spoke topology whereby the Paris site is the hub, and both the London and Zurich sites are the spokes. The London site is linked up to the Paris site via a point-to-point VC #1. Likewise, the Zurich site is linked up to the Paris site via a point-to-point VC #2. In this instance, the layer-3 routing adjacencies are established between the CE routers at the various customer sites, and the service provider is not aware of this routing information at all. As illustrated in Figure 1.2, from the perspective of the CE routers, the service provider infrastructure appears as point-to-point links between Paris–London and Paris–Zurich.

The overlay VPN model has two further constraints. One is the high level of difficulty in sizing the intersite circuit capacities. The other is the requirement

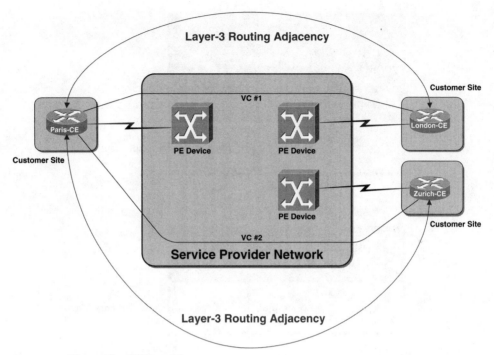

Figure 1.1: The overlay VPN model.

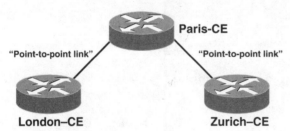

Figure 1.2: Perception of the SP infrastructure from the CE routers.

of a fully meshed deployment of point-to-point links or VCs over the service provider's backbone to attain optimal routing.

1.4.1.1 Layer-1 implementation. Figure 1.3 illustrates the overlay VPN layer-1 implementation, which adopts the traditional time division multiplexing (TDM) solution. In this scenario, the service provider assigns bit pipes and establishes the physical-layer (Layer-1) connectivity between customer sites via ISDN, DS0, T1, E1, SONET, or SDH, and the customer is accountable for implementation of all higher layers, such as PPP, HDLC, and IP.

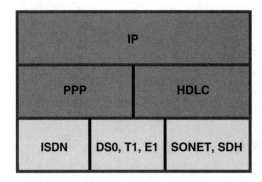

Figure 1.3: Overlay VPN layer-1 implementation.

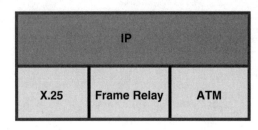

Figure 1.4: Overlay VPN layer-2 implementation.

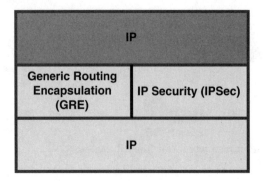

Figure 1.5: Overlay VPN layer-3 implementation.

1.4.1.2 Layer-2 implementation.   Figure 1.4 illustrates the overlay VPN layer-2 implementation, which adopts the traditional switched WAN solution. In this scenario, the service provider is responsible for establishing layer-2 VCs between customer sites via X.25, Frame Relay, or ATM, and the customer is accountable for the IP layer and above.

1.4.1.3 Layer-3 implementation.   Figure 1.5 illustrates the overlay VPN layer-3 implementation, whereby the VPN is implemented with point-to-point IP-over-IP tunnels. This is commonly referred as IP tunneling whereby a destination can be reached transparently without the source having to know the topology specifics. Therefore, virtual networks can be created by tying otherwise unconnected devices or hosts together through a tunnel. Tunnels also enable the use of private network addressing across a service provider's backbone without the need for network address translation (NAT). Tunnels are established with generic routing encapsulation (GRE) or IP security (IPSec). The GRE implementation is simpler and quicker but less secure, whereas the IPSec deployment is more complex and resource-intensive (CPU cycle) but it provides more robust security.

GRE tunnels provide a specific conduit across the shared WAN and encapsulate traffic with new packet headers to ensure delivery to specific destinations. Since traffic can enter a tunnel only at an endpoint, the network is private. GRE tunnels do not provided true confidentiality but can carry encrypted traffic. For better protection, GRE can be used in conjunction with IPSec. When we deploy a layer-3 VPN over a public network, the VPN tunnels are built across a distrusted shared infrastructure. IPSec provides robust security services such as confidentiality, integrity, and authentication to ensure that sensitive information is securely transported across these tunnels and not circumvented accidentally or intentionally.

IPSec is an Internet Engineering Task Force (IETF) standard that enables encrypted communication between users and devices. It can be implemented transparently and seamlessly into the network infrastructure. End users need not have any knowledge that packets are being intercepted and transformed by IPSec. Because it operates at the network layer (layer-3), IPSec is ideally positioned to enforce corporate network security. Since IPSec encryption works only on IP unicast frames, it can be used in conjunction with GRE to alleviate this deficit. This is because GRE is capable of handling the transportation of multiprotocol and IP multicast traffic between two sites, which have only IP unicast connectivity.

1.4.2   Peer-to-peer model

The peer-to-peer model adopts a simple routing scheme for the customer. Both provider and customer network use the same network protocol and all the customer routes are carried within the core network (service provider network). The PE routers exchange routing information with the CE routers, and layer-3 routing adjacencies are established between the CE and PE routers at each site. Because peer-to-peer routing has been implemented, routing between sites is now optimal. Fully meshed deployment of point-to-point links or VCs over the service provider backbone is no longer applicable to attain optimal routing. Since there is no overlay mesh to contend with, the addition of new sites is easier, and circuit capacity sizing is not an issue. Because the service provider now participates in customer routing, provider-assigned or public address space needs to be deployed at the customer's network, so private addressing is no longer an option.

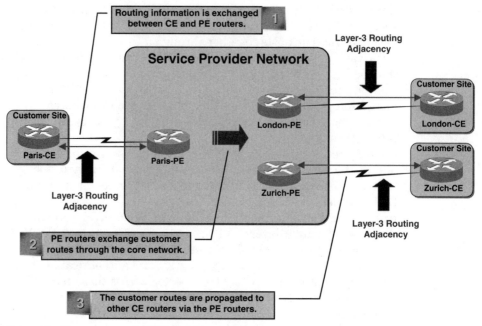

Figure 1.6: The peer-to-peer VPN model.

Figure 1.6 illustrates the deployment of a peer-to-peer VPN. In this scenario, customer routing information is exchanged between Paris-CE and Paris-PE. The customer routes are then broadcast through the core network to London-PE and Zurich-PE, which in turn, propagate these routes to their respective CE routers. The shared router and dedicated router approaches are derivatives of the peer-to-peer model. We discuss these two approaches in Sections 1.4.2.1 and 1.4.2.2.

1.4.2.1 Shared PE router model. In this model, a common PE router that carries customer routes is deployed. Individual customers routes are separated with packet filters on PE-CE interfaces. The packet filters are managed so that information goes to the proper site and different customers are separated. The complexity of these packet filters results in high maintenance costs and a significant impact on performance.

Figure 1.7 illustrates the Shared PE router model. In this scenario, there are three separate VPNs: VPN-101, VPN-201, and VPN-301. These VPNs are deployed over four different customer sites. VPN-101 is deployed for Paris as well as Lyon; VPN-201 is deployed for Brussels; and VPN-301 is deployed for Munich. In Figure 1.7, we can see that London-PE carries all the customer routes for VPN-101, VPN-201, and VPN-301. Isolation between VPNs is achieved with packet filters (access lists) on the PE-CE interfaces: Serial0/0, Serial0/1, Serial0/2, and Serial0/3.

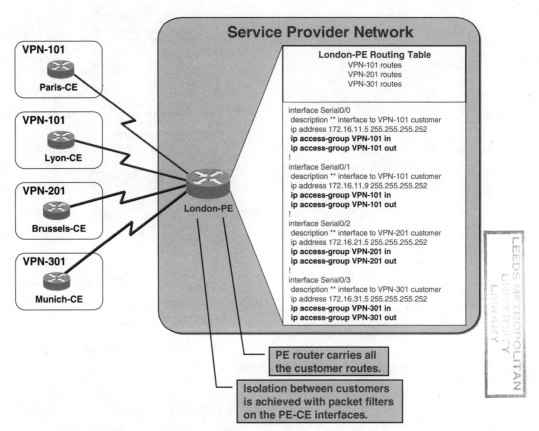

Service Provider Network

VPN-101
Paris-CE

VPN-101
Lyon-CE

VPN-201
Brussels-CE

VPN-301
Munich-CE

London-PE

London-PE Routing Table
VPN-101 routes
VPN-201 routes
VPN-301 routes

interface Serial0/0
 description ** interface to VPN-101 customer
 ip address 172.16.11.5 255.255.255.252
 ip access-group VPN-101 in
 ip access-group VPN-101 out
!
interface Serial0/1
 description ** interface to VPN-101 customer
 ip address 172.16.11.9 255.255.255.252
 ip access-group VPN-101 in
 ip access-group VPN-101 out
!
interface Serial0/2
 description ** interface to VPN-201 customer
 ip address 172.16.21.5 255.255.255.252
 ip access-group VPN-201 in
 ip access-group VPN-201 out
!
interface Serial0/3
 description ** interface to VPN-301 customer
 ip address 172.16.31.5 255.255.255.252
 ip access-group VPN-301 in
 ip access-group VPN-301 out

**PE router carries all
the customer routes.**

**Isolation between customers
is achieved with packet filters
on the PE-CE interfaces.**

Figure 1.7: Shared PE router model.

1.4.2.2 Dedicated PE router model. In this model, each customer has a dedicated PE router that carries only its own routes. Customer segregation is achieved through lack of routing information on the PE router. The P router contains all customer routes and filters routing updates between different PE routers using Border Gateway Protocol (BGP) Communities. Because each customer has a dedicated PE router, this approach is expensive to deploy, and hence it is not a cost-effective solution.

Figure 1.8 illustrates a dedicated PE router model. In this scenario, there are two separate VPNs: VPN-101 and VPN-201. These VPNs are deployed over four different customer sites. VPN-101 is deployed for two locations in Paris and one location in Brussels, and VPN-201 is deployed for one location in London. In Figure 1.8, we can see that the P router in the service provider network contains all the customer routes for VPN-101 and VPN-201. It filters routing updates between Paris-PE, London-PE, and Brussels-PE using BGP Communities. VPN-101 routes are tagged with a community value of 101:1, and VPN-201 routes are tagged with a community value of 201:1. In other words, the P

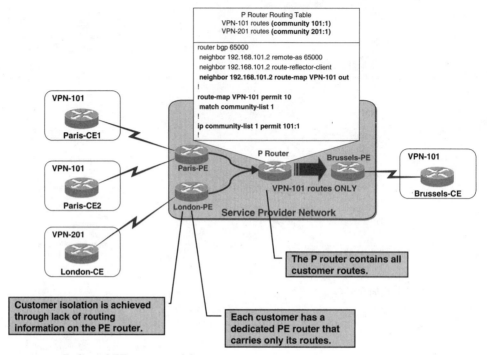

Figure 1.8: Dedicated PE router model.

router will propagate routes with a community value of 101:1 to VPN-101 and routes with a community value of 201:1 to VPN-201. Each customer or VPN has a dedicated PE router that carries only its routes. Customer isolation is simply achieved through the lack of routing information on the PE routers.

In this example, routes from Paris-CE1 and Paris-CE2, which are both in VPN-101, are announced to a dedicated PE router—Paris-PE. Paris-PE in turn advertises the routes to the P router. The P router uses a community list to match routes with a community value of 101:1 and propagates the routes that match this community value to another dedicated PE router, Brussels-PE, which serves the customer site that is in VPN-101 at Brussels.

1.4.3   MPLS VPN model

MPLS VPN is a true peer-to-peer model that combines the best of both worlds. It unites the customer security and segregation features implemented in the overlay model with the simplified customer routing deployed in the traditional peer-to-peer model. The MPLS VPN architecture is very similar to the dedicated PE router model, except the dedicated per customer routers are implemented as virtual routing tables within the PE router. In other words, customer segregation is achieved through the concept of virtual routing and forwarding (VRF) whereby the PE router is subdivided into virtual routers serving differ-

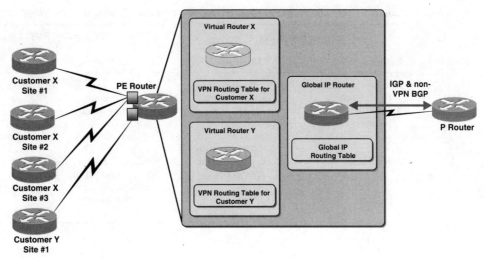

Figure 1.9: MPLS VPN model.

ent VPNs (or customer sites). This establishes overlapping addresses in different customer sites since each customer is now assigned an independent routing table.

The PE routers hold the routing information only for directly connected VPNs. As a result, the size of the PE routing table is significantly reduced. The amount of routing information is proportional to the number of VPNs attached to the PE router. As such, the PE routing table will still grow when the number of directly connected VPNs increases. In addition, the PE routers participate in customer routing, ensuring optimal routing between sites and easy provisioning. Full routing within the service provider backbone is no longer required because multi protocol label switching (MPLS) (see Chapter 7 for more in-depth MPLS concepts), and not traditional IP routing, is used to forward packets. The details of the MPLS VPN architecture are discussed in Chapter 8.

Figure 1.9 illustrates the MPLS VPN model. There are two separate customers—Customer X and Customer Y. Customer X spans three different sites, while Customer Y has only one site. Customer isolation is achieved with a dedicated per customer virtual router. In Figure 1.9, we can see that routing within the PE router is split into two separate planes, one for VPN routing and the other for global IP routing.

In the VPN routing plane, the PE router is subdivided into virtual router X which serves customer X, and virtual router Y which serves customer Y. Each of these virtual routers participates in customer routing and keeps an independent VPN routing table for its respective customers. Meanwhile, the PE router also has a global IP router in the global IP routing plane that take cares of the IGP and non-VPN BGP routing between the various PE and P routers.

TABLE 1.1   Upside and Downside of Overlay VPN versus Peer-to-Peer VPN.

Overlay VPN Model		Peer-to-Peer VPN Model	
Upside	Downside	Upside	Downside
■ Allows replicate IP addressing ■ Full isolation between customers ■ Secure VPN service	■ Difficult to size intersite circuit capacity ■ Fully meshed circuit requirement for optimal routing ■ Layer-3 CE routing adjacencies between sites	■ Routing between sites is optimal ■ Circuit capacity sizing between sites is not an issue ■ Simpler routing configuration for customers (no overlay mesh)	■ All VPN routes are carried in the service provider IGP ■ Replicate IP addressing is no longer an option ■ Complex filters or dedicated devices

TABLE 1.2   Benefits of MPLS VPN

MPLS VPN Model	
Combined benefits of overlay and peer-to-peer VPN models	
■ Routing between sites is optimal ■ Allows replicate IP addressing ■ Secure VPN service	■ PE routers hold only pertinent VPN routers ■ Full isolation between customers ■ No complex filters or dedicated routers

1.5   Comparison Between Various VPN Implementations

This section compares some of the benefits and drawbacks between overlay VPN and peer-to-peer VPN. Table 1.1 gives a quick appraisal of the upside and downside of the overlay VPN model versus the peer-to-peer VPN model.

Table 1.2 illustrates the benefits of the MPLS VPN model, which are combined from the overlay and peer-to-peer VPN models for the best of both worlds.

1.6   IPSec versus MPLS VPNs

Both IPSec and MPLS VPNs are implemented at the network layer of the OSI model. In other words, they are layer-3 VPNs that use the IP protocol as the network layer protocol. However, the deployment of these different VPN architectures is rather controversial because IPSec VPN is based on the overlay model (see Section 1.4.1), while MPLS VPN is based on the peer-to-peer model (see Section 1.4.3) which offers the combined benefits of traditional overlay and peer-to-peer VPN models (see Table 1.2). The controversy between the IPSec and MPLS VPNs is narrowed down to one important factor: who owns the VPN.

In other words, the implementation of these VPN architectures depends on whether the VPN is managed by the customer or the service provider.

Customers who manage their own VPNs usually adopt the overlay approach. Overlay VPNs are connection oriented. They are based on creating point-to-point connections and not networks. In IPSec VPNs, IPSec tunnels are created to provide point-to-point connectivity at the customer's business site. This also implies that IP connectivity is required to establish these point-to-point tunnels. The best way to interconnect the sites is via the Internet through a service provider who is responsible only for providing Internet connectivity between the customer sites. In this case, the customers will have to build and secure their own VPNs across the Internet by deploying IPSec. These site-to-site enterprise-managed VPNs are also known as Internet VPNs. Because site-to-site peering is required when implementing IPSec, scalability issues will arise when the number of customer sites grows.

MPLS is typically offered as a site-to-site VPN service from a service provider. The service provider builds and manages a private IP-based network and offers multiple customers IP connectivity between their sites across this network. This highly scalable connectionless architecture allows individual customers to view the MPLS service as though they had an IP VPN connecting their sites. This setup provides customers the same benefits of a layer-2 private network (see Section 1.4.1.2), but with scalability and the easy management features of an IP (layer-3) network, by eliminating the need for tunnels or VCs. Since MPLS VPN runs across a private IP-based network rather than the Internet, the service provider has the capabilities to offer differentiated levels of services and service level agreements (SLAs) to its customers. A typical MPLS VPN deployment is capable of supporting tens of thousands of VPN groups over the same network. However, because MPLS VPN is based on a service provider's private network, the reach of the service is constrained to the locations at which the service provider operates.

1.7 Summary

There is a growing demand for VPNs. The different types of VPNs such as remote access, site-to-site, and firewall-based VPNs have been examined. The layer-1, layer-2 and layer-3 implementations of overlay VPNs have been explained, along with the shared router and dedicated router approaches for peer-to-peer VPNs. We also briefly examined how the MPLS VPN model operates and compared the benefits and drawbacks of overlay VPNs versus peer-to-peer VPNs and listed the benefits of MPLS VPNs, which are the best of both worlds. The last section of the chapter addressed the controversy on the deployment of IPSec and MPLS VPN architectures. The remainder of this book discusses implementing IPSec VPNs in Enterprises and deploying MPLS VPNs for Service Providers.

Implementing IPSEC
VPN in Enterprises

IPSec and IKE Overview

2.1 Introduction

Before the actual implementation of IPSec VPN, it is important to have a clear understanding of the related algorithms, technologies, and operational processes. This chapter covers the operations of IP security protocol (IPSec) and Internet key exchange (IKE), as well as their associated technologies, in-depth. These associated technologies should be thoroughly understood in order to deploy a robust VPN security policy and to implement a fully operational IPSec VPN.

2.2 IPSec Overview

IPSec is implemented at the network layer. It protects and authentices IP packets between participating IPSec devices (peers) such as firewalls, routers, clients, and other IPSec-compliant products. IPSec is a framework of open standards, which is not tied down to any specific algorithms. As such, IPSec allows for newer and better algorithms to be implemented without patching the existing standards, and it provides data confidentiality, data integrity, and data origin authentication between participating peers at the IP layer. In the following sections, we will look at the definitions for these security services and describe the technologies associated with them.

2.3 IPSec VPN Services

The IPSec framework provides three important security services for VPNs:

- *Confidentiality* is the security service that protects data and the external characteristics of communication from unauthorized disclosure by concealing source and destination addresses, message length, or frequency of communication. With confidentiality, the intended recipients know what was being sent but unintended parties cannot determine what was sent. Encryption algorithms (see Section 2.4) can be used to provide confidentiality.

- *Integrity* is a security service that ensures that any modifications to data are detectable. There are two types of integrity: connectionless and partial sequence integrity. Connectionless integrity is a service that detects the modification of individual IP packets without regard to the ordering of the packet in a traffic stream. Partial sequence integrity is referred to as antireplay integrity, and it detects arrival of duplicate IP packets. With integrity, data is transmitted from source to destination without undetected change. Hashing algorithms (see Section 2.5) can be used to provide integrity. Although authentication and integrity services are often mentioned separately, in practice they are closely related, and almost always offered together.

- *Data Origin Authentication* is a security service that verifies the identity of the claimed source of data. This service is usually bundled with connectionless integrity service. With data origin authentication, we have to ensure that the data received is the same as the data that was sent and that the claimed sender is, in fact, the actual sender. Mechanisms such as the exchange of digital certificates (see Section 2.7.2) can be use to provide data origin authentication.

2.4 Encryption Concepts and Algorithms

An ordinary message is usually in plain text. The process of camouflaging a message in such a way as to conceal its content is known as encryption. An encrypted message is in cipher text. The process of converting cipher text back into plain text is known as decryption.

A cryptographic algorithm, also called a cipher, is the mathematical function used for encryption and decryption. For better security, quality control, and standardization, modern cryptography incorporates the concept of a key. This key can take many values (usually denoted in binary bits), and is analogous to the combination to a lock. Although the concept of a combination lock is well known, it is often difficult to open up a combination lock without knowing the combination. In addition, the more numbers in a given combination, the more time consuming it will be to guess the combination. The same is true for cryptographic keys; hence the more bits in a key, the less susceptible a key is to being compromised. Therefore, the security robustness in modern day ciphers is based on the key rather than the details of the cipher. This means that the cipher can be published and analyzed openly. It doesn't matter if an eavesdropper knows your algorithm; if your particular key is not revealed, your message can't be read by anyone.

2.4.1 Symmetric algorithms

There are two general types of key-based algorithms: symmetric and asymmetric. In symmetric algorithms, the encryption key can be calculated from the

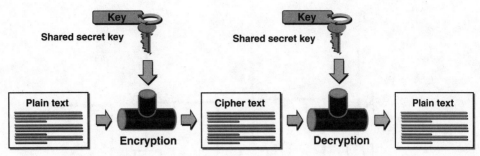

Figure 2.1: Symmetric encryption and decryption.

decryption key and vice versa. In this case, encryption and decryption use the same mathematical functions, and the encryption key and the decryption key are identical. Symmetric algorithms, also known as secret-key algorithms, require that the sender and receiver agree on a key before they can transmit and receive information securely. The security of a symmetric algorithm lies in the key, disclosing the key implies that anyone could encrypt and decrypt messages. Therefore, it is mandatory that the key remains secret. Figure 2.1 illustrates how a symmetric algorithm operates.

Symmetric algorithms are further categorized into stream and block ciphers. Block ciphers operate on data one block (usually in 64-bit block size) at a time. Stream ciphers conversely operate on data one bit (or one byte) at a time. In either case, they are ideal for bulk encryption. As block ciphers are used solely in IPSec, stream ciphers are beyond the scope of this book.

Cipher mode refers to a set of techniques used to apply a block cipher to a data stream. Four basic modes are used:

- Electronic code book (ECB)
- Cipher block chaining (CBC)
- Cipher feedback (CFB)
- Output feedback (OFB)

Of the four different modes, only CBC will be discussed in this book because all the block ciphers currently defined for IPSec are operating in CBC mode. CBC mode as illustrated in Figure 2.2, takes the previous block of cipher text and exclusively-OR (XOR) it with the next block of plain text before encryption. XOR is a logical operation involving binary bits. In other words, "0" XOR with "0" gives you "0"; "0" XOR with "1" gives you "1"; "1" XOR with "0" gives you "1"; and "1" XOR with "1" gives you "0".

The first block is XORed with an initialization vector (IV), which is a block of random bits. The IV is used to ensure that no two messages will yield the same cipher text, even if their plain texts are identical. As for decryption, it is the reverse of encryption. Each block is decrypted and XORed with the IV.

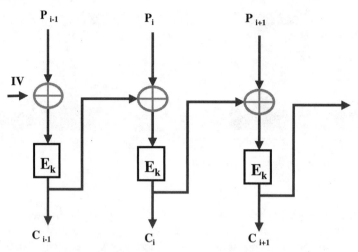

Figure 2.2: Cipher block chaining mode (Encryption).

An example of a symmetric block cipher in CBC mode is the data encryption standard (DES). DES encrypts data in 64-bit blocks and the same algorithm and key are used for both encryption and decryption. The key length is 56 bits. The key can be any 56-bit number and can be altered any time. DES breaks a message into 64-bit blocks, XORs the first block with the IV, and encrypts it with the 56-bit key. The result is a 64-bit block of cipher text that is further XORed with the next 64-bit block. This process is repeated until the entire message has been encrypted.

There is a more secure variant of DES. It is 3DES. In this case, the DES algorithm is applied 3 times to each 64-bit plain text block, using a different 56-bit key each time. A typical approaches uses two dissimilar 56-bit keys, K1 and K2, yielding a length of 112 bits, with K1 being reused for the final (third) encrypt. Other applications use three dissimilar 56-bit keys, K1, K2, and K3, yielding a total key length of 168 bits. Using the same key in all three steps will yield the same result as encrypting with conventional DES. This is because the second step has now become a decrypt operation.

2.4.2 Asymmetric algorithms

Asymmetric algorithms are designed so that the key used for encryption is different from the key used for decryption. In addition, the decryption key cannot be derived from the encryption key. These algorithms are referred to as public-key algorithms because the encryption key is made public. In other words, anyone can use the encryption key (public key) to encrypt a message, but only a specific recipient with the corresponding decryption key (private key) can decrypt the message. Figure 2.3 illustrates how an asymmetric algorithm operates. In other instances the message can be encrypted with the private key and

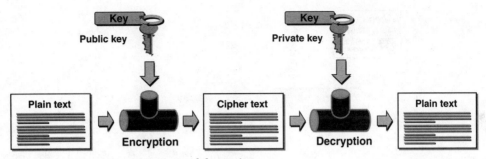

Figure 2.3: Asymmetric encryption and decryption.

decrypted with the public key; this is typically used in digital signatures (see Section 2.7.1).

An example of a public-key cipher is the RSA algorithm developed by Ron Rivest, Adi Shamir, and Leonard Adleman. The public and private keys are functions of a pair of large prime numbers. The security of this approach is based on the fact that it can be relatively easy to multiply these large prime numbers together but extremely difficult to factor the resulting product. In other words, anyone attempting to recover the plain text from the cipher text and the public key will have to go through the knotty task of guessing the decryption (private) key by factoring the resulting product of two large prime numbers. RSA uses a key length of 512 bits, 768 bits, 1024 bits, or longer. A longer key length offers stronger security, but also uses up more CPU processing cycles. This is due to the fact that the mathematical computations in the RSA algorithm are much more complicated than the DES algorithm.

As a result, in hardware, RSA is about 1000 times slower than DES, and in software, DES is up to 100 times faster than RSA. These numbers may change slightly as technology advances, however, RSA will never approach the speed of symmetric algorithms, which makes it an unlikely candidate for bulk data encryption.

Public-key algorithms such as RSA are rarely used for data confidentiality (encryption) because of the performance constraints mentioned earlier. Instead, they are most often used in applications involving authentication via digital signatures and key management. In real life, public-key algorithms are not surrogates for symmetric algorithms. They are typically used to encrypt keys instead of messages.

2.5 Hashing Concepts and Algorithms

A one-way hash function is a mathematical function that is easy to compute in the forward direction, but computationally infeasible to reverse. An input message runs through the mathematical function (the hash function) and results in a string of output bits called a hash (or a message digest). The process is irreversible, that is, the input cannot be derived from the output. It is

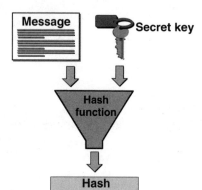

Figure 2.4: A   keyed-hashing function (MAC).

analogous to a coffee grinder. Imagine the message is the coffee beans and the output hash is the resulting ground coffee, it is quite impossible to take the ground coffee and recreate the original coffee beans. In addition, a hash function takes a message of arbitrary length and produces a fixed-length output.

It is important to implement keyed hashing because just performing a one-way hash on some data does not provide any authentication. A hash function alone is like a checksum. Anybody can modify the data and just run the hash algorithm over the modified data. A keyed hash function is a message authentication code (MAC), which, in turn, is a one-way hash function with the addition of a secret key. A MAC is generated through hashing a shared secret key along with the message. The concept is the same as hash functions, except now only the holder of the secret key can verify the hash value. Figure 2.4 illustrates how keyed hashing works.

Keyed hashing can be used to generate a hash (MAC) on each IP packet in an IP data stream. These hashes then become part of the IP packet and are used to verify the integrity of the IP packet when it is received. The recipient runs the same hash (plus shared secret key) on each received IP packet and compares it with the inserted hash value. If nothing in the packet has been modified in transit the results are the same and the integrity of the IP packet remains intact. Common keyed hash functions include:

- Message digest 5 (MD5) is a one-way hash that combines a shared secret and the message (header together with payload), to produce a hash. MD5 processes the input text in 512-bit blocks, divided into 16×32-bit subblocks. The output is a set of 4×32-bit blocks, concatenated to form a single 128-bit hash value.

- Secure Hash Algorithm (SHA)—is similar to MD5 but the algorithm output is a set of 5×32-bit blocks, concatenated to form a single 160-bit hash value.

Note that all message authentications in IPSec use hashed MAC (HMAC) [described in RFC 2104] so MD5 becomes HMAC-MD5 and SHA becomes

HMAC-SHA. HMAC is a variant that provides an additional level of hashing making it cryptographically stronger than the principal hashing function.

2.6   Key Exchanges

Symmetric algorithms and keyed hashing both require a shared key. The security of the encryption and authentication techniques can be completely undermined by an insecure key exchange. With symmetric encryption, the data encryption key is always present and remains fixed. Any intruder who compromises the encryption key can decrypt messages encrypted with it. A common cryptographic technique is to encrypt each individual conversation with a separate key. This is referred to as a session key since it is used for only one particular communication session. The session key is created when it is required to encrypt communications and destroyed when it is no longer needed. As such, the risk of compromising the session (encryption) key is reduced significantly. In the following section, we look at how the Diffie-Hellman key exchange algorithm can be used to ensure the secure exchange of keys.

2.6.1   Diffie-Hellman key exchange

Diffie-Hellman was the first public-key algorithm developed. It is secure because of the complexity of calculating discrete logarithms in a finite field, as compared with the ease of calculating modular exponentiation in the same field. Diffie-Hellman can be used to securely distribute a shared secret key between parties in a key exchange across an insecure communication channel such as the Internet.

All participants in a Diffie-Hellman exchange must first choose a group. This group defines the prime p and the generator (primitive) g to be used. The Diffie-Hellman exchange comprises two parts. In part 1 each side, Alice and Bob, choose a random number (which becomes their private key) and does an exponentiation to produce a public value:

- Alice: public-value-A = g^a MOD p (where a is the random number chosen by Alice).
- Bob: public-value-B = g^b MOD p (where b is the random number chosen by Bob).

In part 2, Alice and Bob exchange their public values. In other words, Alice provides public-value-A to Bob, and Bob provides public-value-B to Alice. Each side then does an exponentiation again, this time using the other party's public value as the generator, which creates the shared secret key, Z:

- Alice: (public-value-B)^a MOD p = g^{ab} MOD p = Z
- Bob: (public-value-A)^b MOD p = g^{ab} MOD p = Z

Once Alice and Bob know the shared secret key Z, they can use it to protect their communications with any conventional secret-key algorithms. Anyone who knows p or g will not be able to calculate the shared secret value Z easily because it is very complicated to compute a discrete logarithm to recover the secret exponents (private keys) Alice and Bob used.

One of the inadequacies with key exchange systems is that the same session key is distributed to numerous sessions. If this key is leaked, it can be used to decrypt the earlier encrypted traffic. Diffie-Hellman can be used to achieve perfect forward secrecy (PFS), which eliminates the need for using long-life session keys. With PFS, a separate Diffie-Hellman session key (secret key) is generated for each session. If any of these keys are recovered, they will only provide access to that specific session's information. The remaining keys that are used to encrypt other sessions are not compromized.

Nevertheless, the Diffie-Hellman exchange still has some limitations. Alice and Bob can use this algorithm to generate and distribute a shared secret key, however, it cannot be used to encrypt and decrypt messages. Another drawback of the Diffie-Hellman exchange is that Alice and Bob have no way to verify that they are talking to each other, so the exchange can be subject to a man-in-the-middle attacks. In other words, Diffie-Hellman provides for confidentiality but not for authentication. In this case, authentication can be achieved via the use of digital signatures (see Section 2.7.1) in the Diffie-Hellman message exchanges.

2.7 Authentication Mechanisms

Confidentiality is mandatory to protect a secret. Without authentication, there is no way of validating and identifying the person with whom you share the secret as who he or she claims to be. In the following sections, we will examine additional authentication mechanisms that can help us identify and validate a sender.

2.7.1 Digital signatures

A digital signature is an encrypted message digest that is appended to a document. It is used to validate the identity of the sender and the integrity of the document. Because the entire document must be known before signature generation, digital signatures are not used to provide the message integrity of an ongoing data stream. HMAC (see Section 2.5) is used in this case.

In practical implementations, digital signatures are created by combining hash functions with public-key algorithms. Instead of signing (encrypting) a document, Alice signs (encrypts) the hash of the document to produce the digital signature and appends it to the original document.

Figure 2.5 shows an example of how a digital signature is created. Alice and Bob first need to come to a consensus on a public-key encryption algorithm, generate public/private key pairs, and exchange their respective public keys.

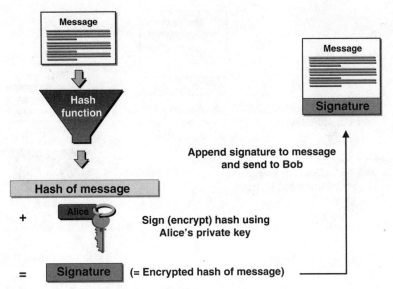

Figure 2.5: Digital signature generation.

They also need to agree on which hash function they will use to generate the digital signature and verify the signature. Alice takes the original document and uses it as input to the hash function to produce an output, which is called the message digest of the document. Alice then signs (encrypts) the message digest with her private key. This encrypted message digest is the digital signature, and it is appended to the original document.

The combination of the document and the digital signature is the message that Alice sends to Bob. Figure 2.6 shows how the digital signature is verified. At the receiving end, Bob separates the received message into the original document and the digital signature. Since the digital signature was encrypted with Alice's private key, Bob can decrypt it with Alice's public key. Bob now has the original hash. Next, Bob takes the document and uses it as input to the same hash function used by Alice. If Bob's calculation of the message digest matches Alice's decrypted message digest, the integrity of the document as well as the authentication of the sender are validated.

Digital signatures require the exchange of public keys beforehand. This exchange of public keys is again susceptible to man-in-the-middle-attack because during the exchange process, there is no way to guarantee that the public keys, which Alice and Bob have received from each other, are genuine. Some imposter (man-in-the-middle) can substitute Alice's public key with his own fabricated public key, therefore, tricking Alice into believing that the public key she has received is actually from Bob. In the next section, we will look at how digital certificates can be used to thwart this kind of attack.

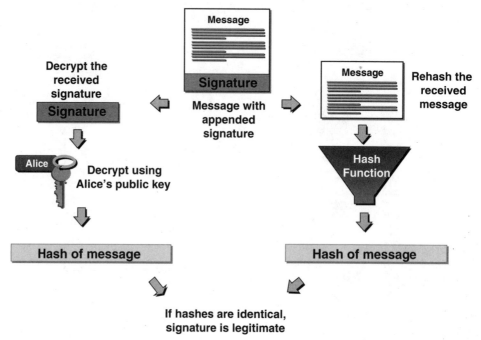

Figure 2.6: Digital signature verification.

2.7.2 Digital certificates

A digital certificate (also commonly referred as public-key certificate) is someone's public key, signed and verified by another trustworthy party. Certificates are used to foil attempts to substitute one's key for another. Typically, a digital certificate uses the X.509 standard format (refer to Chapter 4, Section 4.6 for further details).

Bob's certificate not only contains his public key, it also includes information about him. Bob's certificate is signed and issued by a trusted third-party in whom Alice places absolute trust—a certificate authority (CA). By signing both the key and the information about Bob, the CA certifies that the information about Bob is correct and that the public key truly belongs to Bob.

For instance, Alice wants to verify Bob's public key via a CA before starting to send data. Bob has a valid certificate stored in the CA. Alice requests Bob's digital certificate from the CA. The CA signs the certificate with its private key. Alice has access to the CA's public key and can therefore verify that Bob's certificate signed by the CA is valid. Since Bob's certificate contained his public key, Alice is now certain that the public key is Bob's and no one else's.

2.7.3 Digital envelopes

Digital signature and digital certificate provide proof of identity but they do not provide for privacy. To incorporate confidentiality, we need to put the signed

Figure 2.7: Digital envelope.

document into an envelope and encrypt the envelope. Typically, the envelope contains an encrypted content, and encrypted content-encryption keys for the recipient. The combination of encrypted content and encrypted content-encryption keys for a recipient is referred as a digital envelope (encrypted message) for that recipient.

The process by which enveloped data is constructed by the sender involves the following steps:

- A content-encryption key for a particular content-encryption (symmetric) algorithm is generated at random.
- The content is encrypted with the content-encryption key.
- The content-encryption key is encrypted with the recipient's public key.

The encrypted message and encrypted key are represented together according to the syntax of the public key cryptography standard #7 (PKCS #7). PKCS #7 is a general syntax for data that may be encrypted or signed, such as digital signatures or digital envelopes. The syntax is recursive, so that envelopes can be nested. Digital envelopes are used during CA transactions to ensure the secure distribution of the certificates (PKCS #7 wrapped) to the users.

At the receiving end, the recipient opens the envelope by decrypting the encrypted content-encryption keys with the recipient's private key and decrypting the encrypted content with the recovered content-encryption key. Figure 2.7 shows an example of a digital envelope being sent from Alice to Bob.

2.8 IPSec Protocol Framework

The previous sections discussed encryption, integrity, and authentication in general. The following sections explain how encryption, integrity, and authentication are applied to the IPSec protocol suite.

2.8.1 IPSec

IPSec (RFC 2401) provides security services at the IP layer by enabling a system to select required security protocols, determine the algorithm(s) to use for the service(s), and put in place any cryptographic keys required to provide the requested services. IPSec can be used to protect one or more paths between a pair of hosts, between a pair of security gateways, or between a security gateway and a host. In this case, "security gateway" refers to an intermediate-system that implements IPSec protocols. For instance, a router or a firewall implementing IPSec is a security gateway.

The set of security services that IPSec can provide includes access control, data origin authentication, confidentiality (encryption), and limited traffic flow confidentiality. In addition, IPSec supports two forms of integrity: connectionless and a form of partial sequence integrity. Connectionless integrity is a service that detects modification of an individual IP packet, without regard to the ordering of the packet in a stream of traffic. The form of partial sequence integrity offered in IPSec is referred to as antireplay integrity, and it detects arrival of duplicate IP packets within a constrained window. This is in contrast to connection-oriented integrity, which imposes more stringent sequencing requirements on traffic, for instance, to be able to detect lost or reordered messages. Because these services are provided at the IP layer, they can be used by any higher layer protocol such as TCP, UDP, ICMP, and BGP.

IPSec uses two protocols to provide traffic security: authentication header (AH) and encapsulating security payload (ESP). The AH (RFC 2402) provides connectionless integrity, data origin authentication, and an optional antireplay service. The ESP (RFC 2406) provides confidentiality (encryption), and limited traffic flow confidentiality. It also provides connectionless integrity, data origin authentication, and an antireplay service.

Both AH and ESP are vehicles for access control, based on the distribution of cryptographic keys and the management of traffic flows relative to these security protocols. These protocols may be applied alone or in combination with each other to provide a desired set of security services in IPv4 and IPv6. Each protocol supports two modes of use: transport mode and tunnel mode. In transport mode the protocols provide protection primarily for upper layer protocols; in tunnel mode, the protocols are applied to tunneled IP packets. For both of these protocols, IPSec does not define the specific security algorithms to use, instead it provides an open framework for implementing industry standard algorithms.

2.8.2 Authentication header

The Authentication Header (AH) is used to provide connectionless data integrity and data origin authentication for IP packets. With connectionless data integrity service, the original IP packet is not modified in transit from the source to the destination. Data origin authentication service validates the source of the data. Collectively these joint services are referred as authentication.

The AH is inserted into the IP packet between the IP header and the rest of the packet contents. Immediately preceding the AH, the header will contain the value 51 in its next header (IPv6) or protocol (IPv4) field. AH provides authentication for as much of the IP header as possible, as well as for upper level protocol data. However, some IP header fields may mutate in transit and the value of these fields (such as type of service, fragment offset, time to live, and header checksum) when the packet arrives at the receiver, may not be predictable by the sender. The values of such fields cannot be protected by AH. Thus, the protection provided to the IP header by AH is rather piecemeal.

The AH includes a cryptographic checksum of the IP packet contents, including the parts of the IP header that are immutable (predictable) in transit. The default cryptographic algorithms for computing the checksum are: HMAC-MD5 and HMAC-SHA-1 (see Section 2.5). By taking a received packet, computing the same cryptographic checksum, and comparing it with the checksum value received (found in the AH), the receiver can verify the packet has not been changed in transit.

In addition, AH provides an antireplay service that can be used to defy a denial-of-service (DoS) attack based on an attacker intercepting a sequence of packets and then replaying these intercepted packets. If the arriving packets fall outside the antireplay window, IPSec will reject these packets. This optional antireplay service requires the sender to increment the sequence number used for antireplay, and the service is effective provided that the receiver checks the sequence number.

AH may be used by itself, in combination with the IP encapsulating security payload (ESP), or in a nested fashion through the use of tunnel mode. Security services can be provided between a pair of communicating hosts, between a pair of communicating security gateways, or between a security gateway and a host. ESP may be used to provide the same security services together with a confidentiality (encryption) service. The primary difference for the authentication provided by AH and ESP is the scope of the protection. Specifically, ESP does not protect any IP header fields unless ESP is in the tunnel mode. Note that the AH does not have the capability to keep the contents of the packets confidential. For confidentiality, ESP must be used.

2.8.3 Encapsulating security payload

ESP is used to provide confidentiality, data origin authentication, connectionless integrity, an antireplay service (a form of partial sequence integrity), and limited traffic flow confidentiality. The set of services provided depends on options selected at the time the security association is established, and on the details of its implementation.

Confidentiality is achieved through encryption. ESP supports a wide range of symmetric encryption algorithms for the bulk encryption of data. The default algorithms are: DES and 3DES (see Section 2.4.1). Confidentiality may be selected independent of all other services. However, the application of confidentiality without integrity/authentication (either in ESP or separately in AH) may subject traffic to certain forms of active attacks (such as the header cut-and-paste attack) that could undermine the confidentiality service.

Data origin authentication and connectionless integrity are joint services and are offered as an option together with (optional) confidentiality. The antireplay service may be selected only if data origin authentication is selected, and its selection is solely at the discretion of the receiver. Traffic flow confidentiality requires the selection of tunnel mode, and is most effective if implemented at a security gateway, where traffic aggregation may be able to mask true source-destination patterns. Note that although both confidentiality and authentication are optional, at least one of them must be selected.

ESP may be used by itself, in combination with the IP authentication header (AH), or in a nested fashion through the use of tunnel mode. Security services can be provided between a pair of communicating hosts, between a pair of communicating security gateways, or between a security gateway and a host. The ESP can appear anywhere after the IP header and before the final transport layer (layer-4) protocol. The header immediately preceding the ESP header will contain the value 50 in its next header (IPv6) or protocol (IPv4) field. ESP comprises an unencrypted header followed by the encrypted data payload. The ESP header is inserted after the IP header and before the upper layer protocol header in transport mode or before an encapsulated IP header in tunnel mode. These modes are described in the next section.

2.8.4 IPSec modes

IPSec can operate in one of two modes to accommodate the different types of connections: tunnel mode and transport mode. Figure 2.8 shows the packet format with AH in the transport and tunnel modes. Likewise, Figure 2.9 illustrates the packet format with ESP in the transport and tunnel modes.

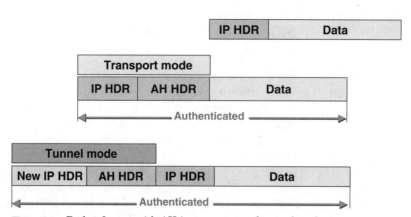

Figure 2.8: Packet format with AH in transport and tunnel modes.

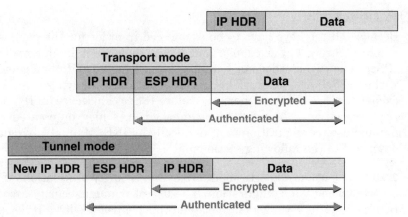

Figure 2.9: Packet format with ESP in transport and tunnel modes.

With tunnel mode, the entire original IP packet is encapsulated with the AH or ESP, and it becomes the payload in a new IP packet with a new unencrypted IP header placed around it. The information in the new IP header is used to route the secure packet from source to destination. This mode allows a secure gateway to act as an IPSec proxy.

Specifically, when implementing tunnel mode ESP, the secure gateway performs the encryption on behalf of the hosts. Over at the host end, no modification is required on the operating system or running applications. The source secure gateway encrypts the packets and forwards them along the IPSec tunnel. The destination-secure gateway decrypts as well as recovers the original IP packet and forwards it on to the destination host.

In other words, tunnel mode is used when one or both sides of the IPSec connection are a secure gateway and the actual destination hosts behind it do not support IPSec. Therefore, the new IP header has the source address of the gateway itself. With tunnel mode operating between two security gateways, the original source and destination addresses can be concealed through the use of encryption (for ESP).

With transport mode, the AH or ESP is placed after the original IP header. In this mode, bandwidth is conserved because there are no encrypted IP headers or IP options and it adds only a few bytes to each packet. In transport mode ESP, only the IP payload is encrypted, and the original IP headers remain intact. This mode also permits devices on the public network to see the final source and destination addresses of an IP packet. This facilitates special processing (for instance, quality of service) in the intermediate network based on the information on the IP header. Nevertheless, the layer-4 header will be encrypted (for ESP), restricting the thorough examination of the packet. Transport mode can be used when both end hosts support IPSec.

2.8.5   IPSec key management

Because all the keys have to be exchanged in order for the parties to communicate securely, key exchange and key management is an important part of IPSec. The two methods of handling key exchange and management specified within IPSec are: manual keying and automated keying with Internet key exchange (IKE). Put another way, before IPSec can secure an IP packet, a security association (SA) must be established. SAs may be created manually by manual user configuration or dynamically by IKE. Manual keying and IKE are described in the following sections.

2.8.5.1   Manual keying.   In manual keying, security associations (SAs) need to be established via configuration. The use of manual security associations is a result of a prior arrangement between the users of the local IPSec peer and the remote IPSec peer. In manual keying, there is no negotiation of security associations, so the configuration information in both peers must be the same in order for traffic to be processed successfully by IPSec. Some of the configuration information includes:

- Defining both the inbound and outbound IPSec session keys for the SAs associated with an AH transform. The AH session key length is 128 bits with MD5, and 160 bits with SHA.

- Defining both the inbound and outbound IPSec session keys for the SAs associated with ESP transform. The ESP session key length is 56 bits with DES, and 168 bits with 3DES.

- Specifying both the inbound and outbound SPI (Security Parameter Index) associated with the SAs for an AH transform. The SPI is an arbitrary number in the range of 256 to 4,294,967,295.

- Specifying both the inbound and outbound SPI associated with the SAs for an ESP transform.

Security associations that are established manually have an infinite lifetime. If a session key is modified, the SA using the key will be deleted and reinitialized. Therefore, the session keys configured at the local peer must match the session keys configured at the remote peer.

For manual keying, it is very difficult and tedious to ensure the SA values match between the peers. Moreover, Diffie-Hellman key exchange has an edge over manual configuration, as it is more secure. Some other disadvantages of using manual keying include:

- Manual keying is not scalable and is generally insecure because of the difficulty in manually generating secure keying material.

- Manually established SAs do not expire.

- Crypto access lists used to determine the interesting traffic in order to trigger the IPSec operation are limited to a single permit entry, and subsequent entries are ignored. Therefore, the SAs established are only for a single data stream.

In the next section, we shall look at a more dynamic and secure way of managing the IPSec keys and SAs using IKE.

2.8.5.2 Internet key exchange. IKE (RFC 2409) automatically negotiates IPSec security associations (SAs) and enables IPSec secure communications without the costly and cumbersome manual pre-configuration discussed in the previous section. It is important to note that SAs are negotiated for both IKE and IPSec, and it is IKE itself that facilitates these SA establishments.

IKE is a hybrid protocol which implements the Oakley key exchange and Skeme key exchange inside the Internet security association and key management protocol (ISAKMP) framework.

The Oakley key exchange describes a series of key exchanges called "modes" and details the services provided by each (for instance, perfect forward secrecy for keys, identity protection, and authentication). IKE borrowed this idea of different modes and incorporated these modes into its exchanges.

The Skeme key exchange describes a versatile key exchange technique, which provides anonymity, repudiation, and quick key refreshment. IKE does not implement the entire SKEME protocol, but only borrows the method of public key encryption for authentication and its concept of fast rekeying using an exchange of nonces (pseudo-random numbers).

ISAKMP (RFC 2408) provides a framework for authentication and key exchange but does not define them. ISAKMP is designed to be key exchange independent. In other words, it is designed to support various key exchanges. ISAKMP uses UDP port 500 for negotiation.

IKE is based on the ISAKMP, Oakley, and Skeme protocols. ISAKMP describes the phases of negotiation, Oakley and Skeme define the method to establish an authenticated key exchange; and IKE presents different exchanges as modes that operate in one of two phases.

Phase 1 is initiated when two peers want to establish a secure, authenticated channel with which to communicate. Either Oakley's main mode or aggressive mode is used in phase 1. The result of main mode or aggressive mode is the authenticated bidirectional IKE SA, which is a secure tunnel between the two peers, and its keying material. Note that both peers must have a shared session key in order to secure (encrypt) the IKE tunnel between them. The Diffie-Hellman key exchange is used to derive the keying material for this common session key. Note that the Oakley protocol incorporated by IKE defines five groups (Oakley groups) to do the Diffie-Hellman exchange. Cisco IOS supports group 1 (a 768-bit key) and group 2 (a 1024-bit key).

Phase 2 uses Oakley's Quick Mode to establish (possibly multiple) SAs on behalf of IPSec, which requires a separate round of key material generation and security parameters negotiation. IPSec uses a different shared session key than IKE. The IPSec session key can be derived by using Diffie-Hellman again to

1. Alice sends interesting traffic to Bob and initiates IPSec process triggered by ACL.

2. Router A and B negotiate an IKE phase 1 session and establish a secure IKE tunnel.

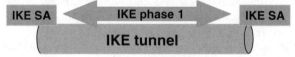

3. Router A and B negotiate an IKE Phase 2 session through the secure IKE tunnel and establish an IPSec tunnel.

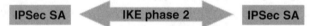

4. Interesting traffic between Alice and Bob is exchanged via the IPSec tunnel.

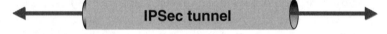

5. The IPSec tunnel is terminated.

Figure 2.10:   Five steps of IPSec operation.

ensure PFS, or by doing a fast rekeying (key refreshing) for the session key derived from the original Diffie-Hellman exchange that created the IKE SA by hashing it with nonces. The result of Quick Mode is two or four (depending on whether AH and/or ESP was used) unidirectional IPSec Security Associations and their keying material. The details of the exchanges for phase 1 and phase 2 are discussed in the following section.

2.9   IPSec Operation

The objective of IPSec is to protect a selected data stream with the desired security services and algorithms. The IPSec's operation can be divided into five primary steps as illustrated in Figure 2.10, and are described in the following sections.

2.9.1   Step 1: Interesting traffic

Interesting traffic initiates the IPSec process. Traffic is considered interesting when it needs to be protected during transit to a specific destination. A security policy is used to determine what traffic needs to be protected and what traffic can be sent in the clear. The policy is then implemented for each particular IPSec peer. Extended access control lists (EACLs) are used to determine the traffic to encrypt. When constructing the (crypto) EACLs, permit statements are used to indicate which selected traffic is to be sent encrypted, and deny statements are used to indicate which selected traffic is to be sent unencrypted. When interesting traffic is generated or transits the local IPSec peer, the peer initiates the next step in the process: negotiating an IKE phase 1 exchange.

2.9.2   Step 2: IKE phase 1

The function of IKE phase 1 is to negotiate IKE policy sets, authenticate the IPSec peers, and set up a secure communications channel between the peers for negotiating IPSec SAs in IKE phase 2. IKE phase 1 happens in two modes: main mode and aggressive mode. Main mode as illustrated in Figure 2.11, has three two-way exchanges (six exchanges) between the initiator and responder.

- First exchange (policy negotiation). The encryption and hash algorithms used to secure the IKE communications are negotiated and agreed upon between the IPSec peers.

- Second exchange (Diffie-Hellman and nonce exchange). A Diffie-Hellman exchange is used to generate shared secret keys and pass nonces (pseudorandom numbers) to the other party, signed, and returned to prove their identity. The shared secret key is used to generate all the other encryption and authentication keys.

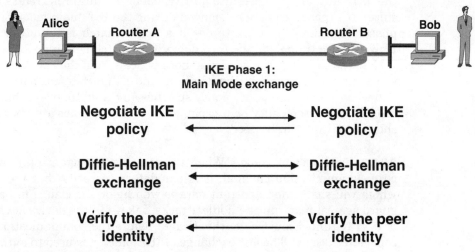

Figure 2.11: IKE phase 1: main mode.

- Third exchange (authentication of exchange and peer). This verifies the other party's identity and is used to authenticate the remote peer. The ultimate objective of main mode is a secure communication channel for subsequent exchanges between the IPSec peers.

The purpose of an aggressive mode is identical to a main mode. The main difference is that aggressive mode establishes the IKE SA in only three exchanges as compared with six exchanges (three two-way exchanges) in main mode. By constricting the number of exchanges, aggressive mode constrains its negotiable options and does not provide identity protection (identities are passed in the clear).

On the first exchange, almost everything is squeezed in: the IKE policy set negotiation; the Diffie-Hellman public key generation; a nonce, which the other party signs; and an identity packet. The responder then sends everything back that is required to complete the exchange. The final thing left is for the initiator to confirm the exchange. Typically, Aggressive Mode is used in remote access environments where the address of the initiator cannot be made known to the responder beforehand, and both parties want to use preshared key authentication.

Note that the outcome of main mode or aggressive mode is the authenticated bidirectional IKE SA and its keying material.

2.9.2.1 IKE policy sets. The encryption, authentication, and other protocols used to secure the IKE communications need to be negotiated in IKE phase 1. The mandatory attributes used by IKE which must be negotiated as part of the IKE SA include: encryption algorithm, hash algorithm, authentication method, and Diffie-Hellman group (Oakley group). Optional attributes such as SA lifetime may also be negotiated. Instead of negotiating each attribute separately, the attributes are grouped into an IKE policy set (also referred as a protection suite). IKE policy sets are exchanged during the IKE main mode first exchange phase. The main mode continues if a policy match is found between peers. However, the tunnel will be torn down if no match is found.

Note that a match is made when both policies from the two peers contain the same encryption, hash, authentication, and Diffie-Hellman parameter values, and when the remote peer's policy specifies an SA lifetime less than or equal to the lifetime in the policy being compared. If the lifetimes are not identical, the shorter lifetime will be used.

2.9.2.2 Peer authentication. When transferring important data over an insecure communication channel, it is mandatory to verify that the person with whom you share the important data is who he or she claims to be in order to guard against man-in-the-middle attacks. Likewise, the device at the other end of a VPN tunnel must be authenticated before the communication channel is considered secure. The last exchange of IKE phase 1 is used to authenticate the remote device (or peer). IKE peer authentication methods such as preshared

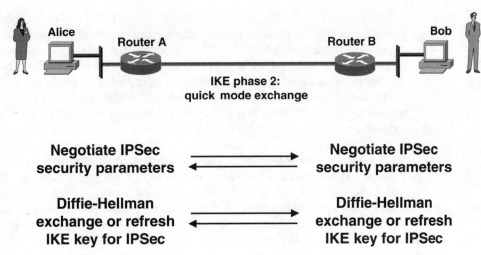

Figure 2.12: IKE phase 2: quick mode.

keys, RSA encrypted nonces, and RSA signatures (digital certificates) are discussed later in chapters 3 and 4.

2.9.3 Step 3: IKE phase 2

The function of IKE phase 2 as illustrated in figure 2.12, is to negotiate the IPSec security parameters used to secure the IPSec tunnel and sets up matching IPSec SAs between the peers. These security parameters are used to protect data and messages exchanged between IPSec endpoints. Specifically, IKE phase 2 performs the following tasks:

- Negotiates IPSec security parameters and transform sets
- Creates IPSec security associations (SAs)
- Ensures security by periodically renegotiating IPSec SAs
- Carries out an additional Diffie-Hellman exchange when necessary

IKE phase 2 has one mode, called quick mode. Quick mode is not a complete exchange itself in that it is bound to an IKE Phase 1 exchange, but is used as part of the SA negotiation process in IKE phase 2. Quick mode occurs after IKE has established the secure tunnel in phase 1. It negotiates a shared IPSec transform, derives shared secret keying material used for the IPSec security algorithms, and establishes IPSec SAs. Quick Mode is also used to renegotiate a new IPSec SA when the IPSec SA lifetime expires.

In addition, quick mode exchanges nonces that are used to generate new shared secret key material and prevent replay attacks from generating bogus SAs. An optional key exchange (KE) payload can be exchanged to allow for an additional Diffie-Hellman exchange and exponentiation per quick mode.

IPSec uses a different shared secret key than IKE. Base quick mode (without the KE payload) is used to refresh the keying material (by hashing it with the nonces) derived from the Diffie-Hellman exchange in IKE phase 1, which is used to create the shared secret key. This does not provide PFS (refer to Section 2.6.1 for more details on PFS). Using the optional KE payload, an additional Diffie-Hellman exchange is performed and PFS is provided for the keying material. Typically, the local IPSec policy dictates whether to use PFS or a key refresh.

2.9.3.1 IPSec transform sets.   The main purpose of IKE phase 2 is to establish a secure IPSec session between endpoints. To achieve this, each pair of endpoints is required to negotiate the level of security required (for instance, the encryption and authentication algorithms for the session as well as the IPSec Mode). Instead of negotiating each protocol separately, the protocols are grouped into an IPSec transform set. IPSec transform sets are exchanged between peers during quick mode. IPSec session-establishment continues if a match is found between sets. However, the session will be torn down if no match is found.

2.9.3.2 Security associations.   Once a transform set is agreed upon between peers, each VPN peer device then populates the information into a SA database (SADB). The stored information includes the encryption as well as the authentication algorithm, IPSec mode, peer (destination) address, shared session key, and SA lifetime (key lifetime). This information is referred to as the security association (SA). The SA is referenced by a 32-bit number, which is known as a security parameter index (SPI).

Instead of sending the individual parameters of the SA across the tunnel, the sending peer inserts the SPI into the IPSec header. The receiving peer uses the tuple ⟨spi, dst, protocol⟩ (where dst is the destination address in the IP header) to uniquely identify the SA. In other words, when the IPSec peer receives the packet, it looks up the destination address and SPI in its database, and then processes the packet according to the protocols listed under the SPI.

Note that the outcome of quick mode is two unidirectional IPSec SAs when either AH or ESP is used, and four unidirectional IPSec SAs when both AH and ESP are used.

2.9.3.3 SA lifetime.   SA lifetime determines the period of time that a security association is valid. In other words, it also determines how often a shared session key is refreshed or recomputed (PFS). Lifetime may be measured in seconds or kilobytes and is applicable to both the IKE SA and IPSec SAs.

2.9.4 Step 4: Data transfer

After IKE phase 2 has completed and quick mode has established the respective IPSec SAs, data is transferred between IPSec peers based on the IPSec security parameters stored in the SA database. In other words, a secure tunnel is established whereby interesting traffic is encrypted and decrypted using the encryption algorithm(s) specified in the IPSec SA. Note that outbound interesting packets are encrypted and inbound interesting packets are decrypted.

2.9.5 Step 5: Tunnel termination

There are two ways IPSec SAs can terminate. One way is through deleting the SAs explicitly and the other is to let the SAs timeout (exhausting its lifetimes). An SA can time out when a specified number of seconds has elapsed or when a specified number of bytes has passed through the tunnel. When the SAs terminate, the associated keys are also discarded. When subsequent IPSec SAs are required for a flow, IKE initiates a new phase 2, and if necessary, a new phase 1 negotiation. In this case, a successful negotiation will produce new SAs and new keys. Note that new SAs are established before the existing SAs expire, so that an existing traffic flow can continue uninterrupted.

2.10 Summary

Chapter 2 first takes the reader through the different types of VPN services that are provided by IPSec. This is followed by an overview of the encryption concepts and the various encryption algorithms for instance DES and 3DES, which are commonly deployed today. Likewise, hashing concepts and algorithms for example MD5 and SHA are also discussed. We then proceed to explain key exchanges such as the Diffie-Hellman algorithm, and enhanced authentication mechanisms such as digital signatures, digital certificates, and digital envelopes. Next, we describe the IPSec framework and look at how to manage IPSec session keys using manual keying and IKE. The concluding section of the chapter identifies the different phases and exchanges that occur during an IPSec operation.

The following chapters—Chapters 3 to 6 are about implementing IPSec from a small-scale point-to-point VPN to a large-scale hub-and-spoke enterprise network. Note that the device configurations found in the case studies from Chapters 3 to 6 have been developed and tested using Cisco IOS Software Release 12.2(10b) and Cisco 2621 routers.

3

Site-to-Site IPSec VPN Using Preshared Keys

3.1 Chapter Overview

In Chapter 2, the emphasis was on the algorithms and technologies associated with IPSec and IKE. Chapter 3 takes a step further by incorporating an actual IPSec site-to-site implementation in conjunction with the IKE preshared keys peer authentication. The chapter includes the actual site-to-site IPSec VPN design and implementation together with various IPSec configuration caveats. The reader will also be given sufficient exposure to using some of the commands and tools meant for troubleshooting, monitoring, and verifying the operation of an IPSec network. Chapter 3 also covers another site-to-site IPSec VPN implementation variation that uses just dynamic crypto maps and dynamic crypto maps accompanied by the tunnel endpoint discovery (TED) mechanism.

3.2 IKE Peer Authentication Overview

In site-to-site IPSec VPNs, it is important that devices are identified in a secure and manageable manner. The device (peer) on the other end of the VPN tunnel must be authenticated before the communication path is considered secure. In addition, we need to authenticate the data flow between devices to ensure that the data has originated from a legitimate party. There are three peer (data origin) authentication methods:

- Preshared keys: A shared secret key, preconfigured into each prospective peer manually, is used to authenticate the peers.

- RSA signatures: Uses the exchange of digital certificates to authenticate the peers.

- RSA encrypted nonces: Nonces (a random number generated by each peer) are encrypted and exchanged between peers. The two nonces are used during the peer authentication process.

This chapter focuses on implementing site-to-site IPSec VPNs using the preshared keys peer-authentication method. RSA signatures and encrypted nonces are discussed later in Chapter 4.

3.3 Preshared Keys Overview and Types

The same preshared key is configured on each IPSec peer. The preshared key is combined with the Diffie-Hellman key, initiator, and responder's nonce and cookie (this is an 8-byte pseudorandom number unique to each peer and the particular exchange in which it is defined) to form the authentication key. The authentication key is combined with device-specific information and sent through a hash algorithm to obtain HASH_I (initiator's hash value) and HASH_R (responder's hash value). IKE peers authenticate each other by exchanging and decoding HASH_I and HASH_R. If the receiving peer is able to independently recreate the same hash, the peer is authenticated. Each local peer must authenticate its remote peer before the tunnel is considered secure.

The preshared key is used during IKE (ISAKMP) phase 1 (main mode) to authenticate the Diffie-Hellman (D-H) shared secret exchange. There are three types of preshared keys: unique, group, and wildcard. Unique preshared keys are fixed to a specific IP address. They are defined for pairs of peers when the IP addresses of the peers are known initally. Group preshared keys are tied to a group name identity; these are applicable only to remote access. Wildcard preshared keys are not coupled with any unique information to determine a peer's identity. They are valid with any other device, as long as that device is also defined to use the same key value. Any devices that have the key will authenticate successfully. When using wildcard preshared keys, every device in the network uses the same key. If a single device in your network is compromised and the wildcard preshared key has been gathered, then all devices are compromised. Readers are strongly advised to avoid deploying wildcard preshared keys for site-to-site device authentication.

Preshared keys are simple to implement if your secured network is small (fewer than ten nodes). However, they do not scale well with a large or growing network (see Chapter 4 for scalability discussion). Depending on how strong the preshared keys are and how often they are replaced, they may not provide strong device authentication. In fact, preshared key is the least secure of the three different authentication methods.

3.4 Preparing for IKE (ISAKMP) and IPSec

Implementing IPSec encryption can be a complex task. Readers are encouraged to plan in advance to minimize erratic configuration. The IPSec security policy should be based on the overall company security policy.

3.4.1 IKE (ISAKMP) policy

When the IKE negotiation begins, IKE looks for an IKE policy that is the same on both peers. The initiator (peer that initiates the negotiation) will send all its policies to the responder (remote peer), and the responder will seek a match by comparing its own policies in order of priority (highest priority first) with the initiator's received policies until a match is determined. If no suitable match is located, IKE rejects negotiation and IPSec will not be established. If a match is found, IKE will complete the negotiation, and IPSec security associations will be created. A match is made when both the initiator and responder agree on a common (shared) IKE policy. IKE policy or security association (SA) includes the following parameters:

- Message encryption algorithm—DES (default), 3DES
- Message integrity (hash) algorithm—SHA-1 (default), MD5
- Peer authentication method—preshared keys, RSA encrypted nonces, RSA signatures (default)
- Diffie-Hellman group identifier—group 1 (768-bit key, default), group 2 (1024-bit key)
- IKE SA lifetime—86,400 seconds (1 day, default)

These parameters (refer to Chapter 2 for more detail discussion) are referred to as a protection suite. Protection suites are negotiated as a transform unit by exchanging ISAKMP SA payloads. In Cisco IOS, if no IKE policy is explicitly specified, a default (implicit) policy 65535 comprising all the default values listed above will be used. If more than one IKE policy is defined, the initiator will offer all the policies to the responder, and at least one of these policies must match. If more than one policy matches, the stronger match (preferably set to a higher priority) should be selected.

3.4.2 IPSec policy

An IPSec policy defines a combination of IPSec parameters used during IKE phase 2 (Quick Mode) and forms the basis of an offer to another device. The initiator may offer more than one transform set (combinations of AH, ESP, modes) to the responder, who must match at least one of these proposals before IPSec SAs can be established. Note there are two IPSec modes: transport and tunnel. The tunnel mode is the default. Refer to Chapter 2 for more details on IPSec modes.

Lifetimes for the IPSec SAs can also be defined. If a value is not explicitly set, the defaults are applied:

- Seconds: 3600 (1 hour, default)
- Kilobytes: 4,608,000 (10 Mbps for 1 hour, default)

In general, if the lifetimes are not identical, the smaller of the values that is exchanged between the peers will be used. In addition, IPSec uses a different shared key from IKE. The IPSec shared key can be derived by using Diffie-Hellman again to ensure perfect forward secrecy (PFS) or by refreshing the shared secret derived from the original Diffie-Hellman exchange generated in the main mode by hashing it with nonces. PFS generates a new key based on fresh seed material altogether by carrying out a Diffie-Hellman key exchange every time a new IPSec SA needs a key generation. The strength of the Diffie-Hellman key exchange is configurable; group 1 (768-bit key) and group 2 (1024-bit key) are supported. Note that by default PFS is not enabled.

3.4.3   Activating IKE (ISAKMP) and IPSec

The process of establishing the security associations and sending protected and/or authenticated data requires an activation mechanism, which is based on matching the IP address and applications of hosts to be protected at the local and remote peers. A combination of a crypto map and access lists are used to implement this trigger mechanism. In addition, the crypto map is used to define other IPSec parameters such as lifetimes and PFS requirements. There are two different types of crypto map:

1. A static crypto map is used when the IP address of the peer is known beforehand. This map is characterized by the inclusion of the "set peer" statement.

2. Dynamic crypto map is used when the identity of the remote peer is not known beforehand. In this case, so long as the local peer can successfully authenticate the remote peer, the IP address of the remote will be used as the IPSec peering endpoint. In other words, the local peer learns the remote peer's IP address dynamically, which suits remote users very well. The only constraint with this map is that the local router cannot be the initiator of the SA establishment process unless tunnel endpoint discovery (see Section 3.7) is configured. The inclusion of an access list is also optional.

3.5   Case Study 3.1: Preshared Keys Example

3.5.1   Case overview and network topology

This case study illustrates the preshared keys peer authentication between a site-to-site IPSec VPN using Cisco routers. The site-to-site VPN spans two different customer locations: Brussels-R1 and London-R2. EIGRP is configured as the IGP between the two routers. The site-to-site VPN topology is shown in Figure 3.1.

Peer authentication via preshared keys is a nonpublic key option. Using this method, each peer shares a secret key that has been preconfigured into the router. The ability for each side to demonstrate knowledge of this shared secret, without explicitly mentioning it, authenticates the exchange. This method is

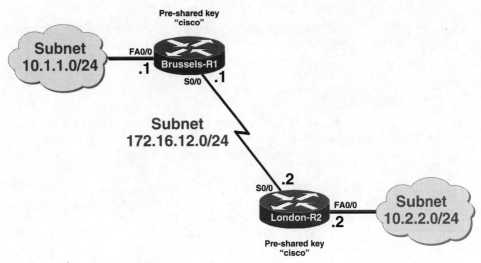

Figure 3.1: Site-to-site VPN topology for Case Study 3.1.

appropriate for small installations, but it will encounter scaling issues when the network grows.

3.5.2 ISAKMP Preshared keys and IPSec configurations

Code Listings 3.1 and 3.2 are the ISAKMP preshared keys and IPSec configurations for Brussels-R1 and London-R2, respectively. A preshared key of "cisco" is used here. In the configurations, comments (in *italics*) precede certain configuration lines to explain them. Note that in Cisco IOS, we can insert comment lines into the configuration by heading the comments with an exclamation mark, which tells the router to ignore these lines.

Code Listing 3.1:   ISAKMP preshared keys and IPSec configuration for Brussels-R1

```
hostname Brussels-R1
!
! Supersede the default policy and use preshared keys for
! peer authentication, and hash algorithm MD5.
! The rest of the IKE SA parameters remain as
! the default values (see Section 3.4.1).
crypto isakmp policy 110
 hash md5
 authentication pre-share
```

```
!
! Specify the preshared key (shared secret key) "cisco".
crypto isakmp key cisco address 172.16.12.2
!
! Specify the ESP settings for IPSec,
! which is later applied to the crypto map.
! The IPSec tunnel (default) mode is used here.
! Note that AH is not implemented in this case study.
! Therefore, AH settings is not defined in this configuration.
crypto ipsec transform-set chapter3 esp-des
!
! Specify the crypto map chapter3 where we define our peer London-R2,
! transform-set chapter3, and our crypto access-list 101.
! Note that PFS is not enabled here,
! and the IPSec SA lifetimes remain as
! the default values (see Section 3.4.2).
crypto map chapter3 10 ipsec-isakmp
 set peer 172.16.12.2
 set transform-set chapter3
 match address 101
!
interface FastEthernet0/0
 ip address 10.1.1.1 255.255.255.0
!
interface Serial0/0
 bandwidth 64
 ip address 172.16.12.1 255.255.255.0
 clockrate 64000
! Applied the crypto map to an interface to activate crypto engine.
 crypto map chapter3
!
! EIGRP is configured as the routing protocol
! between network 10.0.0.0 and 172.16.0.0.
router eigrp 12
 network 10.0.0.0
 network 172.16.0.0
 no auto-summary
!
! This is the crypto access list that we have
! referenced in crypto map chapter3.
! We are encrypting IP traffic between the remote FastEthernet LANs.
access-list 101 permit ip host 10.1.1.1 host 10.2.2.2
```

Code Listing 3.2:   ISAKMP preshared keys and IPSec configuration for London-R2

```
hostname London-R2
!
! Supersede the default policy and use preshared keys for
! peer authentication and hash algorithm MD5.
! The rest of the IKE SA parameters remain as
! the default values (see Section 3.4.1).
crypto isakmp policy 110
 hash md5
 authentication pre-share
!
! Specify the preshared key (shared secret key) "cisco".
crypto isakmp key cisco address 172.16.12.1
!
! Specify the ESP settings for IPSec,
! which is later applied to the crypto map.
! The IPSec tunnel (default) mode is used here.
! Note that AH is not implemented in this case study.
! Therefore, AH settings is not defined in this configuration.
crypto ipsec transform-set chapter3 esp-des
!
! Specify the crypto map chapter3 where we define our peer Brussels-R1,
! transform-set chapter3, and our crypto access-list 102.
! Note that PFS is not enabled here,
! and the IPSec SA lifetimes remain as
! the default values (see Section 3.4.2).
crypto map chapter3 10 ipsec-isakmp
 set peer 172.16.12.1
 set transform-set chapter3
 match address 102
!
interface FastEthernet0/0
 ip address 10.2.2.2 255.255.255.0
!
interface Serial0/0
 bandwidth 64
 ip address 172.16.12.2 255.255.255.0
! Applied the crypto map to an interface to activate crypto engine.
 crypto map chapter3
!
! EIGRP is configured as the routing protocol
! between network 10.0.0.0 and 172.16.0.0.
```

```
router eigrp 12
 network 10.0.0.0
 network 172.16.0.0
 no auto-summary
!
! This is the crypto access-list that we have
! referenced in crypto map chapter3.
! We are encrypting IP traffic between the remote FastEthernet LANs.
access list 102 permit ip host 10.2.2.2 host 10.1.1.1
```

3.5.3 IPSec configuration caveats

- Ensure that your regular access lists (if any) are configured so that ISAKMP (IKE), ESP, and AH traffic are not blocked at interfaces used by IPSec. ISAKMP uses UDP port 500; ESP is assigned IP protocol number 50, and AH is assigned IP protocol number 51.

- Note that encryption takes place only on the output of an interface, and decryption takes place only on input to the interface. The policy for encryption and decryption is symmetrical, which implies that defining one gives you the other automatically.

- With the crypto maps and their associated extended access lists, only the encryption policy is explicitly defined. When the decryption policy matches packets, it uses the same information but reverses source and destination addresses and ports. The "match address X" line in the crypto map is used to depict the encryption of packets leaving an interface. However, packets must also be matched for decryption as they enter the interface. This is done automatically by navigating the access list with the source and destination addresses and ports reversed.

- The access list (X) pointed to by the crypto map should describe traffic in the outbound direction only. IP packets that do not match the access list will still be sent out but not encrypted.

- A "deny" in the access list does not imply that the packet will be dropped. It indicates host or application traffic that should not be encrypted.

- Beware of using "any" in extended access lists, because this causes your traffic to be dropped unless it is headed to the matching decrypting interface.

- You are advised not to use the "permit any any" statement, because this will cause all outbound traffic to be protected and will require protection for all inbound traffic. Otherwise, all inbound packets that lack IPSec protection will be silently dropped. Note that with this statement, IPSec would just drop any packets that do not have IPSec protection. Therefore, you have to be certain which packets you want to protect.

- If you must use the "any" keyword in a permit statement, preamble the statement with a progression of deny statements to filter out any traffic (that would otherwise fall within the permit statement) that you do not want to be protected.

- Ensure IP routing works before attempting to do crypto. If the remote peer does not have IP connectivity to the local peer, you will not be able to have an encryption session with that peer.

- Only two routers are allowed to share a Diffie-Hellman session key. In other words, one router cannot exchange encrypted packets with more than one peer using the same Diffie-Hellman session key.

- Ensure that traffic is set to arrive at an interface that is ready to encrypt or decrypt it. If the encrypted traffic arrives on an interface other than the one with the crypto map applied, it will be dropped silently.

- IP traffic is dropped during key renegotiation each time the key expires. Consider changing the key timeout (IPSec SA lifetime). The one-hour (3600 seconds) default is very short. Try increasing it to one day.

- Choose only the traffic that you want to encrypt as this saves CPU cycles.

- Currently, IPSec does not support encryption of broadcast and multicast traffic. It supports only unicast traffic. If ensuring the integrity of routing updates is a requirement in a network design, use routing protocols such as RIPv2, OSPF, or EIGRP that have a built-in authentication feature. Alternatively, use IPSec in conjunction with GRE (Generic Routing Encapsulation) because GRE supports routing updates, multiprotocol, and multicast traffic. GRE is discussed in more depth in Chapter 6.

3.5.4   Monitoring IP routes between the IPSec routers

EIGRP (in Autonomous System 12) is configured as the routing protocol between Brussels-R1 and London-R2. Since IPSec is running on top of the IP layer, it is important that IP routing is working properly so that the peer and application host addresses are reachable by the IPSec routers. Code listings 3.3 and 3.4 illustrate the IP routing tables for the two IPSec routers.

Code Listing 3.3:   IP routing table for Brussels-R1

```
Brussels-R1#show ip route
<Output Omitted>

     172.16.0.0/24 is subnetted, 1 subnets
C       172.16.12.0 is directly connected, Serial0/0
     10.0.0.0/24 is subnetted, 2 subnets
D       10.2.2.0 [90/40514560] via 172.16.12.2, 03:51:59, Serial0/0
C       10.1.1.0 is directly connected, FastEthernet0/0
```

Code Listing 3.3 indicates that FastEthernet LAN segment (10.2.2.0/24) at London-R2 is reachable from Brussels-R1.

Code Listing 3.4:   IP routing table for London-R2

```
London-R2#show ip route
<Output Omitted>

    172.16.0.0/24 is subnetted, 1 subnets
C       172.16.12.0 is directly connected, Serial0/0
    10.0.0.0/24 is subnetted, 2 subnets
D       10.1.1.0 [90/40514560] via 172.16.12.1, 03:53:48, Serial0/0
C       10.2.2.0 is directly connected, FastEthernet0/0
```

Code Listing 3.4 indicates that FastEthernet LAN segment (10.1.1.0/24) at Brussels-R1 is reachable from London-R2.

3.5.5   Monitoring and verifying IKE (ISAKMP) and IPSec operations

3.5.5.1   Verifying IKE (ISAKMP) and IPSec policies.   Code listings 3.5 to 3.10 look at how you can verify the IKE and IPSec policies that you have configured using the "show crypto isakmp policy," "show crypto map," and "show crypto ipsec transform-set" commands.

Code Listing 3.5:   Brussels-R1 ISAKMP policy

```
Brussels-R1#show crypto isakmp policy
Protection suite of priority 110
   encryption algorithm:   DES - Data Encryption Standard (56 bit keys).
   hash algorithm:       Message Digest 5
   authentication method:  Pre-Shared Key
   Diffie-Hellman group:    #1 (768 bit)
   lifetime:     86,400 seconds, no volume limit
Default protection suite
   encryption algorithm:   DES - Data Encryption Standard (56 bit keys).
   hash algorithm:       Secure Hash Standard
   authentication method:  Rivest-Shamir-Adleman Signature
   Diffie-Hellman group:    #1 (768 bit)
   lifetime:     86,400 seconds, no volume limit
```

Code Listing 3.5 illustrates the ISAKMP policy belonging to Brussels-R1. From the output, we can see that the encryption algorithm, hash algorithm, and authentication method correspond with DES, MD5, and preshared key, respectively, which are in sync with what we have configured for Brussels-R1 in Code Listing 3.1. The Diffie-Hellman group identifier and the ISAKMP lifetime remain as the default values (see Section 3.4.1). Note that a default protection suite, comprising all the default values, is always listed as the last (lowest priority, 65535) policy.

Code Listing 3.6:   Brussels-R1 crypto map

```
Brussels-R1#show crypto map
Crypto Map "chapter3" 10 ipsec-isakmp
   Peer = 172.16.12.2
   Extended IP access list 101
      access list 101 permit ip host 10.1.1.1 host 10.2.2.2
   Current peer: 172.16.12.2
   Security association lifetime: 4,608,000 kilobytes/3600 seconds
   PFS (Y/N): N
 Transform sets={ chapter3, }
```

Code Listing 3.6 illustrates crypto map chapter3 that is implemented in Brussels-R1. From the crypto map, we can see that the remote IPSec peer is London-R2 (172.16.12.2), the crypto access list 101 is used to encrypt IP traffic between the remote FastEthernet LANs, and the IPSec transform set is chapter3 (see Code Listing 3.7). In this instance, PFS is not enabled and the IPSec SA lifetime remains as the default value (see Section 3.4.2).

Code Listing 3.7:   Brussels-R1 IPSec transform set

```
Brussels-R1#show crypto ipsec transform-set chapter3
Transform set chapter3: { esp-des  }
   will negotiate = { Tunnel,   },
```

In Code Listing 3.7, we can see that ESP-DES is the only transform defined for IPSec, and the IPSec tunnel (default) mode is being implemented. Brussels-R1's crypto map (see Code Listing 3.6), together with its IPSec transform, illustrates the overall IPSec policy that is implemented in the router.

Code Listing 3.8:   London-R2 ISAKMP policy

```
London-R2#show crypto isakmp policy
Protection suite of priority 110
    encryption algorithm:   DES - Data Encryption Standard (56 bit keys).
    hash algorithm:       Message Digest 5
```

```
   authentication method: Pre-Shared Key
   Diffie-Hellman group:    #1 (768 bit)
   lifetime:      86400 seconds, no volume limit
Default protection suite
   encryption algorithm:  DES - Data Encryption Standard (56 bit keys).
   hash algorithm:      Secure Hash Standard
   authentication method: Rivest-Shamir-Adleman Signature
   Diffie-Hellman group:    #1 (768 bit)
   lifetime:      86,400 seconds, no volume limit
```

Code Listing 3.8 illustrates the ISAKMP policy belonging to London-R2. From the output, we can see that the encryption algorithm, hash algorithm, and authentication method correspond with DES, MD5, and preshared key, respectively, which are in sync with what we have configured for London-R2 in Code Listing 3.2. The Diffie-Hellman group identifier and the ISAKMP lifetime remain as the default values (see Section 3.4.1). Note that a default protection suite, comprising all the default values, is always listed as the last (lowest priority, 65535) policy.

Code Listing 3.9:   London-R2 crypto map

```
London-R2#show crypto map
Crypto Map "chapter3" 10 ipsec-isakmp
   Peer = 172.16.12.1
   Extended IP access list 102
       access-list 102 permit ip host 10.2.2.2 host 10.1.1.1
   Current peer: 172.16.12.1
   Security association lifetime: 4,608,000 kilobytes/3600 seconds
   PFS (Y/N): N
   Transform sets={ chapter3, }
```

Code Listing 3.9 illustrates crypto map chapter3 which is implemented in London-R2. From the crypto map, we can see that the remote IPSec peer is Brussels-R1 (172.16.12.1), the crypto access list 102 is used to encrypt IP traffic between the remote FastEthernet LANs, and the IPSec transform set is chapter3 (see Code Listing 3.10). In this instance, PFS is not enabled and the IPSec SA lifetime remains as the default value (see Section 3.4.2).

Code Listing 3.10:   London-R2 IPSec transform set

```
London-R2#show crypto ipsec transform-set chapter3
Transform set chapter3: { esp-des   }
   will negotiate = { Tunnel,   },
```

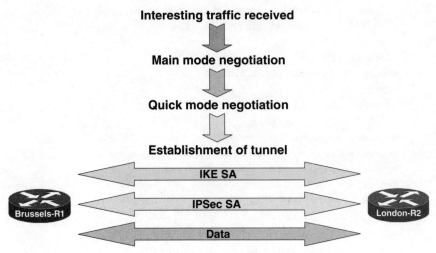

Figure 3.2: Functionality flow chart for IKE and IPSec debugging.

In Code Listing 3.10, we can see that ESP-DES is the only transform defined for IPSec, and the IPSec tunnel (default) mode is being implemented. London-R2's crypto map (see code listing 3.9), together with its IPSec transform, illustrates the overall IPSec policy that is implemented in the router.

3.5.5.2 Monitoring IKE (ISAKMP) and IPSec debug trace. The "debug crypto isakmp" and "debug crypto ipsec" commands are used to observe the IKE and IPSec modes of operation and the security association sequences. The functionality flow chart for these two debug commands is portrayed in Figure 3.2. The flow chart will concur with the actual IKE and IPSec debug trace as illustrated in Code Listing 3.11.

Code Listing 3.11 illustrates the IKE (ISAKMP) and IPSec debug trace when some IP traffic (Ping) is generated from Brussels-R1's end. In the trace output below, italic comments are inserted at specific debugging sequences or "checkpoints" to describe them.

Code Listing 3.11:   IKE and IPSec debug trace at Brussels-R1

```
! Ensure that the "debug crypto isakmp" and
! "debug crypto ipsec" commands are activated.
Brussels-R1#show debug
Cryptographic Subsystem:
  Crypto ISAKMP debugging is on
  Crypto IPSEC debugging is on
```

```
! Generate IP traffic from Brussels-R1 (10.1.1.1) to London-R2 (10.2.2.2)
! using the extended ping command.
Brussels-R1#ping
Protocol [ip]:
Target IP address: 10.2.2.2
Repeat count [5]: 4
Datagram size [100]:
Timeout in seconds [2]:
Extended commands [n]: y
Source address or interface: 10.1.1.1
Type of service [0]:
Set DF bit in IP header? [no]:
Validate reply data? [no]:
Data pattern [0xABCD]:
Loose, Strict, Record, Timestamp, Verbose[none]:
Sweep range of sizes [n]:
Type escape sequence to abort.
Sending 4, 100-byte ICMP Echos to 10.2.2.2, timeout is 2 seconds:
!!!!
Success rate is 100 percent (4/4), round-trip min/avg/max = 40/42/44 ms
! The ping source and destination addresses match access-list 101
! defined in crypto map chapter3.

Brussels-R1#

! Traffic matching an access list specification triggers
! a policy formulation by the sender.
! If more than one policy exists for a particular destination,
! then gather all relevant policies.
! In this case, IPSec requests SAs between 10.1.1.1 and 10.2.2.2,
! on behalf of access list 101, and uses the transform set chapter3.
*Mar  1 04:07:52: IPSEC(sa_request): ,
  (key eng. msg.) src= 172.16.12.1, dest= 172.16.12.2,
! The src is the local crypto endpoint, and the dest is
! the remote crypto endpoint as configured in the crypto map.

    src_proxy= 10.1.1.1/255.255.255.255/0/0 (type=1),
    dest_proxy= 10.2.2.2/255.255.255.255/0/0 (type=1),
! The src_proxy is the source interesting traffic,
! and the dest_proxy is the destination interesting traffic,
! both are defined in access list 101.
    protocol= ESP, transform= esp-des ,
    lifedur= 3600s and 4608000kb,
! The protocol and the transforms are specified by crypto map chapter3,
! which has been triggered, and so are the lifetimes (default values).
```

```
    spi= 0x0(0), conn_id= 0, keysize= 0, flags= 0x4004
! Note that the SPI is still 0.


! IKE negotiate an ISAKMP security association (policy)
! by checking for a matching IKE policy.
! This SA will protect any key and/or
! parameter negotiation required by IPSec.
! This is ISAKMP Phase 1 using Oakley Main Mode.
*Mar  1 04:07:52: ISAKMP (28): beginning Main Mode exchange
! The value 28 (in parenthesis) is the connection ID assigned to IKE.
*Mar 1 04:07:52: ISAKMP (28): sending packet to 172.16.12.2 (I) MM_NO_STATE
*Mar  1 04:07:52:  ISAKMP  (28):  received  packet  from  172.16.12.2  (I)
MM_NO_STATE
*Mar  1 04:07:52: ISAKMP (28): processing SA payload. message ID = 0
*Mar  1 04:07:52:  ISAKMP  (28):  Checking  ISAKMP  transform  1  against
priority 110 policy
! Policy 110 is the only ISAKMP policy configured
! on this local router (initiator).


*Mar  1 04:07:52: ISAKMP:         encryption DES-CBC
*Mar  1 04:07:52: ISAKMP:         hash MD5
*Mar  1 04:07:52: ISAKMP:         default group 1
*Mar  1 04:07:52: ISAKMP:         auth pre-share
! These are the ISAKMP attributes being
! offered by the remote end (responder).


*Mar  1 04:07:52: ISAKMP (28): atts are acceptable. Next payload is 0
! IKE has found a matching policy since policy 110 on this router
! and the atts (attributes) that are offered by the remote end match.


! If the initiator doesn't offer an acceptable proposal,
! that is, there is no matching policy on the responder,
! NO PROPOSAL CHOSEN will result.
! If preshared keys are to be used, but the keys are not the same
! for both parties or a key was not defined, NO PROPOSAL CHOSEN will result.
! The router will attempt to match the default policy 65535
! if the initiator does not match any other ISAKMP policy.


! Preshared key authentication will start now.
! Each peer authenticates the other by using the IKE SA,
! and public/shared keys and nonces are exchanged.
! This is the actual Diffie-Hellman shared secret calculation.
! Process KE, which is the preshared key information,
! then process the nonces and generate the shared key SKEYID,
! which will be used as the actual encryption key.
```

```
*Mar  1 04:07:52: ISAKMP (28): SA is doing preshared key authentication
using id type ID_IPV4_ADDR
*Mar  1 04:07:52: ISAKMP (28): sending packet to 172.16.12.2 (I) MM_SA_SETUP
*Mar  1 04:07:53: ISAKMP (28): received packet from 172.16.12.2 (I)
MM_SA_SETUP
*Mar  1 04:07:53: ISAKMP (28): processing KE payload. message ID = 0
*Mar  1 04:07:53: ISAKMP (28): processing NONCE payload. message ID = 0
*Mar  1 04:07:53: ISAKMP (28): SKEYID state generated

! Next, authenticate the Diffie-Hellman Exchange
! using MD5 as the hash algorithm
! to make sure the payload has not been intercepted and tampered with.
*Mar  1 04:07:53: ISAKMP (28): processing vendor id payload
*Mar  1 04:07:53: ISAKMP (28): speaking to another IOS box!
*Mar  1 04:07:53: ISAKMP (28): ID payload
    next-payload    : 8
    type            : 1
    protocol        : 17
    port            : 500
    length          : 8
*Mar  1 04:07:53  : ISAKMP (28): Total payload length: 12
*Mar  1 04:07:53  : ISAKMP (28): sending packet to 172.16.12.2 (I) MM_KEY_EXCH
*Mar  1 04:07:53  : ISAKMP (28): received packet from 172.16.12.2 (I)
MM_KEY_EXCH
*Mar  1 04:07:53  : ISAKMP (28): processing ID payload. message ID = 0
*Mar  1 04:07:53  : ISAKMP (28): processing HASH payload. message ID = 0
*Mar  1 04:07:53  : ISAKMP (28): SA has been authenticated with 172.16.12.2
! Preshared key authentication has succeeded at this point.
! The ISAKMP SA has been successfully negotiated.

! Now, IKE negotiates to set up the IPSec SA
! by searching for a matching transform set.
! This is ISAKMP phase 2 using Oakley quick mode.
*Mar 1 04:07:53  : ISAKMP (28): beginning quick mode exchange, M-ID of
847079453

! IKE asks for SPIs from IPSec.
! For inbound security associations, IPSec controls its own SPI space.
*Mar  1 04:07:53  : IPSEC(key_engine): got a queue event...
*Mar  1 04:07:53  : IPSEC(spi_response): getting spi 219485627 for SA
    from 172.16.12.2      to 172.16.12.1      for prot 3
! IKE gets the SPI from IPSec to offer to the remote end.

*Mar  1 04:07:53  : ISAKMP (28): sending packet to 172.16.12.2 (I) QM_IDLE
*Mar  1 04:07:53  : ISAKMP (28): received packet from 172.16.12.2 (I) QM_IDLE
```

```
*Mar   1 04:07:53: ISAKMP (28): processing SA payload. message ID = 847079453
*Mar   1 04:07:53: ISAKMP (28): Checking IPSec proposal 1
! Here IKE will process the IPSec attributes offered by the remote end.

*Mar   1 04:07:53: ISAKMP: transform 1, ESP_DES
! This is the protocol offered by the remote end
! according to its transform set.

*Mar   1 04:07:53: ISAKMP:  attributes in transform:
*Mar   1 04:07:53: ISAKMP:      encaps is 1
*Mar   1 04:07:53: ISAKMP:      SA life type in seconds
*Mar   1 04:07:53: ISAKMP:      SA life duration (basic) of 3600
*Mar   1 04:07:53: ISAKMP:      SA life type in kilobytes
*Mar   1 04:07:53: ISAKMP:      SA life duration (VPI) of  0x0 0x46 0x50 0x0
*Mar   1 04:07:53: ISAKMP (28): atts are acceptable.
! A matching IPSec transform set has been found at the two peers.
! Now the IPSec SA can be created (one SA for each direction).

! Here IKE will ask IPSec to validate the IPSec SA proposal
! that it has negotiated with the remote end.
! In this case, it will be validated against
! the one sent initially by the initiator.
*Mar   1 04:07:53: IPSEC(validate_proposal_request): proposal part #1,
   (key eng. msg.) dest= 172.16.12.2, src= 172.16.12.1,
      dest_proxy= 10.2.2.2/255.255.255.255/0/0 (type=1),
      src_proxy= 10.1.1.1/255.255.255.255/0/0 (type=1),
      protocol= ESP, transform= esp-des ,
      lifedur= 0s and 0kb,
      spi= 0x0(0), conn_id= 0, keysize= 0, flags= 0x4

! The initiator must offer at least one matching policy
! relevant  to  the  responder;  otherwise,  NO MATCHING PROPOSAL  will
result.
! The source and destination proxies must equate
! to access lists or policy rules on the parties.
! If the access list rules don't match or overlap,
! INVALID PROXY IDS will result.
! The source and destination proxies are the
! original source and destination hosts/networks.
! Note that the SPI is still 0.

! After the proposal is accepted, IKE generates
! a shared key for encryption for IPSec.
! Generally,  the  original  Diffie-Hellman-generated  shared  secret  key  is
refreshed
```

```
! by combining it with a random value (another nonce) as shown below.
*Mar  1  04:07:53:  ISAKMP (28): processing NONCE payload. message ID =
847079453
*Mar  1 04:07:53 : ISAKMP (28): processing ID payload. message ID = 847079453
*Mar  1 04:07:53 : ISAKMP (28): processing ID payload. message ID = 847079453
*Mar  1 04:07:53 : ISAKMP (28): creating IPSec SAs
*Mar  1 04:07:53 :         inbound SA from 172.16.12.2 to 172.16.12.1
(proxy 10.2.2.2    to 10.1.1.1          )
*Mar  1 04:07:53 :              has spi 219485627 and conn_id 29 and flags 4
*Mar  1 04:07:53 :              lifetime of 3600 seconds
*Mar  1 04:07:53 :              lifetime of 4608000 kilobytes
*Mar  1 04:07:53 :         outbound SA from 172.16.12.1 to 172.16.12.2
(proxy 10.1.1.1    to 10.2.2.2          )
*Mar  1 04:07:53 :              has spi 205267075 and conn_id 30 and flags 4
*Mar  1 04:07:53 :              lifetime of 3600 seconds
*Mar  1 04:07:53 :              lifetime of 4608000 kilobytes
! Two IPSec SAs have been negotiated,
! an incoming SA with the SPI generated by the initiator
! and an outbound SA with the SPI proposed by the responder.
! Crypto engine entries have also been created.

! IKE then notifies IPSec of the new SAs (one SA for each direction).
*Mar  1 04:07:53 : IPSEC(key_engine): got a queue event...

! The following output relates to the inbound SA.
! The conn_id value references an entry in
! the crypto engine connection table.
*Mar  1 04:07:53 : IPSEC(initialize_sas): ,
   (key eng. msg.) dest= 172.16.12.1, src= 172.16.12.2,
     dest_proxy= 10.1.1.1/255.255.255.255/0/0 (type=1),
     src_proxy= 10.2.2.2/255.255.255.255/0/0 (type=1),
     protocol= ESP, transform= esp-des ,
     lifedur= 3600s and 4608000kb,
     spi= 0xD1515BB(219485627), conn_id= 29, keysize= 0, flags= 0x4

! The following output relates to the outbound SA.
! The conn_id value references an entry in
! the crypto engine connection table.
*Mar  1 04:07:53 : IPSEC(initialize_sas): ,
   (key eng. msg.) src= 172.16.12.1, dest= 172.16.12.2,
     src_proxy= 10.1.1.1/255.255.255.255/0/0 (type=1),
     dest_proxy= 10.2.2.2/255.255.255.255/0/0 (type=1),
     protocol= ESP, transform= esp-des ,
     lifedur= 3600s and 4608000kb,
     spi= 0xC3C2083(205267075), conn_id= 30, keysize= 0, flags= 0x4
! IPSec now installs the SA information into its SA database (SADB).
```

```
! Each SA is unidirectional so we need to see
! two SAs created on each participating peer,
! one inbound and one outbound.
*Mar  1 04:07:53: IPSEC(create_sa): sa created,
  (sa) sa_dest= 172.16.12.1, sa_prot= 50,
    sa_spi= 0xD1515BB(219485627),
    sa_trans= esp-des , sa_conn_id= 29
*Mar  1 04:07:53: IPSEC(create_sa): sa created,
  (sa) sa_dest= 172.16.12.2, sa_prot= 50,
    sa_spi= 0xC3C2083(205267075),
    sa_trans= esp-des , sa_conn_id= 30
! The SADB has been updated and the IPSec SAs have been initialized.
! The tunnel is now fully operational.

*Mar  1 04:07:53: ISAKMP (28): sending packet to 172.16.12.2 (I) QM_IDLE
! The IPSec SA has now been successfully negotiated.
! IKE will now go into a state known as QM-IDLE (quick mode idle).
```

3.5.5.3 Verifying IKE (ISAKMP) and IPSec security associations.

In Code Listings 3.12 to 3.17, we conduct a postmodem on the IKE and IPSec SAs after generating IP traffic from Brussels-R1 to London-R2 as illustrated previously in Code Listing 3.11.

Code Listing 3.12:   Verifying IKE SA for Brussels-R1

```
Brussels-R1#show crypto isakmp sa
    dst              src           state       conn-id    slot
172.16.12.2      172.16.12.1       QM_IDLE       28         0
```

Code listing 3.12 lists the current status of the IKE SA for Brussels-R1, which is in the quick mode idle state. The IKE tunnel starts at Brussels-R1 (172.16.12.1) and ends at London-R2 (172.16.12.2). The IKE SA is also assigned a connection ID value of 28 (that references an entry in the crypto engine connection table shown in Code Listing 3.14). All these values coincide with those displayed in the IKE and IPSec debug trace of Code Listing 3.11.

Code Listing 3.13:   Verifying IPSec SA for Brussels-R1

```
Brussels-R1#show crypto ipsec sa

interface: Serial0/0
    Crypto map tag: chapter3, local addr. 172.16.12.1
```

```
local   ident (addr/mask/prot/port): (10.1.1.1/255.255.255.255/0/0)
remote ident (addr/mask/prot/port): (10.2.2.2/255.255.255.255/0/0)
current_peer: 172.16.12.2
  PERMIT, flags={origin_is_acl,}
 #pkts encaps: 4, #pkts encrypt: 4, #pkts digest 0
 #pkts decaps: 4, #pkts decrypt: 4, #pkts verify 0
 #send errors 0, #recv errors 0

  local crypto endpt.: 172.16.12.1, remote crypto endpt.: 172.16.12.2
 path mtu 1500, media mtu 1500
 current outbound spi: C3C2083

 inbound esp sas:
  spi: 0xD1515BB(219485627)
    transform: esp-des ,
    in use settings ={Tunnel, }
    slot: 0, conn id: 29, crypto map: chapter3
    sa timing: remaining key lifetime (k/sec): (4607999/3450)
    IV size: 8 bytes
    replay detection support: N

 inbound ah sas:


 outbound esp sas:
  spi: 0xC3C2083(205267075)
    transform: esp-des ,
    in use settings ={Tunnel, }
    slot: 0, conn id: 30, crypto map: chapter3
    sa timing: remaining key lifetime (k/sec): (4607999/3450)
    IV size: 8 bytes
    replay detection support: N

 outbound ah sas:
```

In Code Listing 3.13, the Brussels-R1's IPSec SA states that four interesting data packets, from local host 10.1.1.1 to remote host 10.2.2.2, have been encrypted, and another four interesting packets, from remote host 10.2.2.2 to local host 10.1.1.1, have been decrypted through interface Serial0/0 where crypto map chapter3 is applied. The IPSec tunnel is established from local crypto endpoint 172.16.12.1 to remote crypto endpoint 172.16.12.2.

In addition, the inbound ESP SA has been populated into the SADB with information such as ESP-DES transform, IPSec tunnel mode, SPI value 219485627, connection ID value 29 (that references an entry in the crypto engine connection table shown in code listing 3.14), and crypto map chapter3. Likewise, outbound ESP SA has been populated into the SADB with information such as ESP-DES transform, IPSec tunnel mode, SPI value 205267075, connection ID value 30 (that references an entry in the crypto engine connection table shown in Code Listing 3.14), and crypto map chapter3. Notice that no inbound or outbound AH information is displayed because the AH protocol is not implemented in this case study.

Once again, all these values coincide with those displayed in the IKE and IPSec debug trace as shown Code Listing 3.11.

Code Listing 3.14:   Brussels-R1 crypto engine connection table

---

```
Brussels-R1#show crypto engine connection active

  ID Interface     IP-Address     State   Algorithm       Encrypt   Decrypt
  28 no idb        no address     set     DES_56_CBC            0         0
  29 Serial0/0     172.16.12.1    set     DES_56_CBC            0         4
  30 Serial0/0     172.16.12.1    set     DES_56_CBC            4         0
```

Code Listing 3.14 illustrates the crypto engine connection table for Brussels-R1. From the table, we can see that connection ID 28 belongs to the IKE SA in Code Listing 3.12, and connection ID 29 and 30 belong to the two ESP SAs in Code Listing 3.13. Note that the outbound SA encrypts four data packets, and the inbound SA decrypts another four, concurring with Code Listing 3.13.

Code Listing 3.15:   Verifying IKE SA for London-R2

```
London-R2#show crypto isakmp sa
     dst              src           state       conn-id     slot
 172.16.12.2      172.16.12.1      QM_IDLE        29          0
```

Code Listing 3.15 lists the current status of the IKE SA for London-R2, which is in the quick mode idle state. The IKE tunnel starts at Brussels-R1 (172.16.12.1) and ends at London-R2 (172.16.12.2). The IKE SA is also assigned a connection ID value of 29 (that references an entry in the crypto engine connection table shown in Code Listing 3.17).

Code Listing 3.16:   Verifying IPSec SA for London-R2

```
London-R2#show crypto ipsec sa

interface: Serial0/0
    Crypto map tag: chapter3, local addr. 172.16.12.2
```

```
local  ident (addr/mask/prot/port): (10.2.2.2/255.255.255.255/0/0)
remote ident (addr/mask/prot/port): (10.1.1.1/255.255.255.255/0/0)
current_peer: 172.16.12.1
  PERMIT, flags={origin_is_acl,}
 #pkts encaps: 4, #pkts encrypt: 4, #pkts digest 0
 #pkts decaps: 4, #pkts decrypt: 4, #pkts verify 0
 #send errors 0, #recv errors 0

 local crypto endpt.: 172.16.12.2, remote crypto endpt.: 172.16.12.1
 path mtu 1500, media mtu 1500
 current outbound spi: D1515BB

 inbound esp sas:
  spi: 0xC3C2083(205267075)
    transform: esp-des ,
    in use settings ={Tunnel, }
    slot: 0, conn id: 30, crypto map: chapter3
    sa timing: remaining key lifetime (k/sec): (4607999/3412)
    IV size: 8 bytes
    replay detection support: N

 inbound ah sas:

 outbound esp sas:
  spi: 0xD1515BB(219485627)
    transform: esp-des ,
    in use settings ={Tunnel, }
    slot: 0, conn id: 31, crypto map: chapter3
    sa timing: remaining key lifetime (k/sec): (4607999/3412)
    IV size: 8 bytes
    replay detection support: N

 outbound ah sas:
```

In Code Listing 3.16, the London-R2's IPSec SA states that four interesting data packets, from local host 10.2.2.2 to remote host 10.1.1.1, have been encrypted, and another four interesting packets, from remote host 10.1.1.1 to local host 10.2.2.2, have been decrypted through interface Serial0/0 where crypto map chapter3 is applied. The IPSec tunnel is established from local crypto endpoint 172.16.12.2 to remote crypto endpoint 172.16.12.1.

In addition, the inbound ESP SA has been populated into the SADB with information such as ESP-DES transform, IPSec tunnel mode, SPI value 205267075, connection ID value 30 (that references an entry in the crypto engine connection table listed in Code Listing 3.17), and crypto map chapter3. Likewise, outbound ESP SA has been populated into the SADB with information such as ESP-DES transform, IPSec tunnel mode, SPI value 219485627, connection ID value 31 (that references an entry in the crypto engine connection table listed in Code Listing 3.17), and crypto map chapter3. Notice that no inbound or outbound AH information is displayed because the AH protocol is not implemented in this case study.

Code Listing 3.17:   London-R2 crypto engine connection table

---

```
London-R2#show crypto engine connection active

ID Interface      IP-Address      State  Algorithm      Encrypt  Decrypt
29 no idb         no address      set    DES_56_CBC          0        0
30 Serial0/0      172.16.12.1     set    DES_56_CBC          0        4
31 Serial0/0      172.16.12.1     set    DES_56_CBC          4        0
```

Code Listing 3.17 illustrates the crypto engine connection table for London-R2. From the table, we can see that connection ID 29 belongs to the IKE SA in Code Listing 3.15, and connection ID 30 and 31 belong to the two ESP SAs in Code Listing 3.16. Note that the outbound SA encrypts four data packets, and the inbound SA decrypts another four, concurring with Code Listing 3.16.

3.6   Case Study 3.2: Dynamic Crypto Map Deployment

3.6.1   Dynamic crypto map primer

Dynamic crypto maps simplify configuration because a crypto map statement is not required for each IP address range. They are deployed in networks where the remote peers are unknown. An example of this is remote clients using dialup, cable, or DSL, who acquire dynamically assigned IP addresses. In this case, the remote clients need to authenticate themselves to the local router's IKE by a mechanism other than an IP address, such as a fully qualified domain name (FQDN). Once authenticated, the security association request can be processed against a dynamic crypto map that is set up to accept requests matching the local policy that is specified in the client's IPSec configuration.

Dynamic crypto maps are available only for use by IKE. A dynamic crypto map entry is essentially a crypto map entry without all the parameters configured. It acts as a policy template where the missing parameters are later dynamically configured (as the result of an IPSec negotiation) to match a remote peer's requirements. This allows remote peers to exchange IPSec traffic with the router even if the router does not have a crypto map entry specifically configured to meet all of the remote peer's requirements.

A dynamic crypto map can be originated only from the remote end. It is used when a remote peer tries to initiate an IPSec security association with the router. In other words, dynamic crypto maps only accept incoming IKE requests and cannot initiate IKE requests; the router cannot use them to initiate new IPSec security associations with remote peers.

As the local router (which is configured with the dynamic crypto map) cannot be the initiator of the SA establishment process, we cannot be certain that a tunnel will always exist between the remote device and the local router. The network time protocol (NTP) can be configured on the local router (NTP server) and remote peers (NTP clients) to alleviate this problem. When a remote peer generates NTP traffic to the local router that is configured as the timeserver, it forces IPSec tunnel establishment from the remote end. This work-around implementation allows the use of dynamic crypto maps, at the same time, ensuring the constant existence of an IPSec tunnel.

In addition, the reader should take into consideration that even though dynamic crypto maps ease configuration management by accepting IKE requests from any IP address, they also weaken rather than reinforce VPN security. For instance, if dynamic crypto maps are used with wildcard preshared keys, you will be unable to verify the initiator's identity. Since static crypto maps are tied to the static IP addresses of the remote peers, they are comparatively more secure than dynamic crypto maps.

3.6.2 Dynamic crypto map configurations

Even though dynamic crypto maps are generally deployed in environments where the remote peers' IP addresses remain unknown, for illustrative purposes in this case study we emulate Brussels-R1 as the remote user and configure the dynamic crypto map on London-R2 as illustrated in Figure 3.3.

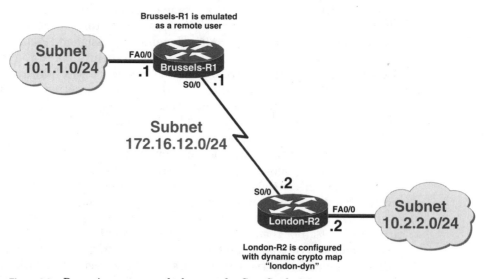

Figure 3.3: Dynamic crypto map deployment for Case Study 3.2.

This case study is derived directly from Case Study 3.1, and the IPSec configuration for Brussels-R1 remains unchanged as before. Only the configuration for London-R2 is modified as illustrated in Code Listing 3.18. Note that dynamic crypto map does not eliminate the need to predefine the remote peer's IP address for IKE authentication when preshared keys are used.

Code Listing 3.18:   London-R2 dynamic crypto map configuration

---

```
hostname London-R2
!
crypto isakmp policy 110
 hash md5
 authentication pre-share
crypto isakmp key cisco address 172.16.12.1
!
! Define IPSec transform sets transform1,
! transform2, and transform3
crypto ipsec transform-set transform1 ah-sha-hmac
crypto ipsec transform-set transform2 esp-null esp-md5-hmac
crypto ipsec transform-set transform3 esp-des
!
! Create a dynamic crypto map london-dyn.
! Specify transform1, transform2, and transform3
! for the crypto map entry.
! These transform sets are listed in order
! of priority (highest priority first).
! Generally, this is the only configuration statement
! required in dynamic crypto map entries.
! Even though access lists are optional, we still
! configure access list 102 in this case.
! This access list determines which traffic
! IPSec should protect and which it should not.
! If this is configured, the data flow identity
! proposed by the IPSec peer must fall
! within a permit statement for this crypto access list.
! If an access list is not configured,
! the router will accept any data flow identity
! proposed by the IPSec peer.
! However, if an access list is configured
! but it does not exist or is empty,
! the router will drop all packets.
crypto dynamic-map london-dyn 10
 set transform-set transform1 transform2 transform3
 match address 102
```

```
!
! Add the dynamic crypto map london-dyn
! into (static) crypto map chapter3.
crypto map chapter3 10 ipsec-isakmp dynamic london-dyn
!
interface FastEthernet0/0
 ip address 10.2.2.2 255.255.255.0
!
interface Serial0/0
 bandwidth 64
 ip address 172.16.12.2 255.255.255.0
! Applied the crypto map to Serial0/0.
 crypto map chapter3
!
router eigrp 12
 network 10.0.0.0
 network 172.16.0.0
 no auto-summary
!
! This is the crypto access list that we have
! referenced in crypto dynamic map london-dyn.
! We are encrypting IP traffic between
! the remote FastEthernet LANs.
access-list 102 permit ip host 10.2.2.2 host 10.1.1.1
```

3.6.3 Verifying dynamic crypto map operations

Code Listing 3.19:   Verifying dynamic crypto map operations

```
! Since London-R2 is only configured with a dynamic crypto map,
! it will not be able to initiate IKE requests. As such,
! we generate the IP traffic at Brussels-R1 end.
! But first we need to do some housekeeping by clearing the SAs
! that are previously established in Case Study 3.1.

! Clear active IKE connections.
Brussels-R1#clear crypto isakmp
London-R2#clear crypto isakmp

! Restart all IPSec SAs so they will use
! the most current configuration settings.
Brussels-R1#clear crypto sa
London-R2#clear crypto sa
```

```
! The crypto engine connection table for Brussels-R1 is now empty.
Brussels-R1#show crypto engine connection active


   ID Interface      IP-Address     State   Algorithm        Encrypt  Decrypt


! The crypto engine connection table for London-R2 is now empty.
London-R2#show crypto engine connection active


   ID Interface      IP-Address     State   Algorithm        Encrypt  Decrypt


! Generate IP traffic from Brussels-R1 (10.1.1.1)
! to London-R2 (10.2.2.2) using the extended ping command.
Brussels-R1#ping
Protocol [ip]:
Target IP address: 10.2.2.2
Repeat count [5]: 4
Datagram size [100]:
Timeout in seconds [2]:
Extended commands [n]: y
Source address or interface: 10.1.1.1
Type of service [0]:
Set DF bit in IP header? [no]:
Validate reply data? [no]:
Data pattern [0xABCD]:
Loose, Strict, Record, Timestamp, Verbose[none]:
Sweep range of sizes [n]:
Type escape sequence to abort.
Sending 4, 100-byte ICMP Echos to 10.2.2.2, timeout is 2 seconds:
!!!!
Success rate is 100 percent (4/4), round-trip min/avg/max = 40/43/44 ms
Brussels-R1#
! Brussels-R1 receives a response (ICMP echo-reply)
! from London-R2, indicating that the dynamic crypto map
! configured for London-R2 is functioning properly.


! A quick glance at Brussels-R1 crypto engine connection table
! indicates that IPSec encryption has indeed taken place.
Brussels-R1#show crypto engine connection active


   ID Interface      IP-Address     State   Algorithm        Encrypt  Decrypt
   33 no idb         no address     set     DES_56_CBC          0        0
   34 Serial0/0      172.16.12.1    set     DES_56_CBC          0        4
   35 Serial0/0      172.16.12.1    set     DES_56_CBC          4        0
```

```
! Likewise, London-R2 crypto engine connection table indicates
! that IPSec encryption is working fine.
London-R2#show crypto engine connection active

  ID  Interface     IP-Address      State   Algorithm      Encrypt   Decrypt
  37  no idb        no address      set     DES_56_CBC        0         0
  38  Serial0/0     172.16.12.1     set     DES_56_CBC        0         4
  39  Serial0/0     172.16.12.1     set     DES_56_CBC        4         0

! From the above results, we can conclude that the dynamic crypto map
! in London-R2 has served its purpose and is working properly.
```

Code Listing 3.19 illustrates the sequence of steps that we have performed to verify whether the dynamic crypto map configuration in Code Listing 3.18 is working properly.

Code Listing 3.20:   London-R2 IPSec transform sets

```
London-R2#show crypto ipsec transform-set
Transform set transform1: { ah-sha-hmac   }
   will negotiate = { Tunnel,  },

Transform set transform2: { esp-null esp-md5-hmac   }
   will negotiate = { Tunnel,  },

Transform set transform3: { esp-des   }
   will negotiate = { Tunnel,  },
```

Code Listing 3.20 illustrates the transform sets that we have defined in London-R2. The reader is encouraged to specify multiple (up to three) transform sets in a crypto map entry to match a broader range of remote peers with diverse requirements (local policies).

Code Listing 3.21:   London-R2 dynamic crypto map template

```
London-R2#show crypto dynamic-map tag london-dyn
Crypto Map Template"london-dyn" 10
    Extended IP access list 102
        access-list 102 permit ip host 10.2.2.2 host 10.1.1.1
    Current peer: 0.0.0.0
    Security association lifetime: 4,608,000 kilobytes/3600 seconds
    PFS (Y/N): N
    Transform sets={ transform1, transform2, transform3, }
```

Code Listing 3.21 illustrates the dynamic crypto map template london-dyn that is configured for London-R2. Transform sets and crypto access list are the only parameters configured for this dynamic crypto map set. A dynamic crypto map entry is essentially a policy template without all the parameters configured. This allows remote peers to exchange IPSec traffic with the router even if the router does not have a crypto map entry specifically configured to meet all of the remote peer's requirements.

Code Listing 3.22:   Verifying London-R2 dynamic crypto map

```
London-R2#show crypto map

Crypto Map "chapter3" 10 ipsec-isakmp
   Dynamic map template tag: london-dyn

Crypto Map "chapter3" 20 ipsec-isakmp
   Peer = 172.16.12.1
   Extended IP access list
       access list  permit ip host 10.2.2.2 host 10.1.1.1
      dynamic (created from dynamic map london-dyn/10)
   Current peer: 172.16.12.1
   Security association lifetime: 4,608,000 kilobytes/3600 seconds
   PFS (Y/N): N
   Transform sets={ transform3, }
```

As illustrated in Code Listing 3.22, a dynamic crypto map set is included by reference as part of a crypto map set. When London-R2 accepts Brussels-R1's request, at the point that it installs the new IPSec SAs, it also installs a temporary crypto map entry created from dynamic map london-dyn. This entry is filled in with the outcome of the negotiation. At this point, London-R2 performs normal processing, using this temporary crypto map entry as a regular entry. The temporary crypto map entry is removed once all the corresponding security associations expire.

Note that when a crypto map entry with the lowest priority (highest sequence numbers) references a dynamic crypto map set, it becomes somewhat like a default crypto map entry that will be examined only when the other (static) map entries are not successfully matched.

3.7   Tunnel endpoint discovery

Tunnel endpoint discovery (TED) is an enhancement to the IPSec feature. The limitation with dynamic crypto map is that the local router can never be the initiator of the SA establishment process. TED eases this constraint. Any device with the TED-based dynamic crypto map can initiate an IKE conversation.

TED, use in conjunction with a dynamic crypto map, allows IPSec to scale to large networks by reducing multiple encryptions, decreasing the setup time, and allowing for simpler configurations on participating peer routers. Each node has a clear-cut configuration that defines the local network that the router is protecting and the IPSec transforms that are required.

TED uses a discovery probe (a special IKE packet that uses UDP port 500), sent from the initiator, to determine which remote IPSec peer (responder) is responsible for the specific host or subnet the original traffic is supposed to reach. The responder will recognize the probe and, instead of forwarding it to the actual destination, will return its address as the destination tunnel endpoint. The responder will also learn the address of the initiator via the probe. Once the address of the responding peer is determined, the initiator will continue with IKE main mode as usual.

London-R2 in Case Study 3.2 will have the capability to initiate IKE requests when TED is enabled using the "crypto map chapter3 10 ipsec-isakmp dynamic london-dyn discover" command. Notice that the "discover" option is added to turn on TED.

3.8 Summary

Chapter 3 gives an overview of the different types of IKE peer authentication, along with how preshared keys and can be used in IKE peer authentication. We then move on to explain the different types of preshared keys and identify some of the IPSec configuration caveats, as well as to prepare and plan for IKE and IPSec before the actual configuration. The configuring and monitoring of preshared keys for a site-to-site IPSec VPN are illustrated in Case Study 3.1.

Next, we explain the functionalities of dynamic crypto maps. The deploying and monitoring of dynamic crypto maps are illustrated in Case Study 3.2. The concluding section of the chapter gives a quick preview of tunnel endpoint discovery (TED) and how it can be used in conjunction with dynamic crypto maps.

Site-to-Site IPSec VPN Using RSA Encrypted Nonces and RSA Signatures

4.1 Chapter Overview

In Chapter 3, we implemented a site-to-site IPSec VPN together with the IKE pre-shared keys peer authentication. Chapter 4 is an extension of Chapter 3 in which we further explore how a site-to-site IPSec VPN can be used in conjunction with the other two IKE peer authentication techniques: RSA encrypted nonces and RSA signatures. The chapter first presents an overview of RSA encrypted nonces and takes the readers step-by-step through the actual configuration and monitoring process. The chapter then proceeds to identify the scalability issues when using pre-shared keys and RSA encrypted nonces peer authentication.

Consequently, more sophisticated authentication components such as digital signatures, digital certificates, and the certificate authority (CA) are brought in collectively to address the scalability issues. Next, we need a standard protocol to manage all these components, as well as to support the secure issuance of certificates to network devices in a scalable way, using existing technology whenever possible. The protocol that can cater to all these functionalities is the Simple Certificate Enrollment Protocol (SCEP). Hence, Chapter 4 gives the readers a better understanding of SCEP by going through the protocol operations incrementally.

Digital certificates, the CA, and SCEP together form the basis for the IKE RSA signatures peer authentication. Last, Chapter 4 progressively describes the CA server setup, certificate enrollment, configuration, monitoring, and verification processes for deploying the IKE RSA signatures peer authentication in a site-to-site IPSec VPN.

4.2 RSA Keys Overview

RSA is the public key cryptography developed by Ron Rivest, Adi Shamir, and Leonard Adleman. The security of RSA is based on the fact that it can be comparatively easy to multiply large prime numbers together but very difficult to factor the resulting product. RSA has become the most popular public key cryptography. The technique literally produces RSA keys, which come in pairs: one public key and one private key. The keys act as complements. If data is encrypted with the public key, then only the holder of the corresponding private key can decrypt the message. If data is encrypted with the private key, then anyone with the public key can decrypt it.

Not only can RSA be used to encrypt or decrypt data, for authentication purposes it can also be used to construct or verify a digital signature. The construction of a digital signature can be envisioned as RSA encryption, and the verification of a digital signature can be envisioned as RSA decryption. In the following sections, we shall look at how RSA encrypted nonces and RSA signatures can be used for peer authentication during IKE phase 1.

4.3 IKE Using RSA Encrypted Nonces

RSA encrypted nonces require that each party generate a nonce, which is a pseudorandom number. The nonces are encrypted in the other party's public key and exchanged. On receipt of the remote peer's nonce, the party at each end formulates an authentication key made up of the Diffie-Hellman key, initiator, and responder's nonce and cookie (this is an 8-byte pseudorandom number unique to each peer and the particular exchange in which it is defined). The authentication key is combined with device-specific information and run through a hash algorithm. The output is HASH_I (initiator's hash value) or HASH_R (responder's hash value), which are encrypted and exchanged. The remote peer's identity is authenticated if the local peer can validate the received hash using the locally generated, nonce-based authentication key.

RSA encrypted nonces provide a powerful method of authenticating the IPSec peers and the Diffie-Hellman key exchange. RSA encrypted nonces provide *repudiation*, a quality that prevents a third party from being able to prove that a communication between two parties ever took place. A shortcoming is that they are not easy to configure and, therefore, are more difficult to scale to a large number of peers. RSA encrypted nonces require that peers possess each other's public keys without the support of a certification authority (CA). There are two alternatives for peers to have each other's public keys:

1. Manually configure and exchange RSA keys.

2. Apply RSA signatures used earlier during a successful ISAKMP negotiation with a remote peer. To make this happen, specify two policies: a higher-priority policy with RSA encrypted nonces and a lower-priority policy with RSA signatures. When IKE negotiations occur, RSA signatures will be used the first time because the peers do not possess each other's public keys yet. Subsequently, future IKE negotiations will be able to use RSA encrypted nonces because the public keys will have been exchanged. This alternative would require the presence of a certification authority.

If RSA signature mode is negotiated and encryption is configured, the peer will request both signature and encryption keys. Generally speaking, the router will apply for as many keys as the configuration will support. If RSA encryption is not configured, it will just request a signature key. Note that RSA encrypted nonces must initially be exchanged via some out-of-band method. This might lead to security susceptibilities.

Implementing IPSec encryption can be a complex task. Readers are advised to plan in advance to minimize improper configuration (refer to the IKE and IPSec pre-configuration notes discussed in Chapter 3).

4.4   Case Study 4.1: RSA Encrypted Nonces Example

Case Study 4.1 is identical to Case Study 3.1 in Chapter 3 except that we are using RSA encrypted nonces as the IKE peer authentication between the site-to-site IPSec VPN. The site-to-site VPN spans two different customer locations: Brussels and London. EIGRP is configured as the IGP between the two routers. The site-to-site VPN topology is shown in Figure 4.1.

In this scenario, a shared secret key is not created. Instead, each router generates its own RSA key pair and needs to configure the peer's RSA public key manually. In other words, a local router needs to have a public RSA key for each remote peer with which it wishes to establish a security association. Since this is a fully manual process, this peer authentication method will encounter scaling limitation in large-scale IPSec deployment.

4.4.1   Generating RSA key pairs

In this example, each router first generates an RSA key pair followed by configuring the remote peer's public RSA key. In other words, we need to

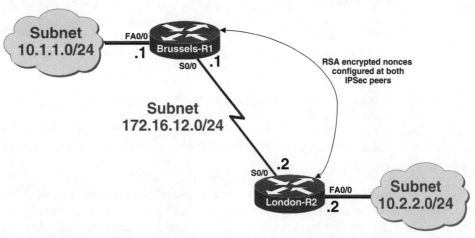

Figure 4.1:   Site-to-site VPN topology for Case Study 4.1.

configure London-R2's public key in Brussels-R1, and Brussels-R1's public key in London-R2. Note that the RSA private key that is generated is saved in the private configuration in the Non-Volatile Random Access Memory (NVRAM) of the router, which is never displayed to the user or backed up to another device.

In Code Listing 4.1, we look at how to generate the RSA key pair for Brussels-R1 using the "crypto key generate rsa usage-keys" command. The "show crypto key mypubkey rsa" command is used to verify the RSA key pair that was created.

Code Listing 4.1:   Generating RSA key pair for Brussels-R1

```
! Use the "crypto key generate rsa" command to generate
! RSA key pairs for Brussels-R1.
! Note that RSA keys are generated in pairs:
! one public and one private.
! In this example, for illustrative purposes,
! we specify the "usage-keys" option, which means
! two pairs of RSA keys will be generated.
! One pair will be used with any IKE policy
! that specifies RSA signatures as the peer
! authentication method, and the other pair
! will be used with any IKE policy that specifies
! RSA encrypted nonces as the authentication method.

! Generally, special-usage keys are generated if both types of
! RSA authentication methods are included in the IKE policies.
! With special-usage keys, each key is not exposed unnecessarily.

! If you generate general-purpose keys
! (that is, without specifying the "usage-keys" option),
! only one pair of RSA keys will be created,
! and this pair will be used with IKE policies
! specifying either RSA signatures or
! RSA encrypted nonces as the authentication method.
! Therefore, a general-purpose key pair is used more often than
! a special-usage key pair, increasing that key's exposure.

Brussels-R1(config)#crypto key generate rsa usage-keys
The name for the keys will be: Brussels-R1.couver.com.sg
Choose the size of the key modulus in the range of 360 to 2048 for your
    Signature Keys. Choosing a key modulus greater than 512 may take
    a few minutes.
```

```
! A longer modulus offers stronger security
! but takes a longer time to generate
! and uses up more CPU processing cycles.
! A length of less than 512 is normally not recommended.

How many bits in the modulus [512]:
Generating RSA keys ...
[OK]
Choose the size of the key modulus in the range of 360 to 2048 for your
   Encryption Keys. Choosing a key modulus greater than 512 may take
   a few minutes.

How many bits in the modulus [512]:
Generating RSA keys ...
[OK]


! Use the "show crypto key mypubkey rsa" to view
! the RSA public keys generated by Brussels-R1.

Brussels-R1#show crypto key mypubkey rsa
% Key pair was generated at: 22:28:09 GMT Mar 1 2002
Key name: Brussels-R1.couver.com.sg
 Usage: Signature Key
 Key Data:
    305C300D 06092A86 4886F70D 01010105 00034B00 30480241 00C27972 7E99C89E
    54235ED1 4350CF74 B7593759 3A439130 3E769ACA 0E378BDE EA894391 B021FEA8
    B0C8EBE2 3D0C876D DBA6EFCB 7E22F53A 06EDECB1 B0A62B2F A3020301 0001
% Key pair was generated at: 22:28:11 GMT Mar 1 2002
Key name: Brussels-R1.couver.com.sg
 Usage: Encryption Key
 Key Data:
    305C300D 06092A86 4886F70D 01010105 00034B00 30480241 00C77365 BC79CABC
    7A0C4EE0 823E08CA B68E4D76 BD27CE4D A38D6829 2671E546 6EBDE7E9 7B9F821A
    6A6488DD 8434449F 7D048473 13A1DE24 4F17B1CC 8C0B40A7 0D020301 0001
```

In Code Listing 4.2, we look at how to generate the RSA key pair for London-R2 using the "crypto key generate rsa usage-keys" command. The "show crypto key mypubkey rsa" command is used to verify the RSA key pair that was created.

Code Listing 4.2:   Generating RSA key pair for London-R2

```
! Use the "crypto key generate rsa" command
! to generate RSA key pairs for London-R2.
```

```
London-R2(config)#crypto key generate rsa usage-keys
The name for the keys will be: London-R2.couver.com.sg
Choose the size of the key modulus in the range of 360 to 2048 for your
    Signature Keys. Choosing a key modulus greater than 512 may take
    a few minutes.

How many bits in the modulus [512]:
Generating RSA keys ...
[OK]
Choose the size of the key modulus in the range of 360 to 2048 for your
    Encryption Keys. Choosing a key modulus greater than 512 may take
    a few minutes.

How many bits in the modulus [512]:
Generating RSA keys ...
[OK]
! Use the "show crypto key mypubkey rsa" to view
! the RSA public keys generated by London-R2.

London-R2#show crypto key mypubkey rsa
% Key pair was generated at: 22:14:07 GMT Mar 1 2002
Key name: London-R2.couver.com.sg
 Usage: Signature Key
 Key Data:
   305C300D  06092A86  4886F70D  01010105  00034B00  30480241  00BD5CE1  487D3682
   7CA5D396  FE4520FE  BA9C8128  4B5E1206  F821E0BF  3196FE77  5762E9C8  85E617A7
   6D552709  1EF22922  C812BCDA  F45BF31F  639F2E22  3C7053B9  41020301  0001
% Key pair was generated at: 22:14:11 GMT Mar 1 2002
Key name: London-R2.couver.com.sg
 Usage: Encryption Key
 Key Data:
   305C300D  06092A86  4886F70D  01010105  00034B00  30480241  00B5CF1C  FCAA5382
   A611F795  7DA3DA23  C46BC0E1  0BECC495  F368AC5C  7FCF975F  0F1C941F  D649430B
   1B759BED  1803A929  D68A1A05  7B637040  FA772DFA  7ED75595  7D020301  0001
```

4.4.2 Specifying and verifying RSA public keys

In Code Listing 4.3 we go through the modus operandi of specifying the RSA public keys of London-R2 (the remote peer) in Brussels-R1 (the local peer).

Code Listing 4.3: Specifying the RSA public keys of London-R2 in Brussels-R1

```
! Use the "crypto key pubkey-chain rsa" command
! to manually specify and store London-R2's
! RSA public keys in Brussels-R1 as follows.
! In this example, London-R2 uses its IP address as its identity.
```

```
Brussels-R1(config)#crypto key pubkey-chain rsa
Brussels-R1(config-pubkey-chain)#addressed-key 172.16.12.2 encryption
Brussels-R1(config-pubkey-key)#key-string

! The RSA encryption key generated for London-R2
! in Code Listing 4.2 is entered here.
Brussels-R1(config-pubkey)# 305C300D 06092A86 4886F70D 01010105
Brussels-R1(config-pubkey)# 00034B00 30480241 00B5CF1C FCAA5382
Brussels-R1(config-pubkey)# A611F795 7DA3DA23 C46BC0E1 0BECC495
Brussels-R1(config-pubkey)# F368AC5C 7FCF975F 0F1C941F D649430B
Brussels-R1(config-pubkey)# 1B759BED 1803A929 D68A1A05 7B637040
Brussels-R1(config-pubkey)# FA772DFA 7ED75595 7D020301 0001
Brussels-R1(config-pubkey)#quit
Brussels-R1(config-pubkey-key)#exit
Brussels-R1(config-pubkey-chain)#addressed-key 172.16.12.2 signature
Brussels-R1(config-pubkey-key)#key-string

! Enter the RSA signature key generated for London-R2
! in Code Listing 4.2 here.
Brussels-R1(config-pubkey)# 305C300D 06092A86 4886F70D 01010105
Brussels-R1(config-pubkey)# 00034B00 30480241 00BD5CE1 487D3682
Brussels-R1(config-pubkey)# 7CA5D396 FE4520FE BA9C8128 4B5E1206
Brussels-R1(config-pubkey)# F821E0BF 3196FE77 5762E9C8 85E617A7
Brussels-R1(config-pubkey)# 6D552709 1EF22922 C812BCDA F45BF31F
Brussels-R1(config-pubkey)# 639F2E22 3C7053B9 41020301 0001
Brussels-R1(config-pubkey)#quit
Brussels-R1(config-pubkey-key)#exit
Brussels-R1(config-pubkey-key)#^Z
Brussels-R1#
```

The "show crypto key pubkey-chain rsa" command executed in Code Listing 4.4 helps us verify the RSA public keys (London-R2's) stored on Brussels-R1.

Code Listing 4.4:   Verifying the RSA public keys stored on Brussels-R1

```
! The "show crypto key pubkey-chain rsa" command
! illustrates the RSA public keys stored on Brussels-R1.
! This includes the remote peers' RSA public keys manually
! configured at the local router and keys received by
! the local router via a certificate
! if certification authority support is configured.
```

```
Brussels-R1#show crypto key pubkey-chain rsa
Codes: M - Manually configured, C - Extracted from certificate


Code  Usage     IP-Address        Name
M     Encrypt   172.16.12.2
M     Signing   172.16.12.2


! This sample shows the manually configured special-usage
! RSA public keys for the remote peer London-R2 (172.16.12.2).
! Note that if certificate support is not configured,
! none of the remote peers' keys would be listed
! as "C" in the code column. Instead all would have to be
! manually configured, and the remote peers' keys
! would be shown as "M".


! We can also use the "show crypto key pubkey-chain rsa"
! with the "address" keyword to display more specific details
! about the RSA public keys (London-R2's) stored on Brussels-R1.

Brussels-R1#show crypto key pubkey-chain rsa address 172.16.12.2

Key address: 172.16.12.2
  Usage: Signature key
  Source: Manual
  Data:
    305C300D  06092A86  4886F70D  01010105  00034B00  30480241  00BD5CE1  487D3682
    7CA5D396  FE4520FE  BA9C8128  4B5E1206  F821E0BF  3196FE77  5762E9C8  85E617A7
    6D552709  1EF22922  C812BCDA  F45BF31F  639F2E22  3C7053B9  41020301  0001

Key address: 172.16.12.2
  Usage: Encryption Key
  Source: Manual
  Data:
    305C300D  06092A86  4886F70D  01010105  00034B00  30480241  00B5CF1C  FCAA5382
    A611F795  7DA3DA23  C46BC0E1  0BECC495  F368AC5C  7FCF975F  0F1C941F  D649430B
    1B759BED  1803A929  D68A1A05  7B637040  FA772DFA  7ED75595  7D020301  0001
```

In Code Listing 4.5 we go through the modus operandi of specifying the RSA public keys of Brussels-R1 (the remote peer) in London-R2 (the local peer).

Code Listing 4.5:   Specifying the RSA public keys of Brussels-R1 in London-R2

```
! Use the "crypto key pubkey-chain rsa" command to manually specify
! and store Brussels-R1's RSA public keys in London-R2 as follows.
! In this example, Brussels-R1 uses its IP address as its identity.
```

```
London-R2(config)#crypto key pubkey-chain rsa
London-R2(config-pubkey-chain)#addressed-key 172.16.12.1 encryption
London-R2(config-pubkey-key)#key-string

! Enter the RSA encryption key generated for Brussels-R1
! in Code Listing 4.1 here.
London-R2(config-pubkey)# 305C300D 06092A86 4886F70D 01010105
London-R2(config-pubkey)# 00034B00 30480241 00C77365 BC79CABC
London-R2(config-pubkey)# 7A0C4EE0 823E08CA B68E4D76 BD27CE4D
London-R2(config-pubkey)# A38D6829 2671E546 6EBDE7E9 7B9F821A
London-R2(config-pubkey)# 6A6488DD 8434449F 7D048473 13A1DE24
London-R2(config-pubkey)# 4F17B1CC 8C0B40A7 0D020301 0001
London-R2(config-pubkey)#quit
London-R2(config-pubkey-key)#exit
London-R2(config-pubkey-chain)#addressed-key 172.16.12.1 signature
London-R2(config-pubkey-key)#key-string

! Enter the RSA signature key generated for Brussels-R1
! in Code Listing 4.1 here.
London-R2(config-pubkey)# 305C300D 06092A86 4886F70D 01010105
London-R2(config-pubkey)# 00034B00 30480241 00C27972 7E99C89E
London-R2(config-pubkey)# 54235ED1 4350CF74 B7593759 3A439130
London-R2(config-pubkey)# 3E769ACA 0E378BDE EA894391 B021FEA8
London-R2(config-pubkey)# B0C8EBE2 3D0C876D DBA6EFCB 7E22F53A
London-R2(config-pubkey)# 06EDECB1 B0A62B2F A3020301 0001
London-R2(config-pubkey)#quit
London-R2(config-pubkey-key)#exit
London-R2(config-pubkey-key)#^Z
London-R2#
```

The "show crypto key pubkey-chain rsa" command executed in Code Listing 4.6 helps us verify the RSA public keys (Brussels-R1's) stored on London-R2.

Code Listing 4.6:   Verifying the RSA public keys stored on London-R2

```
! The "show crypto key pubkey-chain rsa" command
! illustrates the RSA public keys stored on London-R2.

London-R2#show crypto key pubkey-chain rsa
Codes: M - Manually configured, C - Extracted from certificate

Code Usage    IP-Address        Name
M    Encrypt  172.16.12.1
M    Signing  172.16.12.1
```

```
! This sample shows manually configured special usage
! RSA public keys for the remote peer Brussels-R1 (172.16.12.1).


! We can also use the "show crypto key pubkey-chain rsa"
! with the "address" keyword to display more specific details
! about the RSA public keys (Brussels-R1's) stored on London-R2.

London-R2#show crypto key pubkey rsa address 172.16.12.1

Key address: 172.16.12.1
 Usage: Signature Key
 Source: Manual
 Data:
   305C300D  06092A86  4886F70D  01010105  00034B00  30480241  00C27972  7E99C89E
   54235ED1  4350CF74  B7593759  3A439130  3E769ACA  0E378BDE  EA894391  B021FEA8
   B0C8EBE2  3D0C876D  DBA6EFCB  7E22F53A  06EDECB1  B0A62B2F  A3020301  0001

Key address: 172.16.12.1
 Usage: Encryption Key
 Source: Manual
 Data:
   305C300D  06092A86  4886F70D  01010105  00034B00  30480241  00C77365  BC79CABC
   7A0C4EE0  823E08CA  B68E4D76  BD27CE4D  A38D6829  2671E546  6EBDE7E9  7B9F821A
   6A6488DD  8434449F  7D048473  13A1DE24  4F17B1CC  8C0B40A7  0D020301  0001
```

4.4.3 ISAKMP RSA encrypted nonces and IPSec configurations

Code Listings 4.7 and 4.8 are the ISAKMP RSA encrypted nonces and IPSec configurations for Brussels-R1 and London-R2, respectively. In the configurations, comments (in *italics*) precede certain configuration lines to explain them.

Code Listing 4.7: RSA encrypted nonces and IPSec configuration for Brussels-R1

```
hostname Brussels-R1
!
! Define domain name for Brussels-R1.
ip domain-name couver.com.sg
!
! Supersede the default policy and use RSA encrypted nonces
! for peer authentication and hash algorithm MD5.
crypto isakmp policy 110
 hash md5
 authentication rsa-encr
```

```
!
! Specify the ESP settings for IPSec,
! which is later applied to the crypto map.
crypto ipsec transform-set chapter4 esp-des
!
crypto key pubkey-chain rsa
 addressed-key 172.16.12.2 encryption
  address 172.16.12.2
  key-string
! London-R2's RSA public encryption key
    305C300D 06092A86 4886F70D 01010105 00034B00 30480241 00B5CF1C FCAA5382
    A611F795 7DA3DA23 C46BC0E1 0BECC495 F368AC5C 7FCF975F 0F1C941F D649430B
    1B759BED 1803A929 D68A1A05 7B637040 FA772DFA 7ED75595 7D020301 0001
  quit
 addressed-key 172.16.12.2 signature
  address 172.16.12.2
  key-string
! London-R2's RSA public signature key
    305C300D 06092A86 4886F70D 01010105 00034B00 30480241 00BD5CE1 487D3682
    7CA5D396 FE4520FE BA9C8128 4B5E1206 F821E0BF 3196FE77 5762E9C8 85E617A7
    6D552709 1EF22922 C812BCDA F45BF31F 639F2E22 3C7053B9 41020301 0001
  quit
 !
! Specify the crypto map chapter4 where we define
! our peer London-R2, transform-set chapter4,
! and our crypto access-list 101.
 crypto map chapter4 10 ipsec-isakmp
 set peer 172.16.12.2
 set transform-set chapter4
 match address 101
!
interface FastEthernet0/0
 ip address 10.1.1.1 255.255.255.0
!
interface Serial0/0
 bandwidth 64
 ip address 172.16.12.1 255.255.255.0
 clockrate 64000
! Applied the crypto map to an interface
! to activate crypto engine.
 crypto map chapter4
!
! EIGRP is configured as the routing protocol
! between network 10.0.0.0 and 172.16.0.0.
router eigrp 12
```

```
 network 10.0.0.0
 network 172.16.0.0
 no auto-summary
!
! This is the crypto access list that we have
! referenced in crypto map chapter4.
! We are encrypting IP traffic between
! the remote FastEthernet LANs.
access list 101 permit ip host 10.1.1.1 host 10.2.2.2
```

Code Listing 4.8:   RSA Encrypted nonces and IPSec configuration for London-R2

```
hostname London-R2
!
! Define domain name for London-R2.
ip domain-name couver.com.sg
!
! Supersede the default policy and use RSA encrypted nonces
! for peer authentication and hash algorithm MD5.
crypto isakmp policy 110
 hash md5
 authentication rsa-encr
!
! Specify the ESP settings for IPSec,
! which is later applied to the crypto map.
crypto ipsec transform set chapter4 esp-des
!
crypto key pubkey-chain rsa
 addressed-key 172.16.12.1 encryption
  address 172.16.12.1
  key-string
! Brussels-R1's RSA public encryption key
   305C300D 06092A86 4886F70D 01010105 00034B00 30480241 00C77365 BC79CABC
   7A0C4EE0 823E08CA B68E4D76 BD27CE4D A38D6829 2671E546 6EBDE7E9 7B9F821A
   6A6488DD 8434449F 7D048473 13A1DE24 4F17B1CC 8C0B40A7 0D020301 0001
  quit
 addressed-key 172.16.12.1 signature
  address 172.16.12.1
  key-string
! Brussels-R1's RSA public signature key
   305C300D 06092A86 4886F70D 01010105 00034B00 30480241 00C27972 7E99C89E
   54235ED1 4350CF74 B7593759 3A439130 3E769ACA 0E378BDE EA894391 B021FEA8
   B0C8EBE2 3D0C876D DBA6EFCB 7E22F53A 06EDECB1 B0A62B2F A3020301 0001
  quit
```

```
!
! Specify the crypto map chapter4 where we define
! our peer Brussels-R1, transform set chapter4,
! and our crypto access list 102.
 crypto map chapter4 10 ipsec-isakmp
 set peer 172.16.12.1
 set transform-set chapter4
 match address 102
!
interface FastEthernet0/0
 ip address 10.2.2.2 255.255.255.0
!
interface Serial0/0
 bandwidth 64
 ip address 172.16.12.2 255.255.255.0
! Applied the crypto map to an interface
! to activate crypto engine.
 crypto map chapter4
!
! EIGRP is configured as the routing protocol
! between network 10.0.0.0 and 172.16.0.0.
router eigrp 12
 network 10.0.0.0
 network 172.16.0.0
 no auto-summary
!
! This is the crypto access list that we have
! referenced in crypto map chapter4.
! We are encrypting IP traffic between
! the remote FastEthernet LANs.
access list 102 permit ip host 10.2.2.2 host 10.1.1.1
```

4.4.4 Verifying IKE (ISAKMP) and IPSec policies

Code Listings 4.9 to 4.14 look at how you can verify the IKE and IPSec policies that you have configured using the "show crypto isakmp policy," "show crypto map," and "show crypto ipsec transform-set" commands.

Code Listing 4.9 illustrates the ISAKMP policy belonging to Brussels-R1. From the command output, we can see that the encryption algorithm, hash algorithm, and authentication method correspond with DES, MD5, and RSA encrypted nonces, respectively.

Code Listing 4.9:   Brussels-R1 ISAKMP policy

```
Brussels-R1#show crypto isakmp policy
Protection suite of priority 110
    encryption algorithm:   DES - Data Encryption Standard (56 bit keys).
    hash algorithm:      Message Digest 5
    authentication method:  Rivest-Shamir-Adleman Encryption
    Diffie-Hellman group:    #1 (768 bit)
    lifetime:      86,400 seconds, no volume limit
Default protection suite
    encryption algorithm:   DES - Data Encryption Standard (56 bit keys).
    hash algorithm:      Secure Hash Standard
    authentication method:  Rivest-Shamir-Adleman Signature
    Diffie-Hellman group:    #1 (768 bit)
    lifetime:      86,400 seconds, no volume limit
```

Code Listing 4.10 illustrates crypto map chapter4 that is implemented in Brussels-R1. From the crypto map, we can see that the remote IPSec peer is London-R2 (172.16.12.2), the crypto access list 101 is used to encrypt IP traffic between the remote FastEthernet LANs, and the IPSec transform set is chapter4 (see Code Listing 4.11).

Code Listing 4.10:   Brussels-R1 crypto map

```
Brussels-R1#show crypto map
Crypto Map "chapter4" 10 ipsec-isakmp
  Peer = 172.16.12.2
  Extended IP access list 101
      access list 101 permit ip host 10.1.1.1 host 10.2.2.2
  Current peer: 172.16.12.2
  Security association lifetime: 4608000 kilobytes/3600 seconds
  PFS (Y/N): N
  Transform sets={ chapter4, }
```

In Code Listing 4.11, we can see that ESP-DES is the only transform defined for IPSec, and the IPSec tunnel (default) mode is being implemented. Brussels-R1's crypto map (see Code Listing 4.10), together with its IPSec transform, illustrates the overall IPSec policy that is implemented in the router.

Code Listing 4.11:   Brussels-R1 IPSec transform set

```
Brussels-R1#show crypto ipsec transform set
Transform set chapter4: { esp-des  }
  will negotiate = { Tunnel,  },
```

Code Listing 4.12 illustrates the ISAKMP policy belonging to London-R2. From the command output, we can see that the encryption algorithm, hash algorithm, and authentication method correspond with DES, MD5, and RSA encrypted nonces, respectively.

Code Listing 4.12:   London-R2 ISAKMP policy

```
London-R2#show crypto isakmp policy
Protection suite of priority 110
    encryption algorithm:  DES - Data Encryption Standard (56 bit keys).
    hash algorithm:      Message Digest 5
    authentication method: Rivest-Shamir-Adleman Encryption
    Diffie-Hellman group:    #1 (768 bit)
    lifetime:      86,400 seconds, no volume limit
Default protection suite
    encryption algorithm:  DES - Data Encryption Standard (56 bit keys).
    hash algorithm:      Secure Hash Standard
    authentication method: Rivest-Shamir-Adleman Signature
    Diffie-Hellman group:    #1 (768 bit)
    lifetime:          86,400 seconds, no volume limit
```

Code Listing 4.13 illustrates crypto map chapter4 that is implemented in London-R2. From the crypto map, we can see that the remote IPSec peer is Brussels-R1 (172.16.12.1), the crypto access list 102 is used to encrypt IP traffic between the remote FastEthernet LANs, and the IPSec transform set is chapter4 (see Code Listing 4.14).

Code Listing 4.13:   London-R2 crypto map

```
London-R2#show crypto map
Crypto Map "chapter4" 10 ipsec-isakmp
   Peer = 172.16.12.1
   Extended IP access list 102
       access-list 102 permit ip host 10.2.2.2 host 10.1.1.1
   Current peer: 172.16.12.1
   Security association lifetime: 4608000 kilobytes/3600 seconds
   PFS (Y/N): N
  Transform sets={ chapter4, }
```

In Code Listing 4.14, we can see that ESP-DES is the only transform defined for IPSec, and the IPSec tunnel (default) mode is being implemented.

London-R2's crypto map (see Code Listing 4.13), together with its IPSec transform, illustrates the overall IPSec policy that is implemented in the router.

Code Listing 4.14:   London-R2 IPSec Transform Set

```
London-R2#show crypto ipsec transform set
Transform set chapter4: { esp-des   }
   will negotiate = { Tunnel,   },
```

4.4.5   Monitoring IKE (ISAKMP) and IPSec debug trace

The "debug crypto isakmp" and "debug crypto ipsec" commands are used to observe the IKE and IPSec modes of operation and security association sequences. Code Listing 4.15 illustrates the IKE (ISAKMP) and IPSec debug trace when some IP traffic (ping) is generated from London-R2's end. In the trace output below, italic comments are inserted at specific debugging sequence or "checkpoint" to describe them.

Code Listing 4.15:   IKE and IPSec debug trace at London-R2

```
! Ensure that the "debug crypto isakmp" and
! "debug crypto ipsec" commands are activated.
London-R2#show debug
Cryptographic Subsystem:
   Crypto ISAKMP debugging is on
   Crypto IPSEC debugging is on

! Generate IP traffic from London-R2 (10.2.2.2) to
! Brussels-R1 (10.1.1.1) using the extended ping command.
London-R2#ping
Protocol [ip]:
Target IP address: 10.1.1.1
Repeat count [5]: 3
Datagram size [100]:
Timeout in seconds [2]:
Extended commands [n]: y
Source address or interface: 10.2.2.2
Type of service [0]:
Set DF bit in IP header? [no]:
Validate reply data? [no]:
Data pattern [0xABCD]:
Loose, Strict, Record, Timestamp, Verbose[none]:
```

```
Sweep range of sizes [n]:
Type escape sequence to abort.
Sending 3, 100-byte ICMP Echos to 10.1.1.1, timeout is 2 seconds:
!!!
Success rate is 100 percent (3/3), round-trip min/avg/max = 40/42/44 ms
! The ping source and destination addresses match
! access list 102 defined in crypto map chapter4.


London-R2#


! IPSec requests SAs between 10.2.2.2 and 10.1.1.1, on behalf of
! access list 102, and uses the transform set chapter4.
*Mar  1 23:55:17: IPSEC(sa_request): ,
   (key eng. msg.) src= 172.16.12.2, dest= 172.16.12.1,
     src_proxy= 10.2.2.2/255.255.255.255/0/0 (type=1),
     dest_proxy= 10.1.1.1/255.255.255.255/0/0 (type=1),
! The src_proxy is the source interesting traffic,
! and the dest_proxy is the destination interesting traffic,
! both are defined in access-list 102.
      <Output Omitted>
! This is ISAKMP Phase 1 using Oakley Main Mode.
*Mar  1 23:55:17: ISAKMP (81): beginning Main Mode exchange
! The value 81 (in parenthesis) is the connection ID assigned to IKE.
<Output Omitted>
*Mar  1 23:55:17: ISAKMP (81): Checking ISAKMP transform 1 against
priority 110 policy
! Policy 110 is the only ISAKMP policy
! configured on this local router (initiator).
*Mar  1 23:55:17: ISAKMP:      encryption DES-CBC
*Mar  1 23:55:17: ISAKMP:      hash MD5
*Mar  1 23:55:17: ISAKMP:      default group 1
*Mar  1 23:55:17: ISAKMP:      auth RSA encr
! These are the ISAKMP attributes being offered
! by the remote end (responder).
! The authentication method is RSA encrypted nonces.
*Mar  1 23:55:17: ISAKMP (81): atts are acceptable. Next payload is 0
! IKE has found a matching policy since policy 110
! on this router, and the atts (attributes) that are
! offered by the remote end match.


! RSA encrypted nonces authentication will start now.
*Mar  1 23:55:17: ISAKMP (81): SA is doing RSA encryption authentication
using id type ID_IPV4_ADDR
<Output Omitted>
*Mar  1 23:55:19: ISAKMP (81): sending packet to 172.16.12.1 (I) MM_KEY_EXCH
```

```
*Mar   1 23:55:19: ISAKMP (81): received packet from 172.16.12.1 (I)
MM_KEY_EXCH
*Mar  1 23:55:19: ISAKMP (81): processing HASH payload. message ID = 0
*Mar  1 23:55:19: ISAKMP (81): SA has been authenticated with 172.16.12.1
! RSA encrypted nonces authentication has succeeded at this point.
! The ISAKMP SA has been successfully negotiated.

! This is ISAKMP phase 2 using Oakley Quick Mode.
*Mar   1 23:55:19: ISAKMP (81): beginning Quick Mode exchange, M-ID of
1348872416
<Output Omitted>
*Mar  1 23:55:19: ISAKMP (81): Checking IPSec proposal 1
! Here IKE will process the IPSec attributes offered by the remote end.

*Mar  1 23:55:19: ISAKMP: transform 1, ESP_DES
! This is the protocol offered by the remote end
! according to its transform set.

*Mar  1 23:55:19: ISAKMP:    attributes in transform:
*Mar  1 23:55:19: ISAKMP:       encaps is 1
*Mar  1 23:55:19: ISAKMP:       SA life type in seconds
*Mar  1 23:55:19: ISAKMP:       SA life duration (basic) of 3600
*Mar  1 23:55:19: ISAKMP:       SA life type in kilobytes
*Mar  1 23:55:19: ISAKMP:       SA life duration (VPI) of  0x0 0x46 0x50
0x0
*Mar  1 23:55:19: ISAKMP (81): atts are acceptable.
! A matching IPSec transform set has been found at the two peers.
! Now the IPSec SA can be created (one SA for each direction).
<Output Omitted>
*Mar  1 23:55:19: ISAKMP (81): Creating IPSec SAs
*Mar  1 23:55:19:          inbound SA from 172.16.12.1      to 172.16.12.2
(proxy 10.1.1.1        to 10.2.2.2         )
*Mar  1 23:55:19:          has spi 183832119 and conn_id 82 and flags 4
*Mar  1 23:55:19:          lifetime of 3600 seconds
*Mar  1 23:55:19:          lifetime of 4,608,000 kilobytes
*Mar  1 23:55:19:          outbound SA from 172.16.12.2      to 172.16.12.1
(proxy 10.2.2.2        to 10.1.1.1         )
*Mar  1 23:55:19:          has spi 306842314 and conn_id 83 and flags 4
*Mar  1 23:55:19:          lifetime of 3600 seconds
*Mar  1 23:55:19:          lifetime of 4608000 kilobytes
! Two IPSec SAs have been negotiated,
! an incoming SA with the SPI generated by the initiator,
! and an outbound SA with the SPI proposed by the responder.
! Crypto engine entries have also been created.
<Output Omitted>
```

```
! IPSec now installs the SA information into its SA database (SADB).
*Mar  1 23:55:19: IPSEC(create_sa): sa created,
  (sa) sa_dest= 172.16.12.2, sa_prot= 50,
    sa_spi= 0xAF50E37(183832119),
    sa_trans= esp-des , sa_conn_id= 82
*Mar  1 23:55:19: IPSEC(create_sa): sa created,
  (sa) sa_dest= 172.16.12.1, sa_prot= 50,
    sa_spi= 0x124A0ACA(306842314),
    sa_trans= esp-des , sa_conn_id= 83
! The SADB has been updated and the IPSec SAs have been initialized.
! The tunnel is now fully operational.

*Mar  1 23:55:19: ISAKMP (81): sending packet to 172.16.12.1 (I) QM_IDLE
! The IPSec SA has now been successfully negotiated.
! IKE will now go into a state known as QM-IDLE (Quick Mode Idle).
```

4.4.6 Verifying IKE (ISAKMP) and IPSec security associations

In Code Listings 4.16 to 4.21, we conduct a postmodem on the IKE and IPSec SAs after generating IP traffic from London-R2 to Brussels-R1 as illustrated previously in Code Listing 4.15.

Code Listing 4.16 lists the current status of the IKE SA for Brussels-R1, which is in the Quick Mode Idle state. The IKE tunnel starts at London-R2 (172.16.12.2) and ends at Brussels-R1 (172.16.12.1). The IKE SA is also assigned a connection ID value of 80 (that references an entry in the crypto engine connection table shown in Code Listing 4.18).

Code Listing 4.16: Verifying IKE SA for Brussels-R1

```
Brussels-R1#show crypto isakmp sa
    dst              src            state        conn-id    slot
172.16.12.1      172.16.12.2        QM_IDLE        80         0
```

Code Listing 4.17: Verifying IPSec SA for Brussels-R1

```
Brussels-R1#show crypto ipsec sa

interface: Serial0/0
    Crypto map tag: chapter4, local addr. 172.16.12.1

  local  ident (addr/mask/prot/port): (10.1.1.1/255.255.255.255/0/0)
  remote ident (addr/mask/prot/port): (10.2.2.2/255.255.255.255/0/0)
  current_peer: 172.16.12.2
    PERMIT, flags={origin_is_acl,}
```

```
#pkts encaps: 3, #pkts encrypt: 3, #pkts digest 0
#pkts decaps: 3, #pkts decrypt: 3, #pkts verify 0
#send errors 0, #recv errors 0

local crypto endpt.: 172.16.12.1, remote crypto endpt.: 172.16.12.2
path mtu 1500, media mtu 1500
current outbound spi: AF50E37

inbound esp sas:
 spi: 0x124A0ACA(306842314)
    transform: esp-des ,
    in use settings ={Tunnel, }
    slot: 0, conn id: 81, crypto map: chapter4
    sa timing: remaining key lifetime (k/sec): (4607999/2671)
    IV size: 8 bytes
    replay detection support: N

inbound ah sas:

outbound esp sas:
 spi: 0xAF50E37(183832119)
    transform: esp-des ,
    in use settings ={Tunnel, }
    slot: 0, conn id: 82, crypto map: chapter4
    sa timing: remaining key lifetime (k/sec): (4607999/2671)
    IV size: 8 bytes
    replay detection support: N
outbound ah sas:
```

In Code Listing 4.17, the Brussels-R1's IPSec SA states that three interesting data packets, from local host 10.1.1.1 to remote host 10.2.2.2, have been encrypted, and another three interesting packets, from remote host 10.2.2.2 to local host 10.1.1.1, have been decrypted through interface Serial0/0 where crypto map chapter4 is applied. The IPSec tunnel is established from local crypto endpoint 172.16.12.1 to remote crypto endpoint 172.16.12.2.

In addition, the inbound ESP SA has been populated into the SADB with information such as ESP-DES transform, IPSec tunnel mode, SPI value 306842314, connection ID value 81 (that references an entry in the crypto engine connection table listed in Code Listing 4.18), and crypto map chapter4. Likewise, outbound ESP SA has been populated into the SADB with information such as ESP-DES transform, IPSec tunnel mode, SPI value 183832119, connection ID value 82 (that references an entry in the crypto engine connection table listed in Code Listing 4.18), and crypto map chapter4. Notice that no inbound or outbound AH information is displayed because the AH protocol is not implemented in this case study.

Code Listing 4.18 illustrates the crypto engine connection table for Brussels-R1. From the table, we can see that connection ID 80 belongs to the IKE SA in Code Listing 4.16, and connection ID 81 and 82 belong to the two ESP SAs in Code Listing 4.17. Note that the outbound SA encrypts three data packets, and the inbound SA decrypts another three, concurring with Code Listing 4.17.

Code Listing 4.18: Brussels-R1 crypto engine connection table

---

```
Brussels-R1#show crypto engine connection active

  ID Interface      IP-Address      State  Algorithm      Encrypt  Decrypt
  80 no idb         no address      set    DES_56_CBC           0        0
  81 Serial0/0      172.16.12.1     set    DES_56_CBC           0        3
  82 Serial0/0      172.16.12.1     set    DES_56_CBC           3        0
```

Code Listing 4.19 lists the current status of the IKE SA for London-R2, which is in the quick mode idle state. The IKE tunnel starts at London-R2 (172.16.12.2) and ends at Brussels-R1 (172.16.12.1). The IKE SA is also assigned a connection ID value of 81 (that references an entry in the crypto engine connection table listed in Code Listing 4.21).

Code Listing 4.19: Verifying IKE SA for London-R2

```
London-R2#show crypto ipsec sa
     dst               src              state       conn-id   slot
172.16.12.1       172.16.12.2          QM_IDLE         81       0
```

Code Listing 4.20: Verifying IPSec SA for London-R2

```
London-R2#show crypto ipsec sa

interface: Serial0/0
    Crypto map tag: chapter4, local addr. 172.16.12.2

    local  ident (addr/mask/prot/port): (10.2.2.2/255.255.255.255/0/0)
    remote ident (addr/mask/prot/port): (10.1.1.1/255.255.255.255/0/0)
    current_peer: 172.16.12.1
      PERMIT, flags={origin_is_acl,}
     #pkts encaps: 3, #pkts encrypt: 3, #pkts digest 0
     #pkts decaps: 3, #pkts decrypt: 3, #pkts verify 0
     #send errors 0, #recv errors 0
```

```
      local crypto endpt.: 172.16.12.2, remote crypto endpt.: 172.16.12.1
      path mtu 1500, media mtu 1500
      current outbound spi: 124A0ACA

      inbound esp sas:
       spi: 0xAF50E37(183832119)
          transform: esp-des ,
          in use settings ={Tunnel, }
          slot: 0, conn id: 82, crypto map: chapter4
          sa timing: remaining key lifetime (k/sec): (4607999/2993)
          IV size: 8 bytes
          replay detection support: N

      inbound ah sas:

      outbound esp sas:
       spi: 0x124A0ACA(306842314)
          transform: esp-des ,
          in use settings ={Tunnel, }
          slot: 0, conn id: 83, crypto map: chapter4
          sa timing: remaining key lifetime (k/sec): (4607999/2993)
          IV size: 8 bytes
          replay detection support: N

      outbound ah sas:
```

In Code Listing 4.20, the London-R2's IPSec SA states that three interesting data packets, from local host 10.2.2.2 to remote host 10.1.1.1, have been encrypted, and another three interesting packets, from remote host 10.1.1.1 to local host 10.2.2.2, have been decrypted through interface Serial0/0 where crypto map chapter4 is applied. The IPSec tunnel is established from local crypto endpoint 172.16.12.2 to remote crypto endpoint 172.16.12.1.

In addition, the inbound ESP SA has been populated into the SADB with information such as ESP-DES transform, IPSec tunnel mode, SPI value 183832119, connection ID value 82 (that references an entry in the crypto engine connection table listed in Code Listing 4.21), and crypto map chapter4. Likewise, outbound ESP SA has been populated into the SADB with information such as ESP-DES transform, IPSec tunnel mode, SPI value 306842314, connection ID value 83 (that references an entry in the crypto engine connection table listed in Code Listing 4.21), and crypto map chapter4. Notice that no inbound or outbound AH information is displayed because the AH protocol is not implemented in this case study.

Code Listing 4.21 illustrates the crypto engine connection table for London-R2. From the table, we can see that connection ID 81 belongs to the IKE SA in Code Listing 4.19, and connection ID 82 and 83 belong to the two ESP SAs in

Code Listing 4.20. Note that the outbound SA encrypts three data packets, and the inbound SA decrypts another three, concurring with Code Listing 4.20.

Code Listing 4.21:   London-R2 crypto engine connection table

---

```
London-R2#show crypto engine active

    ID Interface     IP-Address     State   Algorithm      Encrypt   Decrypt
    81 no idb        no address     set     DES_56_CBC        0        0
    82 Serial0/0     172.16.12.1    set     DES_56_CBC        0        3
    83 Serial0/0     172.16.12.1    set     DES_56_CBC        3        0
```

4.5   Scalability Issues

So far we have discussed the two peer authentication methods: pre-shared keys (see Chapter 3) and RSA encrypted nonces. Both methods require each router to have the other router's key, which is manually performed as follows:

- For pre-shared keys authentication, at each router, specify a shared key to be used between the routers.

- For RSA encrypted nonces authentication, at each router, enter the other router's RSA public key.

Each router uses the other router's key to authenticate the identity of the other router; this authentication always occurs whenever IPSec traffic is exchanged between the two routers. In a mesh topology involving multiple routers, to exchange IPSec traffic among all these routers, shared keys or RSA public keys must first be configured between all the routers. For example, in Figure 4.2, six 2-part key configurations are required for four fully meshed IPSec routers. Note that in a fully meshed network topology, the number of 2-part key configurations can be calculated using the formula: $n * (n - 1)/2$ where n is the total number of IPSec routers.

Every time a new router is added to the IPSec network, the keys between the new router and each of the existing routers must be configured. If a new IPSec router were added to the existing mesh topology in Figure 4.2, four additional 2-part key configurations would be required. Therefore, as the number of devices that require IPSec services grows, the key administration burden also increases. Obviously, pre-shared keys and RSA encrypted nonces do not scale well for bigger and more complex network topologies. For such implementation, certificate-based authentication is preferred (see Section 4.9).

4.6   Digital Certificates

Digital signatures, enabled by public key cryptography, provide another alternative to digitally authenticate devices and individual users. In public key

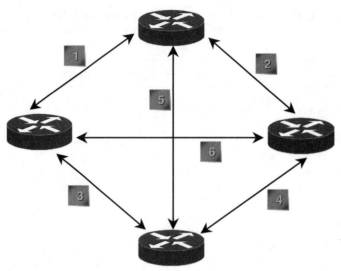

Figure 4.2: Six 2-part key configurations required for four IPSec routers.

cryptography, such as the RSA encryption system, each user has a key pair containing both a public and a private key.

A signature is created when data is encrypted with a user's private key. The receiver verifies the signature by decrypting the message with the sender's public key. The fact that the message could be decrypted using the sender's public key indicates that the holder of the private key, the sender, must have created the message. This process relies on the receiver having a copy of the sender's public key and knowing with a high degree of certainty that it really does belong to the sender, and not to an imposter doubling as the sender. Digital certificates provide this association.

A digital certificate contains information to identify a user or device, such as the name, serial number, company, department, or IP address. It also contains a copy of the entity's public key. In other words, it is a data structure that binds a public key value to a subject. Public key cryptography has been recommended for use with the ISO authentication framework, also known as the X.509 protocols. The most important part of X.509 is its structure for public key certificates. The X.509 certificate consists of the following specific fields:

- Certificate format version—Currently it is X.509 version 1, 2, or 3.
- Certificate serial number—Unique certificate numerical identifier.
- Signature algorithm—Identifies the algorithms (hashing and encryption) used to sign the certificate.
- Issuer—The X.500 distinguished name (DN) of the CA.

- Validity period—Specifies the start and expiration dates for the certificate.
- Subject—The X.500 distinguished name (DN) of the entity (user).
- Subject's public key—Specifies the subject's public key and signature algorithm.
- Extensions—Extends the certificate to allow for additional information.
- CRL distribution points—Specifies the locations of the CRL list for this certificate. CRL will be discussed in Section 4.11.3.
- CA signature—Specifies the CA signature.

Digital certificates simplify authentication and enhance VPN performance. Digital certificates also provide more key entropy (more bits for seeding function), public-private key pair aging, and nonrepudiation—meaning that a third party can prove that a communication between two parties took place. Nevertheless, digital certificates require additional administrative resources to deploy and manage, given their feature complexity. Therefore, the administrative burden for deploying digital certificates to remote access clients is deemed significant.

4.7 Certificate Authority

A certificate authority (CA) is a third party that is explicitly trusted by the receiver to validate identities and to create digital certificates. It signs a digital certificate (containing a device's public key) with its private key. CAs are responsible for managing certificate requests and issuing certificates to participating IPSec network devices. In other words, CAs provide centralized key management for the participating devices and simplify the administration of IPSec network devices. Third-party CA vendors include Microsoft, Verisign, Baltimore, and Entrust. Note that CA forms a part (component) of the public key infrastructure (PKI). PKI is a pervasive security infrastructure whose services are implemented and delivered using public key cryptography. PKI provides efficient and trusted public key certificate management.

The CA can be classified into two trust models in the PKI:

1. Central authority—A flat network design. A single authority root CA signs all certificates.

2. Hierarchical authority—A tiered approach. The ability to sign a certificate is delegated through a hierarchy. The top of the hierarchy is the root CA. It signs certificates for subordinate CAs. The subordinate CAs then sign certificates for lower-level CAs. Eventually, a subordinate CA will sign the user's identity certificate. To validate the identity certificate, the certificate must be validated through the chain of authority.

The CA also revokes certificates with the help of a certificate revocation list (CRL). This is a list of certificates (by certificate serial number) that are no longer valid. This is similar to the revoked credit cards lists that business owners used to verify when processing a credit card purchase.

4.8 Registration Authority

Some CAs have a registration authority (RA) as part of their implementation to offload its registration function. The RA acts as an interface or proxy between the CA and the network devices. It performs administrative tasks on behalf of the CA. The primary function of the RA is to enhance scalability and decrease operational costs by offloading one or more of the following functions from the CA:

- Establish and validate the identity of an end entity (an end-entity is an entity whose name is defined in a certificate subject name field) as part of the initialization process.
- Start the certification process with a CA on behalf of individual end entity.
- Generate keying material on behalf of an end entity.
- Perform key or certification life cycle management functions, for instance to initiate a revocation request from an end entity.

It should be noted that an RA does not issue certificates or CRLs. These functions rest solely on the CA. Note also that there can be more than one RA per CA.

4.9 Implementing IPSec with CA

With a CA, you need not manually configure keys among all the IPSec peers. Instead, you need only enroll each peer with the CA as illustrated in Figure 4.3, requesting a certificate for the peer. When this has been done, each participating peer can dynamically validate all the other participating peers.

To add a new IPSec peer to the network, you only need to configure the new peer to request a certificate from the CA. None of the other peers need any modification. When the new peer attempts an IPSec connection, certificates are automatically exchanged and the peer can be authenticated. Therefore, certificate-based authentication (using RSA signatures) is a more scalable solution than are pre-shared keys or RSA encrypted nonces authentication. Readers can consider deploying CA if the size of the IPSec VPN is more than 20 encrypting devices or if there are design requirements for more robust device authentication.

4.10 Certificate Generation

To participate in a certificate exchange, an end entity must obtain a digital certificate from the CA. This is known as the enrollment process. There are two types of enrollment: file-based and network-based.

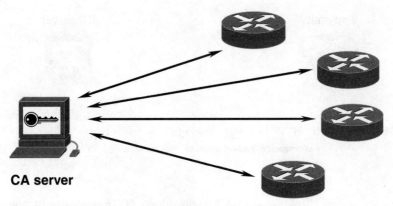

CA server

Figure 4.3:   Each IPSec peer individually enrolls with the CA.

4.10.1   File-based enrollment

This is a manual file transfer-intensive process. The first step is to create a request file (certificate request) known using Public Key Cryptography Standards (PKCS) #10 *(PKCS #10 describes a standard syntax for certification requests. The request is formatted as an Abstract Syntax Notation [ASN.1 {as defined in X.208}] and sent to the CA, which then transforms the request into an X.509 certificate. PKCS #10 is described in RFC 2986.)* After the request file is created, you can either email it to the CA and receive a certificate back or access the CA's website and cut and paste the enrollment request in the space that the CA provides. The CA generates the root and identity certificates. These certificates must then be downloaded and imported into the end entity involved.

4.10.2   Network-based enrollment

This is an automated process that enables the end entity to connect directly to a CA via simple certificate enrollment protocol (SCEP). SCEP provides a standard way of managing the certificate lifecycle and is interoperable with many vendors' devices. The objective of SCEP is to support the secure issuance of certificates to network devices in a scalable way, using existing technology whenever possible. For network-based enrollment to function properly, both the end entity and the CA must support SCEP. SCEP is described in the IETF draft filename draft-nourse-scep-06.txt. The protocol supports the following operations:

- CA and RA public key distribution
- Certificate enrollment
- Certificate revocation

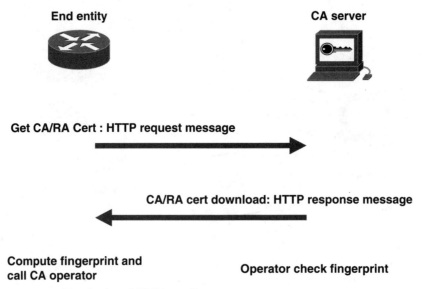

End entity

CA server

Get CA/RA Cert : HTTP request message

CA/RA cert download: HTTP response message

**Compute fingerprint and
call CA operator**

Operator check fingerprint

Figure 4.4: Distribution of CA/RA certificates.

- Certificate query
- CRL query

We examine these operations in the following sections.

4.10.2.1 CA distribution /RA certificate.   To validate the CA's signature, the end entity (receiver) must first know the CA's public key (through its root certificate). Therefore, before any PKI operation can be started, the end entity needs to get the CA/RA certificates. The CA/RA certificate distribution is implemented as a plain HTTP Get operation. After the end entity retrieves the CA certificate, it has to authenticate the CA certificate by comparing the fingerprint with the CA/RA operator. Since the CA signs the RA certificates, there is no need to authenticate the RA certificates. This operation is defined as a transaction consisting of one HTTP Get message and one HTTP Response message as illustrated in Figure 4.4.

In an environment where an RA is present, an end entity performs enrollment through the RA. To setup a secure channel with RA using PKCS #7, the RA certificate(s) have to be obtained by the end entity in addition to the CA certificate(s). Note that PKCS #7 (described in RFC 2315) is a general enveloping syntax that enables both signed and encrypted transmission of arbitrary data. The syntax is recursive, so that envelopes can be nested. This allows multiple certificates to be enveloped within one message (adopts the same concept as PKZIP storing multiple files in a .zip file). When a certificate is sent between a CA and an end entity, the ASN.1 formatted message is encoded. There are two

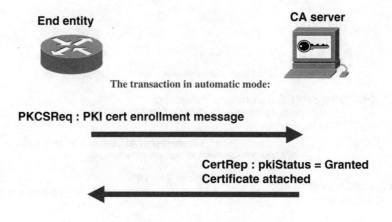

Receive issued certificate

Figure 4.5: Certificate enrollment in automatic mode.

types of digital certificate encoding: distinguished encoding rules (DER
[as defined in X.509]) format (raw binary format, produces octets for transmission), or privacy enhanced mail (PEM [as defined in RFCs 1421–1424])
format (binary converted to base 64 format, converts octets to ASCII for
display).

If an RA is in use, a PKCS #7 with a certificate chain consisting of both RA
and CA certificates is sent back to the end entity. Otherwise, the CA certificate
is directly sent back as the HTTP response payload.

4.10.2.2 Certificate enrollment.   An end entity starts an enrollment transaction
by creating a certificate request using PKCS #10 and sends it to the CA/RA
enveloped using the PKCS #7. After the CA/RA receives the request, it will
automatically approve the request and send the certificate back, or it will
require the end entity to wait until the operator can manually authenticate the
identity of the requesting end entity. Two attributes are included in the PKCS
#10 certificate request: a challenge password attribute and an optional exten-
sionReq attribute, which is a sequence of extensions the end entity would like
to be included in its version 3 certificate extensions. The challenge password is
used for revocation and may be used (at the option of the CA/RA) additionally
as a one-time password for automatic enrollment.

In the automatic mode as illustrated in Figure 4.5, the transaction consists
of one PKCSReq (Request) PKI Message and one CertRep (Reply) PKI message.

In the manual mode as illustrated in Figure 4.6, the end entity enters into
polling mode by periodically sending a GetCertInitial PKI message to the server
until the server operator completes the manual identity authentication, after
which the CA will respond to GetCertInitial by returning the issued certificate.

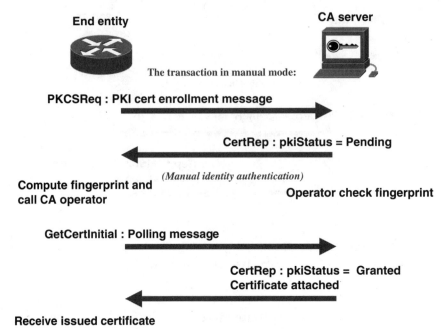

Figure 4.6: Certificate enrollment in manual mode.

This authentication is done through the comparison of MD5/SHA fingerprints (hashes) that are calculated over the data exchanged. Without manual authentication, it may be possible for a "man-in-the-middle" to impersonate one or both parties by substituting his public key for the rest.

4.10.2.3 Certificate revocation.   An end entity should be able to revoke its own certificate. Currently the revocation is implemented as a manual process. To revoke a certificate, the end entity makes a phone call to the CA server operator. The operator will ask for the challenge password, which was sent to the server as an attribute of the PKCS #10 during the certificate request transaction. If the challenge password matches, the certificate is revoked.

4.10.2.4 Certificate access.   A certificate query message is defined to retrieve the certificates from CA. To query a certificate from the CA, an end entity sends a request consisting of the certificate's issuer name and the serial number. This presumes that the end entity has saved the issuer name and the serial number of the issued certificate from the previous enrollment transaction. The transaction to query a certificate consists of one GetCert PKI message and one CertRep PKI message as illustrated in Figure 4.7.

4.10.2.5 CRL distribution.   A CRL query is composed by creating a message that consists of the CA issuer name and the CA's certificate serial number. This

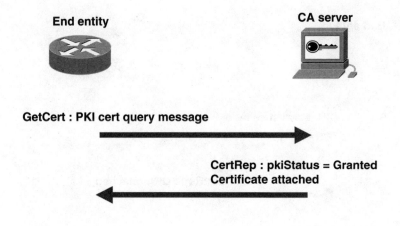

GetCert : PKI cert query message

CertRep : pkiStatus = Granted
Certificate attached

Receive the certificate

Figure 4.7: Certificate Access.

method has some drawbacks because it does not scale well and requires the CA to be constantly available. The message is sent to the CA in the same way as the other SCEP requests. The transaction to query CRL consists of one GetCRL PKI message and one CertRep PKI message, which contain no certificates but CRL. This is illustrated in Figure 4.8.

4.11 Certificate Validation

To validate a digital certificate, the end entity verifies the following:

- Authenticity of the CA signature
- Validity period
- Certificate revocation list

4.11.1 Signature validation

At the CA end, the original identity certificate sent by the end entity is put through a hash algorithm. The output is encrypted (signed) by the CA's private key, and the encrypted hash, which is the CA signature, is appended to the last line of the certificate. The CA-signed certificate is then returned to the end entity.

When a local end entity receives an identity certificate from a remote end entity, it uses the CA's public key (derived from the stored root [CA] certificate that was obtained during the CA/RA certification distribution as discussed in Section 4.10.2.1) to decrypt the hash, which yields the original hash value. The received identity certificate is sent through the same hash algorithm (used by

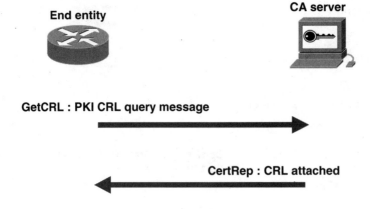

Figure 4.8: CRL distribution.

the CA) to produce a second hash. The CA-generated hash and the end entity–generated hash are compared:

- If they match, the identity certificate is genuine.
- If they do not match, the certificate is invalid, and this implies that there is an invalid signature or identity certificate.

4.11.2 Validity period

A certificate is valid for a specific period of time. The validity period (time range) is set by the CA and consists of "valid from" and "valid to" fields. When a local end entity receives an identity certificate from a remote end entity, the validity range listed in the identity certificate is compared against the system clock, and an error message will be generated if the system clock is not within the validity range.

4.11.3 Certificate revocation list

A certificate revocation list (CRL) is a list issued by the CA that contains certificates that are no longer valid. The certificates are listed by certificate serial number and revocation date. CRLs are signed and issued by the CA. They are valid for a specific period of time and are published (update) periodically or on demand.

Some reasons for revocation include:

- User data changes, for instance the subject name
- A (private) key has been compromised or tampered with

- Violation of CA policy
- An employee has left the organization

Note that the CRL will not be pushed to the end entities by the CA/RA. The query of the CRL can be initialized only by the end entity itself.

One way the CRL may be retrieved is via a simple HTTP GET. If the CA supports this method, it should encode the URL into a CRL distribution point extension in the certificates it issues. CRL distribution points (sometimes referred to as partitioned CRLs) allow revocation information within a single CA domain to be published in multiple CRLs. The certificate can point to the location of the CRL distribution point, so there is no need for the relying end entity to have prior knowledge of where the revocation information for a particular certificate might be located.

Another alternative is to query CRL using LDAP (lightweight directory access protocol). This presumes the CA server supports CRL LDAP publishing and issues the CRL distribution point in the certificate. In this case, the CRL distribution point is encoded as an X.500 distinguished name (DN).

The third method is implemented for the CA, which does not support LDAP CRL publishing or does not implement the CRL distribution point. This method is discussed in Section 4.10.2.5.

Even though verifying CRLs can be configured as optional, it should always be enabled on remote and local devices when digital certificates are deployed, since this is the only revocation scheme for digital certificates.

4.12 IKE Using RSA Signature

RSA signatures use the exchange of digital certificates to authenticate peers. With RSA signatures, HASH_I (initiator's hash value) and HASH_R (responder's hash value) are not just authenticated, but they are also digitally signed (encrypted) with their respective RSA private keys to form digital signatures (encrypted hash).

Identity certificates and digital signatures are exchanged between the initiator and responder during IKE phase 1 negotiation to authenticate the peers. The RSA public keys for decrypting the signatures are included in the digital certificates exchanged between the peers.

For successful peer authentication, the initiator and responder must validate each other's identity certificate (see Section 4.11), decrypt each other's digital signature with the public key found in the identity certificate, and match the decrypted hashes with recomputed ones.

If the certificate binds the peer's FQDN (fully qualified domain name) to its public key, the knowledge of the peer's IP address is not required in advance, which makes RSA signatures ideal for remote users. The use of RSA signatures also provides nonrepudiation of a transaction (a third party can prove that a transaction between two parties took place).

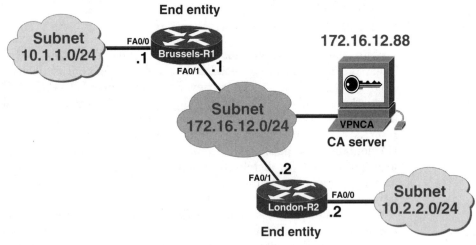

Figure 4.9: Site-to-site VPN topology for Case Study 4.2.

4.13   Case Study 4.2: RSA Signatures Example

Case Study 4.2 is derived directly from Case Study 4.1. In this case, we make
use of a FastEthernet segment (shared media) to emulate the Internet, and the
IKE peer authentication between the site-to-site IPSec VPN is implemented
with RSA signatures, supported by a stand-alone root CA. The site-to-site VPN
spans two different customer locations: Brussels, and London. Brussels-R1 and
London-R2 are both end entities and IPSec peers. EIGRP is configured as
the IGP between the two routers. The site-to-site VPN topology is shown in
Figure 4.9.

This example concerns RSA signatures, which require the use of a CA server.
Each peer is required to obtain certificates from the CA server (VPNCA). When
both peers have valid CA certificates, they automatically exchange identity certificates (RSA public keys) with each other as part of ISAKMP negotiation. All
that is required in this scenario is for each peer to enroll with a CA and obtain
the respective certificates. In this case, a peer no longer needs to retain the
public RSA keys of all the peers in the IPSec VPN.

4.13.1   Setting up the CA server

In this scenario, simple certificate enrollment protocol (SCEP) is used for the
exchange of information between the CA and the end entities—Cisco routers.
SCEP is a PKI communication protocol that leverages existing technology by
using PKCS #7 and PKCS #10. In this case, PKCS #7 is applied in signing and
enveloping certificates, and PKCS #10 is used as the syntax for certificate
request. In addition, HTTP and LDAP are used for transporting the certificates.
The operation aspects of SCEP are covered in Section 4.10.2.

Microsoft has integrated SCEP support into the Windows 2000 CA server through the Server Resource Kit for Windows 2000. This CA requires that the end entity transact with a registration authority (RA), which then forwards the requests through to the CA itself. In this instance, the end entity must receive the CA's public key as well as the public keys (signature and encryption) of the RA (three certificates total).

The router will receive four certificates if a general-purpose key pair (which constitutes one certificate) is generated, and five certificates if special-usage key pairs (which constitute two certificates, one for signature and the other for encryption) are generated. Note that Cisco IOS releases 12.0(5) T and above are required to support the Windows 2000 CA server. The following illustrates the configuration process for the Windows 2000 CA server deployed in this case study.

Set up the Microsoft CA server as a stand-alone root CA:

- Go to Windows Component Wizards: Start → Settings → Control Panel → Windows Components Wizard
- From Windows Component Wizards:
  - Check Certificate Services box
  - Under Certification Authority Type, select Stand-alone root CA radio button
  - Under CA Identifying Information:
    - Enter CA name: *VPNCA*
    - Enter Organization: *GKN*
    - Enter Country/Region: *MY*
    - Enter Valid for: *2 Years*
  - Under Data Storage Location: (*leave as default*)

Next, install Simple Certificate Enrolment Protocol (SCEP) Add-on for Certificate Services from the Windows 2000 Server Resource Kit:

- Click on "cepsetup" found under the /apps/cep/directory in the Windows 2000 Server Resource Kit.
- When the SCEP Add-on for Certificate Services Setup Wizard appears:
  - Click Next>
  - Uncheck Require SCEP Challenge Phrase to Enroll
  - Under SCEP RA Certificate Enrolment:
    - Enter Name: *VPNRA*
    - Enter Company: *GKN*
    - Enter Country/Region: *MY*
    - Click Next>
    - Click Finish

Finally, modify some of the properties of the CA server:

- Go to Certificate Authority Manager: Start → Programs → Administrative Tools → Certificate Authority (*Certificate Authority Manager appears*)
- Right-click on VPNCA icon → select Properties
  - Under VPNCA select Policy Module tab
  - Under Policy Module select Configure . . . tab → select the Default Action Tab:
    - Under Default action: select Always issue the certificate radio button → click Apply (*Note that the Certificate Services must be restarted for these changes to take effect.*)
    - After this select the Security tab:
      - Under Name: click on Everyone
      - Under Permissions: check on Allow box → click Apply → close existing window (*The permission attribute is used by the CA server to manage enroll and read rights. In this case, "Everyone" is given both the enroll and read rights.*)
- Go back to Certification Authority Manager to restart Certificate Services:
  - Right-click on VPNCA icon → All Tasks → Stop Service
  - Right-click on VPNCA icon → All Tasks → Start Service

For simplicity and illustrative purposes, the CA server is configured as a stand-alone root CA, the SCEP challenge phrase for enrollment is disabled, the certificate is always issued, and everyone is given both enroll and read rights. The reader is advised to exercise discretion on which properties to adjust or enable to suit their particular network environments.

4.13.2   Retrieving certificates for end entities

Each end entity (IPSec peer) will have to go through the following steps when RSA signatures are used for IKE authentication:

Step 1. Generate a public-private key pair that will be used for signing—the end-entity will sign any data with its private key. In other words, the receiver may validate that this information really came from a sender by using that sender's public-key to decrypt the digital signature (encrypted hash) attached to the sent message. In Code Listing 4.22, we generate the RSA public-private key pair for Brussels-R1 and London-R2 successively.

Code Listing 4.22:   Generate RSA key pair for Brussels-R1 and London-R2

```
! Ensure that the router's time and date have been
! accurately set with the "clock set" command.
! The clock must be set correctly before generating
! RSA key pairs and enrolling with the CA server
! because the keys and certificates are time-sensitive.
```

```
! When a local end entity receives an identity certificate
! from a remote end entity, the validity range listed in
! the identity certificate is compared against the system clock,
! and an error message will be generated if the system clock
! is not within the valid range.

! To specify the router's time zone, use
! the "clock timezone" command, which sets
! the time zone and an offset from Universal Time Code (UTC).
! The following sets the time zone for Brussels-R1
! to Greenwich Mean Time (GMT).
Brussels-R1(config)#clock timezone GMT 0

! Use the "crypto key generate rsa" command
! to generate the RSA key pair for Brussels-R1.
Brussels-R1(config)#crypto key generate rsa
The name for the keys will be: Brussels-R1.couver.com.sg
Choose the size of the key modulus in the range of 360 to 2048 for your
   general-purpose keys. Choosing a key modulus greater than 512 may take
   a few minutes.

How many bits in the modulus [512]:
Generating RSA keys ...
[OK]

! Use the "show crypto key mypubkey rsa"
! to view the RSA public key generated by Brussels-R1.
Brussels-R1#show crypto key mypubkey rsa
% Key pair was generated at: 09:22:04 GMT Apr 27 2002
Key name: Brussels-R1.couver.com.sg
 Usage: General-Purpose Key
 Key Data:
  305C300D 06092A86 4886F70D 01010105 00034B00 30480241 00B71DF4 E27BB0AB
  45701674 F048D3FB BEA242A4 FC972CA8 00B82B65 76836F18 B368D8A5 84C8C6ED
  EAAA56E3 E93BF179 821413D6 2F44FE97 E57F3267 20279A3D 2D020301 0001

! The following sets the time zone for London-R2
! to Greenwich Mean Time (GMT).
London-R2(config)#clock timezone GMT 0

! Use the "crypto key generate rsa" command to
! generate the RSA key pair for London-R2.
London-R2(config)#crypto key generate rsa
The name for the keys will be: London-R2.couver.com.sg
Choose the size of the key modulus in the range of 360 to 2048 for your
   general-purpose keys. Choosing a key modulus greater than 512 may take
   a few minutes.
```

```
How many bits in the modulus [512]:
Generating RSA keys ...
[OK]


! Use the "show crypto key mypubkey rsa" to view
! the RSA public key generated by London-R2.
London-R2#show crypto key mypubkey rsa
% Key pair was generated at: 09:14:53 GMT Apr 27 2002
Key name: London-R2.couver.com.sg
  Usage: General-Purpose Key
  Key Data:
    305C300D  06092A86  4886F70D  01010105  00034B00  30480241  00E47A82  12EF0470
    0EF1D421  37D386F9  A513FA7C  F226D813  02B5E9BE  7BCADFD4  5BFFA2B9  94FF5BE3
    410CED5B  3C975085  76597372  5918CCF7  CEF3B870  14F3081F  7F020301  0001
```

Step 2. Retrieve the CA/RA certificates and then derive the respective public keys so that any information sent by the CA/RA can be validated before being accepted by the end entity. In Code Listing 4.23, we retrieve the CA/RA certificates from VPNCA for Brussels-R1 and London-R2.

Code Listing 4.23:   Retrieve CA/RA certificates for Brussels-R1 and London-R2

```
! Turn on PKI debugging with the "debug crypto pki messages"
! and "debug crypto pki transactions" commands so that
! we can observe the debug messages for the CA operation.
Brussels-R1#show debug
Cryptographic Subsystem:
  Crypto PKI Msg debugging is on
  Crypto PKI Trans debugging is on


! The router's hostname and domain name must be configured
! for CA support to function properly.
! Define our CA server's hostname as vpnca.
Brussels-R1(config)#ip host vpnca 172.16.12.88

! Specify a unique domain name for the router.
Brussels-R1(config)#ip domain-name couver.com.sg

! Specify the CA server name as vpnca and enter ca-identity mode.
Brussels-R1(config)#crypto ca identity vpnca

! Specify the RA mode since the CA requires the use of an RA.
Brussels-R1(ca-identity)#enrollment mode ra
```

```
! Specify the enrollment URL of the CA server.
Brussels-R1(ca-identity)#enrollment url
http://vpnca/certsrv/mscep/mscep.dll

! Specify that the router can still accept
! other peers' certificates if the CRL is inaccessible.
Brussels-R1(ca-identity)#crl optional

! The router needs to authenticate the CA
! to verify that it is valid.
! The router does this by obtaining
! the CA's self-signed certificate,
! which contains the CA's public key.
! Since the CA signs its own certificate,
! we should manually authenticate the CA's
! public key by contacting the CA administrator
! to verify the CA certificate's fingerprint.

! Authenticate the CA server.
Brussels-R1(config)#crypto ca authenticate vpnca

! Verify the fingerprint of the CA server with the CA administrator.
Certificate has the following attributes:
Fingerprint: 3CA9053E 3F48A7EB 203E9F42 ED046B98

! SCEP uses HTTP for transport.
Apr 27 09:24:07: CRYPTO_PKI: http connection opened

! CA self-signed certificate is received by the end entity.
Apr 27 09:24:08: CRYPTO_PKI: transaction GetCACert completed
Apr 27 09:24:08: CRYPTO_PKI: CA certificate received.

% Please answer 'yes' or 'no'.
% Do you accept this certificate? [yes/no]: yes

! Since the CA is using an RA, the RA signature
! and encryption certificates are retrieved from
! the CA together with the CA (root) certificate.
Apr 27 09:24:11: CRYPTO_PKI: Name: CN = VPNCA, O = GKN, C = MY
Apr 27 09:24:11: CRYPTO_PKI: Name: CN = VPNRA, O = GKN, C = MY
Apr 27 09:24:12: CRYPTO_PKI: Name: CN = VPNRA, O = GKN, C = MY

! The "show crypto ca certificates" command illustrates
! the CA/RA certificates retrieved by the router from the CA.
Brussels-R1#show crypto ca certificates
```

```
CA Certificate
  Status: Available
  Certificate Serial Number: 5A1A8D98B485FCBC4C51E13CDE3460E5
  Key Usage: Not Set
  DN Name
    CN = VPNCA
     O = GKN
     C = MY

RA Signature Certificate
  Status: Available
  Certificate Serial Number: 61598DDB000000000002
  Key Usage: Signature
  DN Name
    CN = VPNRA
     O = GKN
     C = MY

RA KeyEncipher Certificate
  Status: Available
  Certificate Serial Number: 61598FB1000000000003
  Key Usage: Encryption
  DN Name
    CN = VPNRA
     O = GKN
     C = MY


! The procedures for retrieving the CA/RA certificates
! for London-R2 are identical to those for Brussels-R1.


London-R2#show debug
Cryptographic Subsystem:
  Crypto PKI Msg debugging is on
  Crypto PKI Trans debugging is on


London-R2(config)#ip host vpnca 172.16.12.88
London-R2(config)#ip domain-name couver.com.sg
London-R2(config)#crypto ca identity vpnca
London-R2(ca-identity)#enrollment mode ra
London-R2(ca-identity)#enrollment url http://vpnca/certsrv/mscep/mscep.dll
London-R2(ca-identity)#crl optional


London-R2(config)#crypto ca authenticate vpnca
Certificate has the following attributes:
Fingerprint: 3CA9053E 3F48A7EB 203E9F42 ED046B98
```

```
Apr 27 09:16:57: CRYPTO_PKI: http connection opened
Apr 27 09:16:58: CRYPTO_PKI: transaction GetCACert completed
Apr 27 09:16:58: CRYPTO_PKI: CA certificate received.


% Please answer 'yes' or 'no'.
% Do you accept this certificate? [yes/no]: yes


Apr 27 09:17:04: CRYPTO_PKI: Name: CN = VPNCA, O = GKN, C = MY
Apr 27 09:17:04: CRYPTO_PKI: Name: CN = VPNRA, O = GKN, C = MY
Apr 27 09:17:04: CRYPTO_PKI: Name: CN = VPNRA, O = GKN, C = MY


London-R2#show crypto ca certificates
CA Certificate
  Status: Available
  Certificate Serial Number: 5A1A8D98B485FCBC4C51E13CDE3460E5
  Key Usage: Not Set
  DN Name
    CN = VPNCA
     O = GKN
     C = MY


RA Signature Certificate
  Status: Available
  Certificate Serial Number: 61598DDB000000000002
  Key Usage: Signature
  DN Name
    CN = VPNRA
     O = GKN
     C = MY


RA KeyEncipher Certificate
  Status: Available
  Certificate Serial Number: 61598FB1000000000003
  Key Usage: Encryption
  DN Name
    CN = VPNRA
     O = GKN
     C = MY
```

Step 3. Send an enrollment/certificate request to the CA to officially bind the end entity's identity to its public key. Then accept the certificate containing the public key and validate the certificate by verifying that it has been issued and signed by the CA through decrypting a hash value sent with the certificate, which was encrypted (signed) by the CA with its private key. The hash decryption is accomplished with the help of the CA's public key retrieved previously in Step 2. Code Listings 4.24 and 4.25 illustrate how Brussels-R1 and London-R2 enroll with VPNCA.

Code Listing 4.24:   Enroll Brussels-R1 with VPNCA

```
! To obtain certificates from the CA for
! the router's RSA key pair, use the
! "crypto ca enroll" command. This process
! is known as enrolling with the CA.
! Enrolling and obtaining certificates are
! actually two separate processes, but they
! both occur when this command is executed.
! If general-purpose keys were generated previously,
! this command obtains one certificate corresponding
! to the general-purpose RSA key pair.
! If special-usage keys are generated, this command
! obtains two certificates corresponding to each of
! the special-usage RSA key pairs.
! If you already have a certificate for your keys,
! you will not be able to complete this command
! unless you remove existing certificates with
! the "no certificate" command.
Brussels-R1(config)#crypto ca enroll vpnca
%
% Start certificate enrollment ..

! The router administrator has to provide the password
! entered in the script below to the CA administrator
! before certificate revocation can take place.
% Create a challenge password. You will need to verbally provide this
    password to the CA administrator to revoke your certificate.
    For security reasons your password will not be saved in the configuration.
    Please make note of it.

Password:
Re-enter password:

! Format certificate request based on PKCS #10.
% The subject name in the certificate will be: Brussels-R1.couver.com.sg
% Include the router serial number in the subject name? [yes/no]: n
% Include an IP address in the subject name? [yes/no]: n
Request certificate from CA? [yes/no]: y
% Certificate request sent to Certificate Authority
```

```
% The certificate request fingerprint will be displayed.
% The 'show crypto ca certificate' command will also show the fingerprint.
! Note that the "show crypto ca certificate" command will list
! the fingerprint when the PKI status is in pending state.

Brussels-R1(config)#^Z
Brussels-R1#
! For identity validation, the computed MD5/SHA fingerprint
! of the previously created general-purpose key that the
! router sends to the CA should be the same as the fingerprint
! the CA administrator receives through the CA server console.
Brussels-R1#    Fingerprint:  2A565970 7EF82DC2 DDE56F2B 99AD8F64


! PKI certificate enrollment message.
Apr 27 09:25:28: CRYPTO_PKI: transaction PKCSReq completed


! As the certificate request was just initiated,
! there is no PKI status.
Apr 27 09:25:28: CRYPTO_PKI: status:


! Certificate request format is based on PKCS #10.
Apr 27 09:25:28: Write out pkcs#10 content:286
Apr 27 09:25:28: 30 82 01 1A 30 81 C5 02 01 00 30 2A 31 28 30 26 06 09 2A 86
Apr 27 09:25:28: 48 86 F7 0D 01 09 02 16 19 42 72 75 73 73 65 6C 73 2D 52 31
Apr 27 09:25:28: 2E 63 6F 75 76 65 72 2E 63 6F 6D 2E 73 67 30 5C 30 0D 06 09
<Output Omitted>
Apr 27 09:25:28: 69 EB 6C 2A 94 3E 40 C5 1F 53 10 1C AB CC EE B3 52 67 59 1D
Apr 27 09:25:28: 1D 75 63 43 DE B6 22 19 84 E1 98 50 62 66 53 2A CD 37 C2 2A
Apr 27 09:25:28: BF 51 B8 95 43 9B


! Certificate request is enveloped using PKCS #7.
Apr 27 09:25:28: Enveloped Data ...


Apr 27 09:25:28: 30 80 06 09 2A 86 48 86 F7 0D 01 07 03 A0 80 30 80 02 01 00
Apr 27 09:25:28: 31 80 30 81 D0 02 01 00 30 39 30 2B 31 0B 30 09 06 03 55 04
Apr 27 09:25:28: 06 13 02 4D 59 31 0C 30 0A 06 03 55 04 0A 13 03 47 4B 4E 31
<Output Omitted>
Apr 27 09:25:28: 7F 48 62 16 61 F0 F0 C7 F9 E4 65 BA 80 7A 8E AA E6 1F A3 65
Apr 27 09:25:28: F8 D4 33 41 07 22 B1 E9 BE 39 BD 26 A9 46 C8 4C 20 25 AD E4
Apr 27 09:25:28: 12 87 D8 00 00 00 00 00 00 00 00


! CA issues duly signed identity certificate to the router.
! SCEP uses HTTP for transport.
Apr 27 09:25:28: CRYPTO_PKI: http connection opened
Apr 27 09:25:29: CRYPTO_PKI:  received msg of 1709 bytes
```

```
Apr 27 09:25:29: Received pki message: 1563 types
Apr 27 09:25:29: 30 82 06 17 06 09 2A 86 48 86 F7 0D 01 07 02 A0 82 06 08 30
Apr 27 09:25:29: 82 06 04 02 01 01 31 0E 30 0C 06 08 2A 86 48 86 F7 0D 02 05
Apr 27 09:25:29: 05 00 30 82 04 2D 06 09 2A 86 48 86 F7 0D 01 07 01 A0 82 04
<Output Omitted>
Apr 27 09:25:29: 83 01 AB F8 D6 5C BD CA 9F 87 1C B5 3A 62 79 91 06 C7 82 B5
Apr 27 09:25:29: 7C A2 00 E2 56 6D A8 35 8C 81 23 6D 9C 02 67 0E 79 93 B5 B1
Apr 27 09:25:29: 29 D2 67 5E 13 1D BB 0B 9D 0A 3A 99 E1 6E 98 56 78 8B 95 A3
<Output Omitted>

! The router's identity certificate is granted and issued by the CA.
Apr 27 09:25:29: CRYPTO_PKI: status = 100: certificate is granted

! Validate the certificate by verifying that
! it has been issued and signed by the CA.
Apr 27 09:25:29: Verified signed data 1050 bytes:
Apr 27 09:25:29: 30 82 04 16 06 09 2A 86 48 86 F7 0D 01 07 03 A0 82 04 07 30
Apr 27 09:25:29: 82 04 03 02 01 00 31 81 A7 30 81 A4 02 01 00 30 4E 30 2A 31
Apr 27 09:25:29: 28 30 26 06 09 2A 86 48 86 F7 0D 01 09 02 16 19 42 72 75 73
<Output Omitted>
Apr 27 09:25:29: 95 8F 5D 6F 8E 06 8C B1 33 10 4A DC 5D B8 5F FD 80 A5 63 F4
Apr 27 09:25:29: D3 E3 65 2D 24 5D 7E B9 3C 74 96 82 C6 A7 9A 9F 11 7F 98 C1
Apr 27 09:25:29: 7F 9F 3E D9 9F 25 48 31 00

! All enrollment requests are fulfilled,
! and the identity certificate issued from
! the CA is received by the router.
Apr 27 09:25:29: CRYPTO__PKI: All enrollment requests completed.
Apr 27 09:25:29: %CRYPTO-6-CERTRET: Certificate received from Certificate
Authority

! The "show crypto ca certificates" command illustrates the
! identity certificate (requested earlier  from the  CA) and
! the  CA/RA certificates retrieved by the router from the CA.
Brussels-R1#show crypto ca certificates
Certificate
  Status: Available
  Certificate Serial Number: 611E3942000000000007
  Key Usage: General Purpose
  Subject Name
    Name: Brussels-R1.couver.com.sg
  DN Name
    OID.1.2.840.113549.1.9.2 = Brussels-R1.couver.com.sg

RA Signature Certificate
  Status: Available
```

```
Certificate Serial Number: 61598DDB000000000002
Key Usage: Signature
DN Name
   CN = VPNRA
    O = GKN
    C = MY
```

CA Certificate
```
Status: Available
Certificate Serial Number: 5A1A8D98B485FCBC4C51E13CDE3460E5
Key Usage: Not Set
DN Name
   CN = VPNCA
    O = GKN
    C = MY
```

RA KeyEncipher Certificate
```
Status: Available
Certificate Serial Number: 61598FB1000000000003
Key Usage: Encryption
DN Name
   CN = VPNRA
    O = GKN
    C = MY
```

Code Listing 4.25:   Enroll London-R2 with VPNCA

```
! The procedures for enrolling London-R2 with the CA
! are identical to those for Brussels-R1.


London-R2(config)#crypto ca enroll vpnca
%
% Start certificate enrollment ..
% Create a challenge password. You will need to verbally provide this
   password to the CA administrator to revoke your certificate.
   For security reasons your password will not be saved in the configuration.
   Please make note of it.

Password:
Re-enter password:
% The subject name in the certificate will be: London-R2.couver.com.sg
% Include the router serial number in the subject name? [yes/no]: n
% Include an IP address in the subject name? [yes/no]: n
Request certificate from CA? [yes/no]: y
```

```
% Certificate request sent to Certificate Authority
% The certificate request fingerprint will be displayed.
% The 'show crypto ca certificate' command will also show the fingerprint.


London-R2(config)#^Z
London-R2#
London-R2#      Fingerprint:   7991775A 1AA25F8B D3DE4913 987AB624


Apr 27 09:18:05: CRYPTO_PKI: transaction PKCSReq completed
Apr 27 09:18:05: CRYPTO_PKI: status:
Apr 27 09:18:05: Write out pkcs#10 content:284
Apr 27 09:18:05: 30 82 01 18 30 81 C3 02 01 00 30 28 31 26 30 24 06 09 2A 86
Apr 27 09:18:05: 48 86 F7 0D 01 09 02 16 17 4C 6F 6E 64 6F 6E 2D 52 32 2E 63
Apr 27 09:18:05: 6F 75 76 65 72 2E 63 6F 6D 2E 73 67 30 5C 30 0D 06 09 2A 86
<Output Omitted>
Apr 27 09:18:05: BB C4 76 2B B1 18 C9 91 37 87 FA 11 96 52 09 8C 95 8A 8E EC
Apr 27 09:18:05: 01 53 DE D3 0F 3F 66 0B 8E 5B D1 F9 A1 AD 54 3D EA 5D 48 24
Apr 27 09:18:05: E8 52 5C A2
Apr 27 09:18:06: Enveloped Data ...


Apr 27 09:18:06: 30 80 06 09 2A 86 48 86 F7 0D 01 07 03 A0 80 30 80 02 01 00
Apr 27 09:18:06: 31 80 30 81 D0 02 01 00 30 39 30 2B 31 0B 30 09 06 03 55 04
Apr 27 09:18:06: 06 13 02 4D 59 31 0C 30 0A 06 03 55 04 0A 13 03 47 4B 4E 31
<Output Omitted>
Apr 27 09:18:06: 98 D4 48 45 FF 02 A1 86 6D AD A3 FA B6 F2 E0 57 83 94 CF FF
Apr 27 09:18:06: D1 26 D8 3D 8A 9C 8D C9 B4 66 F5 7F A9 F2 99 0E 99 00 00 00
Apr 27 09:18:06: 00 00 00 00 00
Apr 27 09:18:06: CRYPTO_PKI: http connection opened
Apr 27 09:18:07: CRYPTO_PKI:  received msg of 1699 bytes
Apr 27 09:18:07: Received pki message: 1553 types
Apr 27 09:18:07: 30 82 06 0D 06 09 2A 86 48 86 F7 0D 01 07 02 A0 82 05 FE 30
Apr 27 09:18:07: 82 05 FA 02 01 01 31 0E 30 0C 06 08 2A 86 48 86 F7 0D 02 05
Apr 27 09:18:07: 05 00 30 82 04 23 06 09 2A 86 48 86 F7 0D 01 07 01 A0 82 04
<Output Omitted>
Apr 27 09:18:07: CRYPTO_PKI: status = 100: certificate is granted
Apr 27 09:18:07: Verified signed data 1040 bytes:
Apr 27 09:18:07: 30 82 04 0C 06 09 2A 86 48 86 F7 0D 01 07 03 A0 82 03 FD 30
Apr 27 09:18:07: 82 03 F9 02 01 00 31 81 A5 30 81 A2 02 01 00 30 4C 30 28 31
Apr 27 09:18:07: 26 30 24 06 09 2A 86 48 86 F7 0D 01 09 02 16 17 4C 6F 6E 64
<Output Omitted>
Apr 27 09:18:07: 17 63 E2 59 B7 87 9B 67 F9 DF C1 30 BA AF C8 A3 81 25 80 DC
Apr 27 09:18:07: 31 88 B5 84 E5 58 12 BD B0 52 40 D6 DA BC 23 85 A4 D1 CA 8F
Apr 27 09:18:07: 73 C4 3D 31 00
Apr 27 09:18:07: CRYPTO__PKI: All enrollment requests completed.
```

```
Apr 27 09:18:07: %CRYPTO-6-CERTRET: Certificate received from Certificate
Authority

London-R2#show crypto ca certificates
Certificate
  Status: Available
  Certificate Serial Number: 611736FF000000000006
  Key Usage: General Purpose
  Subject Name
    Name: London-R2.couver.com.sg
  DN Name
    OID.1.2.840.113549.1.9.2 = London-R2.couver.com.sg

RA Signature Certificate
  Status: Available
  Certificate Serial Number: 61598DDB000000000002
  Key Usage: Signature
  DN Name
    CN = VPNRA
     O = GKN
     C = MY

CA Certificate
  Status: Available
  Certificate Serial Number: 5A1A8D98B485FCBC4C51E13CDE3460E5
  Key Usage: Not Set
  DN Name
    CN = VPNCA
     O = GKN
     C = MY

RA KeyEncipher Certificate
  Status: Available
  Certificate Serial Number: 61598FB1000000000003
  Key Usage: Encryption
  DN Name
    CN = VPNRA
     O = GKN
     C = MY
```

Step 4. End entity receives the certificate signed and issued by the CA. The end entity's public key appears in the end entity's configuration together with the CA and RA public keys. Code Listings 4.26 and 4.27 illustrate the RSA signatures and IPSec configurations for Brussels-R1 and London-R2. These end entities have previously obtained its identity certificate and the CA/RA certificates from VPNCA, and the certificates are all listed in the configuration.

Code Listing 4.26:   RSA signatures and IPSec configuration for Brussels-R1

```
hostname Brussels-R1
!
! Define time zone for Brussels-R1.
clock timezone GMT 0
! Define hostname for CA server.
ip host vpnca 172.16.12.88
! Define domain name for Brussels-R1.
ip domain-name couver.com.sg
!
! Specify the CA server name as vpnca
crypto ca identity vpnca
! Specify the RA mode since the CA requires the use of an RA.
 enrollment mode ra
! Specify the enrollment URL of the CA server.
 enrollment url http://vpnca:80/certsrv/mscep/mscep.dll
! Specify that the router can still accept other peers'
! certificates if the CRL is inaccessible.
 crl optional
!
! VPNCA's certificate chain
crypto ca certificate chain vpnca
! Brussels-R1's identity certificate retrieved from the CA server
 certificate 611E3942000000000007
  308202F6 308202A0 A0030201 02020A61 1E394200 00000000 07300D06 092A8648
  86F70D01 01050500 302B310B 30090603 55040613 024D5931 0C300A06 0355040A
  1303474B 4E310E30 0C060355 04031305 56504E43 41301E17 0D303230 34323830
  39313835 395A170D 30333034 32383039 32383539 5A302A31 28302606 092A8648
  86F70D01 09021319 42727573 73656C73 2D52312E 636F7576 65722E63 6F6D2E73
  <Output Omitted>
  6E30325F 56504E43 412E6372 74303906 082B0601 05050730 02862D66 696C653A
  2F2F5C5C 74726169 6E30325C 43657274 456E726F 6C6C5C74 7261696E 30325F56
  504E4341 2E637274 300D0609 2A864886 F70D0101 05050003 41009071 AFA1048A
  81888F70 6E4E512F 591F8E95 8F5D6F8E 068CB133 104ADC5D B85FFD80 A563F4D3
  E3652D24 5D7EB93C 749682C6 A79A9F11 7F98C17F 9F3ED99F 2548
  quit
! The RA's signature certificate retrieved from the CA server
 certificate ra-sign 61598DDB000000000002
  3082032C 308202D6 A0030201 02020A61 598DDB00 00000000 02300D06 092A8648
  86F70D01 01050500 302B310B 30090603 55040613 024D5931 0C300A06 0355040A
```

```
    1303474B 4E310E30 0C060355 04031305 56504E43 41301E17 0D303230 34323830
    38343432 335A170D 30333034 32383038 35343233 5A302B31 0B300906 03550406
    13024D59 310C300A 06035504 0A130347 4B4E310E 300C0603 55040313 0556504E
    <Output Omitted>
    63727430 3906082B 06010505 07300286 2D66696C 653A2F2F 5C5C7472 61696E30
    325C4365 7274456E 726F6C6C 5C747261 696E3032 5F56504E 43412E63 7274300D
    06092A86 4886F70D 01010505 00034100 73B78B86 745C7C7C D396AFC2 117FEE56
    9B790A7C AF9BF5D7 48098FFA F9C0DC6E 8F7B31F1 23D60250 7FCF4305 394C8CCA
    FCA33116 1AD06E6A D0197616 FD2941F8
    quit
! The root (CA) certificate retrieved from the CA server
 certificate ca 5A1A8D98B485FCBC4C51E13CDE3460E5
    3082020C 308201B6 A0030201 0202105A 1A8D98B4 85FCBC4C 51E13CDE 3460E530
    0D06092A 864886F7 0D010105 0500302B 310B3009 06035504 0613024D 59310C30
    0A060355 040A1303 474B4E31 0E300C06 03550403 13055650 4E434130 1E170D30
    32303432 38303834 3234385A 170D3034 30343238 30383531 32365A30 2B310B30
    09060355 04061302 4D59310C 300A0603 55040A13 03474B4E 310E300C 06035504
    <Output Omitted>
    2BA029A0 27862566 696C653A 2F2F5C5C 74726169 6E30325C 43657274 456E726F
    6C6C5C56 504E4341 2E63726C 30100609 2B060104 01823715 01040302 0100300D
    06092A86 4886F70D 01010505 00034100 5154CAC8 A01A01F8 41B9EFC3 ADB61DF2
    7AD3D136 9EA22883 30B2E6FD BB27A015 0C81F4DC EEDA3E62 B23563F3 3A10B073
    140ED68E F9A99977 BDE9A4EA 13C045D3
    quit
! The RA's encryption certificate retrieved from the CA server
 certificate ra-encrypt 61598FB1000000000003
    3082032C 308202D6 A0030201 02020A61 598FB100 00000000 03300D06 092A8648
    86F70D01 01050500 302B310B 30090603 55040613 024D5931 0C300A06 0355040A
    1303474B 4E310E30 0C060355 04031305 56504E43 41301E17 0D303230 34323830
    38343432 335A170D 30333034 32383038 35343233 5A302B31 0B300906 03550406
    13024D59 310C300A 06035504 0A130347 4B4E310E 300C0603 55040313 0556504E
    <Output Omitted>
    63727430 3906082B 06010505 07300286 2D66696C 653A2F2F 5C5C7472 61696E30
    325C4365 7274456E 726F6C6C 5C747261 696E3032 5F56504E 43412E63 7274300D
    06092A86 4886F70D 01010505 00034100 32EACF8A F98257C5 F95D7328 877F9200
    6B66B6A7 B7DF3D04 EBD17827 4A3FC6E3 07D371CD 68FC924A 09FB6E53 79DCB644
    1B726241 CB417A04 3A055E3B D4B241C2
    quit
!
! Supersede the default policy and use hash algorithm MD5.
! Notice that peer authentication is not set here because we are
! using the default peer authentication, RSA signatures.
crypto isakmp policy 110
 hash md5
!
```

```
! Specify the ESP settings for IPSec,
! which is later applied to the crypto map.
crypto ipsec transform-set chapter4 esp-des
!
! Specify the crypto map chapter4 where we define
! our peer London-R2, transform set chapter4,
! and our crypto access list 101.
crypto map chapter4 10 ipsec-isakmp
 set peer 172.16.12.2
 set transform-set chapter4
 match address 101
!
interface FastEthernet0/0
 ip address 10.1.1.1 255.255.255.0
!
interface FastEthernet0/1
 ip address 172.16.12.1 255.255.255.0
! Applied the crypto map to an interface
! to activate crypto engine.
 crypto map chapter4
!
! EIGRP is configured as the routing protocol
! between network 10.0.0.0 and 172.16.0.0.
router eigrp 12
 network 10.0.0.0
 network 172.16.0.0
 no auto-summary
!
! This is the crypto access list that we have
! referenced in crypto map chapter4.
! We are encrypting IP traffic between
! the remote FastEthernet LANs.
access-list 101 permit ip host 10.1.1.1 host 10.2.2.2
```

Code Listing 4.27:   RSA signatures and IPSec configuration for London-R2

```
hostname London-R2
!
! Define time zone for London-R2.
clock timezone GMT 0
! Define hostname for CA server.
ip host vpnca 172.16.12.88
! Define domain name for London-R2.
ip domain-name couver.com.sg
```

```
!
! Specify the CA server name as vpnca
crypto ca identity vpnca
! Specify the RA mode since the CA requires the use of an RA.
 enrollment mode ra
! Specify the enrollment URL of the CA server.
 enrollment url http://vpnca:80/certsrv/mscep/mscep.dll
! Specify that the router can still accept other peers'
! certificates if the CRL is inaccessible.
 crl optional
!
crypto ca certificate chain vpnca
! London-R2's identity certificate retrieved from the CA server
 certificate 611736FF000000000006
  308202F2 3082029C A0030201 02020A61 1736FF00 00000000 06300D06 092A8648
  86F70D01 01050500 302B310B 30090603 55040613 024D5931 0C300A06 0355040A
  1303474B 4E310E30 0C060355 04031305 56504E43 41301E17 0D303230 34323830
  39313132 305A170D 30333034 32383039 32313230 5A302831 26302406 092A8648
  86F70D01 09021317 4C6F6E64 6F6E2D52 322E636F 75766572 2E636F6D 2E736730
  <Output Omitted>
  56504E43 412E6372 74303906 082B0601 05050730 02862D66 696C653A 2F2F5C5C
  74726169 6E30325C 43657274 456E726F 6C6C5C74 7261696E 30325F56 504E4341
  2E637274 300D0609 2A864886 F70D0101 05050003 4100BC43 2CE626A4 C904DAE2
  8DD12ED4 82BC0C8C F0BD5C17 63E259B7 879B67F9 DFC130BA AFC8A381 2580DC31
  88B584E5 5812BDB0 5240D6DA BC2385A4 D1CA8F73 C43D
  quit
! The RA's signature certificate retrieved from the CA server
 certificate ra-sign 61598DDB000000000002
  3082032C 308202D6 A0030201 02020A61 598DDB00 00000000 02300D06 092A8648
  86F70D01 01050500 302B310B 30090603 55040613 024D5931 0C300A06 0355040A
  1303474B 4E310E30 0C060355 04031305 56504E43 41301E17 0D303230 34323830
  38343432 335A170D 30333034 32383038 35343233 5A302B31 0B300906 03550406
  13024D59 310C300A 06035504 0A130347 4B4E310E 300C0603 55040313 0556504E
<Output Omitted>
  63727430 3906082B 06010505 07300286 2D66696C 653A2F2F 5C5C7472 61696E30
  325C4365 7274456E 726F6C6C 5C747261 696E3032 5F56504E 43412E63 7274300D
  06092A86 4886F70D 01010505 00034100 73B78B86 745C7C7C D396AFC2 117FEE56
  9B790A7C AF9BF5D7 48098FFA F9C0DC6E 8F7B31F1 23D60250 7FCF4305 394C8CCA
  FCA33116 1AD06E6A D0197616 FD2941F8
  quit
! The root (CA) certificate retrieved from the CA server
 certificate ca 5A1A8D98B485FCBC4C51E13CDE3460E5
  3082020C 308201B6 A0030201 0202105A 1A8D98B4 85FCBC4C 51E13CDE 3460E530
  0D06092A 864886F7 0D010105 0500302B 310B3009 06035504 0613024D 59310C30
  0A060355 040A1303 474B4E31 0E300C06 03550403 13055650 4E434130 1E170D30
```

```
    32303432 38303834 3234385A 170D3034 30343238 30383531 32365A30 2B310B30
    09060355 04061302 4D59310C 300A0603 55040A13 03474B4E 310E300C 06035504
    <Output Omitted>
    2BA029A0 27862566 696C653A 2F2F5C5C 74726169 6E30325C 43657274 456E726F
    6C6C6C56 504E4341 2E63726C 30100609 2B060104 01823715 01040302 0100300D
    06092A86 4886F70D 01010505 00034100 5154CAC8 A01A01F8 41B9EFC3 ADB61DF2
    7AD3D136 9EA22883 30B2E6FD BB27A015 0C81F4DC EEDA3E62 B23563F3 3A10B073
    140ED68E F9A99977 BDE9A4EA 13C045D3
    quit
! The RA's encryption certificate retrieved from the CA server
certificate ra-encrypt 61598FB1000000000003
    3082032C 308202D6 A0030201 02020A61 598FB100 00000000 03300D06 092A8648
    86F70D01 01050500 302B310B 30090603 55040613 024D5931 0C300A06 0355040A
    1303474B 4E310E30 0C060355 04031305 56504E43 41301E17 0D303230 34323830
    38343432 335A170D 30333034 32383038 35343233 5A302B31 0B300906 03550406
    13024D59 310C300A 06035504 0A130347 4B4E310E 300C0603 55040313 0556504E
    <Output Omitted>
    63727430 3906082B 06010505 07300286 2D66696C 653A2F2F 5C5C7472 61696E30
    325C4365 7274456E 726F6C6C 5C747261 696E3032 5F56504E 43412E63 7274300D
    06092A86 4886F70D 01010505 00034100 32EACF8A F98257C5 F95D7328 877F9200
    6B66B6A7 B7DF3D04 EBD17827 4A3FC6E3 07D371CD 68FC924A 09FB6E53 79DCB644
    1B726241 CB417A04 3A055E3B D4B241C2
    quit
!
! Supersede the default policy and use hash algorithm MD5.
! Notice that peer authentication is not set here because we are
! using the default peer authentication, RSA signatures.
crypto isakmp policy 110
 hash md5
!
! Specify the ESP settings for IPSec,
! which is later applied to the crypto map.
crypto ipsec transform-set chapter4 esp-des
!
! Specify the crypto map chapter4 where we define
! our peer Brussels-R1, transform set chapter4,
! and our crypto access list 102.
crypto map chapter4 10 ipsec-isakmp
 set peer 172.16.12.1
 set transform-set chapter4
 match address 102
!
interface FastEthernet0/0
 ip address 10.2.2.2 255.255.255.0
 !
```

```
interface FastEthernet0/1
 ip address 172.16.12.2 255.255.255.0
! Applied the crypto map to an interface
! to activate crypto engine.
 crypto map chapter4
!
! EIGRP is configured as the routing protocol
! between network 10.0.0.0 and 172.16.0.0.
router eigrp 12
 network 10.0.0.0
 network 172.16.0.0
 no auto-summary
!
! This is the crypto access list that we have
! referenced in crypto map chapter4.
! We are encrypting IP traffic between
! the remote FastEthernet LANs.
access-list 102 permit ip host 10.2.2.2 host 10.1.1.1
```

4.13.3   Verifying IKE (ISAKMP) and IPSec policies

Code Listings 4.28 to 4.33 look at how you can verify the IKE and IPSec policies that you have configured using the "show crypto isakmp policy," "show crypto map," and "show crypto ipsec transform-set" commands.

Code Listing 4.28 illustrates the ISAKMP policy belonging to Brussels-R1. From the command output, we can see that the encryption algorithm, hash algorithm, and authentication method correspond with DES, MD5, and RSA signatures, respectively.

Code Listing 4.28:   Brussels-R1 ISAKMP policy

```
Brussels-R1#show crypto isakmp policy
Protection suite of priority 110
    encryption algorithm:    DES - Data Encryption Standard (56 bit keys).
    hash algorithm:       Message Digest 5
    authentication method:    Rivest-Shamir-Adleman Signature
    Diffie-Hellman group:     #1 (768 bit)
    lifetime:          86400 seconds, no volume limit
Default protection suite
    encryption algorithm:    DES - Data Encryption Standard (56 bit keys).
    hash algorithm:       Secure Hash Standard
    authentication method:    Rivest-Shamir-Adleman Signature
    Diffie-Hellman group:     #1 (768 bit)
    lifetime:          86400 seconds, no volume limit
```

Code Listing 4.29 illustrates crypto map chapter4 that is implemented in Brussels-R1. From the crypto map, we can see that the remote IPSec peer is London-R2 (172.16.12.2), the crypto access list 101 is used to encrypt IP traffic between the remote FastEthernet LANs, and the IPSec transform set is chapter4 (see Code Listing 4.30).

Code Listing 4.29:   Brussels-R1 crypto map

```
Brussels-R1#show crypto map
Crypto Map "chapter4" 10 ipsec-isakmp
   Peer = 172.16.12.2
   Extended IP access list 101
      access-list 101 permit ip host 10.1.1.1 host 10.2.2.2
   Current peer: 172.16.12.2
   Security association lifetime: 4608000 kilobytes/3600 seconds
   PFS (Y/N): N
   Transform sets={ chapter4, }
```

In Code Listing 4.30, we can see that ESP-DES is the only transform defined for IPSec, and the IPSec tunnel (default) mode is being implemented. Brussels-R1's crypto map (see Code Listing 4.29), together with its IPSec transform, illustrates the overall IPSec policy that is implemented in the router.

Code Listing 4.30:   Brussels-R1 IPSec transform set

```
Brussels-R1#show crypto ipsec transform-set
Transform set chapter4: { esp-des   }
   will negotiate = { Tunnel,   },
```

Code Listing 4.31 illustrates the ISAKMP policy belonging to London-R2. From the command output, we can see that the encryption algorithm, hash algorithm, and authentication method correspond with DES, MD5, and RSA signatures, respectively.

Code Listing 4.31:   London-R2 ISAKMP policy

```
London-R2#show crypto isakmp policy
Protection suite of priority 110
   encryption algorithm:   DES - Data Encryption Standard (56 bit keys).
   hash algorithm:      Message Digest 5
```

```
    authentication method:    Rivest-Shamir-Adleman Signature
    Diffie-Hellman group:     #1 (768 bit)
    lifetime:                 86400 seconds, no volume limit
Default protection suite
    encryption algorithm:     DES - Data Encryption Standard (56 bit keys).
    hash algorithm:           Secure Hash Standard
    authentication method:    Rivest-Shamir-Adleman Signature
    Diffie-Hellman group:     #1 (768 bit)
    lifetime:       86,400 seconds, no volume limit
```

Code Listing 4.32 illustrates crypto map chapter4 that is implemented in London-R2. From the crypto map, we can see that the remote IPSec peer is Brussels-R1 (172.16.12.1), the crypto access list 102 is used to encrypt IP traffic between the remote FastEthernet LANs, and the IPSec transform set is chapter4 (see Code Listing 4.33).

Code Listing 4.32:　London-R2 crypto map

```
London-R2#show crypto map
Crypto Map "chapter4" 10 ipsec-isakmp
    Peer = 172.16.12.1
    Extended IP access list 102
        access-list 102 permit ip host 10.2.2.2 host 10.1.1.1
    Current peer: 172.16.12.1
    Security association lifetime: 4608000 kilobytes/3600 seconds
    PFS (Y/N): N
    Transform sets={ chapter4, }
```

In Code Listing 4.33, we can see that ESP-DES is the only transform defined for IPSec, and the IPSec tunnel (default) mode is being implemented. London-R2's crypto map (see Code Listing 4.32), together with its IPSec transform, illustrates the overall IPSec policy that is implemented in the router.

Code Listing 4.33:　London-R2 IPSec transform set

```
London-R2#show crypto ipsec transform-set
Transform set chapter4: { esp-des   }
    will negotiate = { Tunnel,   },
```

4.13.4　Monitoring IKE (ISAKMP) and IPSec debug trace

The "debug crypto isakmp" and "debug crypto ipsec" commands are used to observe the IKE and IPSec modes of operation and security association sequences. Code Listing 4.34 illustrates the IKE (ISAKMP) and IPSec debug trace when some IP traffic (ping) is generated from Brussels-R1's end. In the trace output below, italic comments are inserted at specific debugging sequence or "checkpoint" to describe them.

Code Listing 4.34:   IKE and IPSec debug trace at Brussels-R1

```
! Ensure that the "debug crypto isakmp"
! and "debug crypto ipsec" commands are activated.
Brussels-R1#show debug
Cryptographic Subsystem:
   Crypto ISAKMP debugging is on
   Crypto IPSEC debugging is on

! Generate IP traffic from Brussels-R1 (10.1.1.1)
! to London-R2 (10.2.2.2) using the extended ping command.
Brussels-R1#ping
Protocol [ip]:
Target IP address: 10.2.2.2
Repeat count [5]: 3
Datagram size [100]:
Timeout in seconds [2]:
Extended commands [n]: y
Source address or interface: 10.1.1.1
Type of service [0]:
Set DF bit in IP header? [no]:
Validate reply data? [no]:
Data pattern [0xABCD]:
Loose, Strict, Record, Timestamp, Verbose[none]:
Sweep range of sizes [n]:
Type escape sequence to abort.
Sending 3, 100-byte ICMP Echos to 10.2.2.2, timeout is 2 seconds:
Translating "London-R2.couver.com.sg"
!!!
Success rate is 100 percent (3/3), round-trip min/avg/max = 4/6/8 ms
! The ping source and destination addresses match
! access-list 101 defined in crypto map chapter4.


Brussels-R1#

! IPSec requests SAs between 10.1.1.1 and 10.2.2.2,
! on behalf of access list 101, and uses
! the transform set chapter4.
```

```
Apr 27 06:41:18: IPSEC(sa_request): ,
  (key eng. msg.) src= 172.16.12.1, dest= 172.16.12.2,
      src_proxy= 10.1.1.1/255.255.255.255/0/0 (type=1),
      dest_proxy= 10.2.2.2/255.255.255.255/0/0 (type=1),
! The src_proxy is the source interesting traffic,
! and the dest_proxy is the destination interesting traffic,
! both are defined in access list 101.
    <Output Omitted>
! This is ISAKMP Phase 1 using Oakley Main Mode.
Apr 27 06:41:18: ISAKMP (27): beginning Main Mode exchange
! The value 27 (in parentheses) is the connection ID assigned to IKE.
<Output Omitted>
Apr 27 06:41:18: ISAKMP (27): Checking ISAKMP transform 1 against priority
110 policy
! Policy 110 is the only ISAKMP policy
! configured on this local router (initiator).
Apr 27 06:41:18: ISAKMP:      encryption DES-CBC
Apr 27 06:41:18: ISAKMP:      hash MD5
Apr 27 06:41:18: ISAKMP:      default group 1
Apr 27 06:41:18: ISAKMP:      auth RSA sig
! These are the ISAKMP attributes being
! offered by the remote end (responder).
! The authentication method is RSA signatures.
Apr 27 06:41:18: ISAKMP (27): atts are acceptable. Next payload is 0
! IKE has found a matching policy since
! policy 110 on this router and the atts (attributes)
! that are offered by the remote end match.

! RSA signatures authentication will start now.
! The router will use its FQDN as the certificate ID.
Apr 27 06:41:18: ISAKMP (27): SA is doing RSA signature authentication using
id type ID_FQDN
<Output Omitted>
! Generate the Diffie-Hellman (D-H) shared secret
! to encrypt the rest of the IKE transaction..
Apr 27 06:41:18: ISAKMP (27): processing KE payload. message ID = 0
Apr 27 06:41:18: ISAKMP (27): processing NONCE payload. message ID = 0
Apr 27 06:41:18: ISAKMP (27): SKEYID state generated

! The responder's (London-R2's) request for
! the initiator's (Brussels-R1's) identity certificate.
Apr 27 06:41:18: ISAKMP (27): processing CERT_REQ payload. message ID = 0
Apr 27 06:41:18: ISAKMP (27): peer wants a CT_X509_SIGNATURE cert
<Output Omitted>
```

```
! Authenticate the D-H exchange and both parties by exchanging
! identity certificates (public keys). The D-H value and
! the identities are hashed, and the hash is signed with
! the local party private key. The receiver can decrypt the hash
! using the public key of the sender, which is obtained
! during the certificate exchange.
Apr 27 06:41:19: ISAKMP (27): sending packet to 172.16.12.2 (I) MM_KEY_EXCH
Apr 27 06:41:19: ISAKMP (27): received packet from 172.16.12.2 (I) MM_KEY_EXCH
Apr 27 06:41:19: ISAKMP (27): processing ID payload. message ID = 0
Apr 27 06:41:19: ISAKMP (27): processing CERT payload. message ID = 0
Apr 27 06:41:19: ISAKMP (27): processing a CT_X509_SIGNATURE cert
! The initiator has received the identity certificate
! from the responder and is validating the certificate.
! Once the certificate is validated, the responder's
! public key is obtained from the certificate.

Apr 27 06:41:19: ISAKMP (27): cert approved with warning
! This message appears when CRL is optional.
Apr 27 06:41:19: ISAKMP (27): processing SIG payload. message ID = 0
! Decrypt the digital signature (encrypted hash) sent
! from the responder using the public key of
! the responder retrieved  earlier, and
! validate the decrypted hash.

Apr 27 06:41:19: ISAKMP (27): sa->peer.name = 172.16.12.2, sa->peer_id.id.
id_fqdn.fqdn = London-R2.couver.com.sg
! The responder is identified by its FQDN,
! which is also its certificate ID.
Apr 27 06:41:19: ISAKMP (27): SA has been authenticated with 172.16.12.2
! RSA signatures authentication has succeeded at this point.
! The ISAKMP SA has been successfully negotiated.

! This is ISAKMP phase 2 using Oakley Quick Mode.
Apr 27 06:41:19: ISAKMP (27): beginning Quick Mode exchange, M-ID of 664498375
<Output Omitted>
Apr 27 06:41:20: ISAKMP (27): Checking IPSec proposal 1
! Here IKE will process the IPSec attributes
! offered by the remote end.

Apr 27 06:41:20: ISAKMP: transform 1, ESP_DES
! This is the protocol offered by the remote end
! according to its transform set.

Apr 27 06:41:20: ISAKMP:   attributes in transform:
Apr 27 06:41:20: ISAKMP:      encaps is 1
```

```
Apr 27 06:41:20: ISAKMP:        SA life type in seconds
Apr 27 06:41:20: ISAKMP:        SA life duration (basic) of 3600
Apr 27 06:41:20: ISAKMP:        SA life type in kilobytes
Apr 27 06:41:20: ISAKMP:        SA life duration (VPI) of  0x0 0x46 0x50 0x0
Apr 27 06:41:20: ISAKMP (27): atts are acceptable.
! A matching IPSec transform set has been found at the two peers.
! Now the IPSec SA can be created (one SA for each direction).
<Output Omitted>
Apr 27 06:41:20: ISAKMP (27): Creating IPSec SAs
Apr 27 06:41:20:        inbound SA from 172.16.12.2      to 172.16.12.1
(proxy 10.2.2.2        to 10.1.1.1        )
Apr 27 06:41:20:        has spi 229711572 and conn_id 28 and flags 4
Apr 27 06:41:20:        lifetime of 3600 seconds
Apr 27 06:41:20:        lifetime of 4608000 kilobytes
Apr 27 06:41:20:        outbound SA from 172.16.12.1     to 172.16.12.2
(proxy 10.1.1.1        to 10.2.2.2        )
Apr 27 06:41:20:        has spi 327816216 and conn_id 29 and flags 4
Apr 27 06:41:20:        lifetime of 3600 seconds
Apr 27 06:41:20:        lifetime of 4608000 kilobytes
! Two IPSec SAs have been negotiated,
! an incoming SA with the SPI generated by the initiator
! and an outbound SA with the SPI proposed by the responder.
! Crypto engine entries have also been created.
<Output Omitted>

! IPSec now installs the SA information into its SA database (SADB).
Apr 27 06:41:20: IPSEC(create_sa): sa created,
   (sa) sa_dest= 172.16.12.1, sa_prot= 50,
     sa_spi= 0xDB11ED4(229711572),
     sa_trans= esp-des , sa_conn_id= 28
Apr 27 06:41:20: IPSEC(create_sa): sa created,
   (sa) sa_dest= 172.16.12.2, sa_prot= 50,
     sa_spi= 0x138A1418(327816216),
     sa_trans= esp-des , sa_conn_id= 29
! The SADB has been updated and the IPSec SAs have been initialized.
! The tunnel is now fully operational.

Apr 27 06:41:20: ISAKMP (27): sending packet to 172.16.12.2 (I) QM_IDLE
! The IPSec SA has now been successfully negotiated.
! IKE will now go into a state known as QM-IDLE (quick mode idle).
```

4.13.5   Verifying IKE (ISAKMP) and IPSec security associations

In Code Listings 4.35 to 4.40, we conduct a postmodem on the IKE and IPSec
SAs after generating IP traffic from Brussels-R1 to London-R2 as illustrated in
Code Listing 4.34.

Code Listing 4.35 lists the current status of the IKE SA for Brussels-R1, which is in the quick mode idle state. The IKE tunnel starts at Brussels-R1 (172.16.12.1) and ends at London-R2 (172.16.12.2). The IKE SA is also assigned a connection ID value of 27 (that references an entry in the crypto engine connection table listed in Code Listing 4.37).

Code Listing 4.35:   Verifying IKE SA for Brussels-R1

```
Brussels-R1#show crypto isakmp sa
     dst               src              state           conn-id   slot
172.16.12.2       172.16.12.1          QM_IDLE            27       0
```

Code Listing 4.36:   Verifying IPSec SA for Brussels-R1

```
Brussels-R1#show crypto ipsec sa

interface: FastEthernet0/1
    Crypto map tag: chapter4, local addr. 172.16.12.1

   local   ident (addr/mask/prot/port): (10.1.1.1/255.255.255.255/0/0)
   remote ident (addr/mask/prot/port): (10.2.2.2/255.255.255.255/0/0)
   current_peer: 172.16.12.2
     PERMIT, flags={origin_is_acl,}
    #pkts encaps: 3, #pkts encrypt: 3, #pkts digest 0
    #pkts decaps: 3, #pkts decrypt: 3, #pkts verify 0
    #send errors 0, #recv errors 0

    local crypto endpt.: 172.16.12.1, remote crypto endpt.: 172.16.12.2
    path mtu 1500, media mtu 1500
    current outbound spi: 138A1418

    inbound esp sas:
     spi: 0xDB11ED4(229711572)
        transform: esp-des ,
        in use settings ={Tunnel, }
        slot: 0, conn id: 28, crypto map: chapter4
        sa timing: remaining key lifetime (k/sec): (4607999/3506)
        IV size: 8 bytes
        replay detection support: N

    inbound ah sas:
```

```
outbound esp sas:
  spi: 0x138A1418(327816216)
    transform: esp-des ,
    in use settings ={Tunnel, }
    slot: 0, conn id: 29, crypto map: chapter4
    sa timing: remaining key lifetime (k/sec): (4607999/3506)
    IV size: 8 bytes
    replay detection support: N

  outbound ah sas:
```

In Code Listing 4.36, the Brussels-R1's IPSec SA states that three interesting data packets, from local host 10.1.1.1 to remote host 10.2.2.2, have been encrypted, and another three interesting packets, from remote host 10.2.2.2 to local host 10.1.1.1, have been decrypted through interface FastEthernet0/1 where crypto map chapter4 is applied. The IPSec tunnel is established from local crypto endpoint 172.16.12.1 to remote crypto endpoint 172.16.12.2.

In addition, the inbound ESP SA has been populated into the SADB with information such as ESP-DES transform, IPSec tunnel mode, SPI value 229711572, connection ID value 28 (that references an entry in the crypto engine connection table listed in Code Listing 4.37), and crypto map chapter4. Likewise, outbound ESP SA has been populated into the SADB with information such as ESP-DES transform, IPSec tunnel mode, SPI value 327816216, connection ID value 29 (that references an entry in the crypto engine connection table listed in Code Listing 4.37), and crypto map chapter4. Notice that no inbound or outbound AH information is displayed because the AH protocol is not implemented in this case study.

Code Listing 4.37 illustrates the crypto engine connection table for Brussels-R1. From the table, we can see that connection ID 27 belongs to the IKE SA in Code Listing 4.35, and connection ID 28 and 29 belong to the two ESP SAs in Code Listing 4.36. Note that the outbound SA encrypts three data packets, and the inbound SA decrypts another three, concurring with Code Listing 4.36.

Code Listing 4.37: Brussels-R1 crypto engine connection table

---

```
Brussels-R1#show crypto engine connection active

ID  Interface       IP-Address      State   Algorithm     Encrypt     Decrypt
27  no idb          no address      set     DES_56_CBC          0           0
28  FastEthernet0/1 172.16.12.1     set     DES_56_CBC          0           3
29  FastEthernet0/1 172.16.12.1     set     DES_56_CBC          3           0
```

Code Listing 4.38 lists the current status of the IKE SA for London-R2, which is in the quick mode idle state. The IKE tunnel starts at Brussels-R1 (172.16.12.1) and ends at London-R2 (172.16.12.2). The IKE SA is also assigned a connection ID value of 28 (that references an entry in the crypto engine connection table listed in Code Listing 4.40).

Code Listing 4.38:   Verifying IKE SA for London-R2

```
London-R2#show crypto ipsec sa
    dst              src             state          conn-id   slot
172.16.12.2      172.16.12.1       QM_IDLE           28        0
```

Code Listing 4.39:   Verifying IPSec SA for London-R2

```
London-R2#show crypto ipsec sa

interface: FastEthernet0/1
    Crypto map tag: chapter4, local addr. 172.16.12.2

   local  ident (addr/mask/prot/port): (10.2.2.2/255.255.255.255/0/0)
   remote ident (addr/mask/prot/port): (10.1.1.1/255.255.255.255/0/0)
   current_peer: 172.16.12.1
     PERMIT, flags={origin_is_acl,}
    #pkts encaps: 3, #pkts encrypt: 3, #pkts digest 0
    #pkts decaps: 3, #pkts decrypt: 3, #pkts verify 0
    #send errors 0, #recv errors 0

    local crypto endpt.: 172.16.12.2, remote crypto endpt.: 172.16.12.1
    path mtu 1500, media mtu 1500
    current outbound spi: DB11ED4

    inbound esp sas:
     spi: 0x138A1418(327816216)
        transform: esp-des ,
        in use settings ={Tunnel, }
        slot: 0, conn id: 29, crypto map: chapter4
        sa timing: remaining key lifetime (k/sec): (4607999/3555)
        IV size: 8 bytes
        replay detection support: N

    inbound ah sas:

    outbound esp sas:
     spi: 0xDB11ED4(229711572)
```

```
transform: esp-des ,
in use settings ={Tunnel, }
slot: 0, conn id: 30, crypto map: chapter4
sa timing: remaining key lifetime (k/sec): (4607999/3555)
IV size: 8 bytes
replay detection support: N


outbound ah sas:
```

In Code Listing 4.39, the London-R2's IPSec SA states that three interesting data packets, from local host 10.2.2.2 to remote host 10.1.1.1, have been encrypted, and another three interesting packets, from remote host 10.1.1.1 to local host 10.2.2.2, have been decrypted through interface FastEthernet0/1 where crypto map chapter4 is applied. The IPSec tunnel is established from local crypto endpoint 172.16.12.2 to remote crypto endpoint 172.16.12.1.

In addition, the inbound ESP SA has been populated into the SADB with information such as ESP-DES transform, IPSec tunnel mode, SPI value 327816216, connection ID value 29 (that references an entry in the crypto engine connection table listed in Code Listing 4.40), and crypto map chapter4. Likewise, outbound ESP SA has been populated into the SADB with information such as ESP-DES transform, IPSec tunnel mode, SPI value 229711572, connection ID value 30 (that references an entry in the crypto engine connection table listed in Code Listing 4.40), and crypto map chapter4. Notice that no inbound or outbound AH information is displayed because the AH protocol is not implemented in this case study.

Code Listing 4.40 illustrates the crypto engine connection table for London-R2. From the table, we can see that connection ID 28 belongs to the IKE SA in Code Listing 4.38, and connection ID 29 and 30 belong to the two ESP SAs in Code Listing 4.39. Note that the outbound SA encrypts three data packets, and the inbound SA decrypts another three, concurring with Code Listing 4.39.

Code Listing 4.40:   London-R2 crypto engine connection Table

---

```
London-R2#show crypto engine active
```

ID	Interface	IP-Address	State	Algorithm	Encrypt	Decrypt
28	no idb	no address	set	DES_56_CBC	0	0
29	FastEthernet0/1	172.16.12.1	set	DES_56_CBC	0	**3**
30	FastEthernet0/1	172.16.12.1	set	DES_56_CBC	**3**	0

4.14 Summary

Chapter 4 gives an overview of the basic concepts of the RSA algorithm and RSA keys as well as with how RSA encrypted nonces can be used in IKE peer authentication. The configuring and monitoring of RSA encrypted nonces for a site-to-site IPSec VPN are illustrated in Case Study 4.1. We then move on to identify the scalability issues when using pre-shared keys and RSA encrypted nonces, as well as to explain the functionalities of digital signatures and digital certificates. Besides describing how the CA deployment in an IPSec VPN environment can improve scalability, the roles of the certificate authority and registration authority are also further discussed.

Next, we identify the different transactions that occur during certificate generation and explain the differences between file-based enrollment and network-based enrollment. We then proceed to describe the simple certificate enrollment protocol and identify the criteria to validate a digital certificate. We also describe what a certificate revocation list is, explain how RSA signatures can be used in IKE peer authentication, and look at how to set up a stand-alone CA server. The configuring and monitoring of RSA signatures for a site-to-site IPSec VPN are illustrated in Case Study 4.2.

Deploying IPSec VPN in the Enterprise

5.1 Chapter Overview

In Chapters 3 and 4, the focus was on implementing a single site-to-site IPSec VPN and the different IKE peer authentication techniques. Chapter 5 is an extension of those chapters, and the emphasis is now on deploying multiple IPSec VPN sites in large enterprise networks.

When deploying an IPSec VPN in large enterprises, many factors affect the scalability, reliability, interoperability, and performance of the network. Some of these factors include network topology design, such as fully meshed and hub-and-spoke networks; network resiliency techniques, such as HSRP (hot standby router protocol); IP addressing services, such as NAT (network address translation); and performance parameters, such as fragmentation, IKE SA lifetimes, and IKE keepalives. In this chapter, the focus is on the design of scalable IPSec VPNs using the hub-and-spoke topology. Network resiliency, NAT, and performance optimization with IPSec are covered in Chapter 6.

5.2 Meshed Versus Hub-and-Spoke Networks

In a fully meshed IPSec network, every device in the network communicates with every other device via a unique IPSec tunnel. This becomes a scalability issue when the number of devices (or nodes) in the network increases. For instance, a 60-node network will require $n*(n - 1)/2$ tunnels (where n is the number of nodes), or 1770 tunnels to be exact!

Furthermore, with the growing number of nodes in the network, the configuration burden, as well as the complexity, becomes mammoth, and at a certain point it will not be possible to grow the size of the mesh anymore. Keeping track of so many tunnels also creates performance impacts such as heavy CPU utilization. In other words, the limiting factor in this topology is the number of tunnels that the devices can support at a reasonable CPU utilization.

Hub-and-spoke IPSec networks scale better because the hub site can expand to meet growing spoke capacity requirements. In this case, local spoke sites that require connectivity to other remote spoke sites are connected via the hub site, and this reduces the number of IPSec tunnels required for spoke-to-spoke communications. Instead of direct spoke-to-spoke communication, information is now exchanged indirectly between the spoke sites via the hub site. In this instance, a 60-node network will need only (n-1) tunnels, or 59 tunnels, this time.

The limiting factor in this topology is the significant bandwidth requirement for all the traffic that flows through the hub site, which includes all spoke-to-spoke traffic as well as spoke-to-hub traffic. In addition, not all VPN devices support spoke-to-spoke intercommunication via a hub site, and this could contribute an additional constraint. For Cisco routers, the minimum IOS software version to support the IPSec hub and spoke topology is 12.2(5).

5.3   Case Study 5.1: IPSec VPN in a Hub-and-Spoke Topology

5.3.1   Case overview and network topology

This case study demonstrates a hub-and-spoke IPSec network design that spans three routers: Brussels-R1 (spoke 1), Brussels-R2 (spoke 2), and Brussels-R3 (hub). In this instance, information is exchanged between the spoke sites by traversing through the hub. In other words, as illustrated in Figure 5.1, no direct IPSec tunnel exists between the two spoke routers. All data packets are

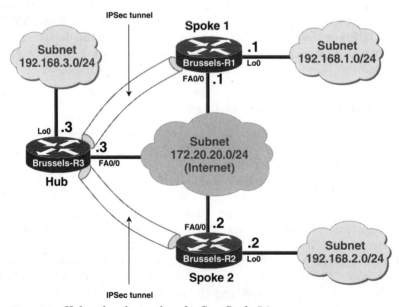

Figure 5.1: Hub-and-spoke topology for Case Study 5.1.

sent across the tunnel to the hub router where it redistributes them through the IPSec tunnel that is shared with the other spoke router.

In the case study, encryption is implemented from:

- Spoke 1 (subnet 192.168.1.0/24) to hub (subnet 192.168.3.0/24) and vice versa.
- Spoke 2 (subnet 192.168.2.0/24) to hub (subnet 192.168.3.0/24) and vice versa.
- Spoke 1 (subnet 192.168.1.0/24) to spoke 2 (subnet 192.168.2.0/24) and vice versa.

5.3.2 Hub-and-spoke IPSec configurations

Code Listings 5.1 to 5.3 are the hub-and-spoke IPSec configurations for Brussels-R1, Brussels-R2, and Brussels-R3, respectively. In the configurations, comments (in *italics*) precede certain configuration lines to explain them.

Code Listing 5.1:   IPSec spoke configuration for Brussels-R1

```
hostname Brussels-R1
!
! Supersede the default policy and use pre-shared keys
! for peer authentication and hash algorithm MD5.
crypto isakmp policy 10
 hash md5
 authentication pre-share
!
! Specify the pre-shared key "key31" to be used
! for the IPSec tunnel between Brussels-R1 and Brussels-R3
crypto isakmp key key31 address 172.20.20.3
!
! Specify the ESP transform settings for IPSec,
! which is later applied to the crypto map.
crypto ipsec transform-set chapter5 esp-des esp-md5-hmac
!
! Specify the crypto map chapter5 where we define
! our hub peer Brussels-R3, transform set chapter5,
! and our crypto access list 110.
crypto map chapter5 10 ipsec isakmp
 set peer 172.20.20.3
 set transform-set chapter5
 match address 110
!
! Emulate subnet 192.168.1.0/24 with a loop-back interface.
interface Loopback0
 ip address 192.168.1.1 255.255.255.0
```

```
!
interface FastEthernet0/0
 ip address 172.20.20.1 255.255.255.0
! Apply the crypto map to an interface
! to activate crypto engine.
 crypto map chapter5
!
! Static routes are configured for
! subnet 192.168.2.0/24 and 192.168.3.0/24.
ip route 192.168.2.0 255.255.255.0 FastEthernet0/0
ip route 192.168.3.0 255.255.255.0 FastEthernet0/0
!
! This is the crypto access list that we
! referenced in crypto map chapter5.
! We are encrypting IP traffic between subnets
! 192.168.1.0/24 and 192.168.3.0/24 (between spoke 1 and hub),
! as well as between subnets 192.168.1.0/24
! and 192.168.2.0/24 (between spoke 1 and spoke 2).
access-list 110 permit ip 192.168.1.0 0.0.0.255 192.168.3.0 0.0.0.255
access-list 110 permit ip 192.168.1.0 0.0.0.255 192.168.2.0 0.0.0.255
```

Code Listing 5.1 illustrates the IPSec spoke-end configuration for Brussels-R1 (spoke 1). Take note of the crypto access list, which is used to define the IP traffic flows that Brussels-R1 is encrypting. In this case, the traffic flows are from spoke 1 to hub and from spoke 1 to spoke 2.

Code Listing 5.2: IPSec spoke configuration for Brussels-R2

```
hostname Brussels-R2
!
! Supersede the default policy and use pre-shared keys
! for peer authentication and hash algorithm MD5.
crypto isakmp policy 10
 hash md5
 authentication pre-share
!
! Specify the pre-shared key "key32" to be used
! for the IPSec tunnel between Brussels-R2 and Brussels-R3.
crypto isakmp key key32 address 172.20.20.3
!
! Specify the ESP transform settings for IPSec,
! which is later applied to the crypto map.
crypto ipsec transform-set chapter5 esp-des esp-md5-hmac
```

```
!
! Specify the crypto map chapter5 where we define
! our hub peer Brussels-R3, transform set chapter5,
! and our crypto access list 120.
crypto map chapter5 10 ipsec-isakmp
 set peer 172.20.20.3
 set transform-set chapter5
 match address 120
!
! Emulate subnet 192.168.2.0/24 with a loop-back interface.
interface Loopback0
 ip address 192.168.2.2 255.255.255.0
!
interface FastEthernet0/0
 ip address 172.20.20.2 255.255.255.0
! Apply the crypto map to an interface
! to activate crypto engine.
 crypto map chapter5
!
! Static routes are configured for
! subnet 192.168.1.0/24 and 192.168.3.0/24.
ip route 192.168.1.0 255.255.255.0 FastEthernet0/0
ip route 192.168.3.0 255.255.255.0 FastEthernet0/0
!
! This is the crypto access list that we have
! referenced in crypto map chapter5.
! We are encrypting IP traffic between subnets 192.168.2.0/24
! and 192.168.3.0/24 (between spoke 2 and hub), as well as
! between subnets 192.168.2.0/24
! and 192.168.1.0/24 (between spoke 2 and spoke 1).
access-list 120 permit ip 192.168.2.0 0.0.0.255 192.168.3.0 0.0.0.255
access-list 120 permit ip 192.168.2.0 0.0.0.255 192.168.1.0 0.0.0.255
```

Code Listing 5.2 illustrates the IPSec spoke-end configuration for Brussels-R2 (spoke 2). Note the crypto access list, which is used to define the IP traffic flows that Brussels-R2 is encrypting. In this case, the traffic flows are from spoke 2 to hub, and from spoke 2 to spoke 1.

Code Listing 5.3:   IPSec Hub Configuration for Brussels-R3

```
hostname Brussels-R3
!
! Supersede the default policy and use pre-shared keys
! for peer authentication and hash algorithm MD5.
```

```
crypto isakmp policy 10
 hash md5
 authentication pre-share
!
! Specify the pre-shared key "key32" to be used for the
! IPSec tunnel between Brussels-R3 and Brussels-R2.
crypto isakmp key key32 address 172.20.20.2
!
! Specify the pre-shared key "key31" to be used for
! the IPSec tunnel between Brussels-R3 and Brussels-R1
crypto isakmp key key31 address 172.20.20.1
!
! Specify the ESP transform settings for IPSec,
! which is later applied to the crypto map.
crypto ipsec transform-set chapter5 esp-des esp-md5-hmac
!
! Specify the first crypto map instance where we define
! our spoke peer Brussels-R1, transform set chapter5,
! and a unique crypto access list 110.
crypto map chapter5 10 ipsec-isakmp
 set peer 172.20.20.1
 set transform-set chapter5
 match address 110
!
! Specify the first crypto map instance where we define
! our spoke peer Brussels-R2, transform set chapter5,
! and a unique crypto access list 120.
crypto map chapter5 20 ipsec-isakmp
 set peer 172.20.20.2
 set transform-set chapter5
 match address 120
!
! Emulate subnet 192.168.3.0/24 with a loop-back interface.
interface Loopback0
 ip address 192.168.3.3 255.255.255.0
!
interface FastEthernet0/0
 ip address 172.20.20.3 255.255.255.0
! Apply the crypto map to an interface
! to activate crypto engine.
 crypto map chapter5
!
! Static routes are configured for
! subnet 192.168.1.0/24 and 192.168.2.0/24.
ip route 192.168.1.0 255.255.255.0 FastEthernet0/0
ip route 192.168.2.0 255.255.255.0 FastEthernet0/0
```

```
!
! In access list 110, we are encrypting IP traffic between subnets
! 192.168.3.0/24 and 192.168.1.0/24 (between hub and spoke 1),
! as well as between subnets 192.168.2.0/24 and
! 192.168.1.0/24 (between spoke 2 and spoke 1).
access-list 110 permit ip 192.168.3.0 0.0.0.255 192.168.1.0 0.0.0.255
access-list 110 permit ip 192.168.2.0 0.0.0.255 192.168.1.0 0.0.0.255
!
! In access list 120, we are encrypting IP traffic between subnets
! 192.168.3.0/24 and 192.168.2.0/24 (between hub and spoke 2),
! as well as between subnets 192.168.1.0/24
! and 192.168.2.0/24 (between spoke 1 and spoke 2).
access-list 120 permit ip 192.168.3.0 0.0.0.255 192.168.2.0 0.0.0.255
access-list 120 permit ip 192.168.1.0 0.0.0.255 192.168.2.0 0.0.0.255
```

Code Listing 5.3 illustrates the IPSec hub-end configuration for Brussels-R3 (hub). Note the two crypto access lists, which are used to define the IP traffic flows that Brussels-R3 is encrypting. In this case, the traffic flows are from hub to spoke 1 and from spoke 2 to spoke 1 (via hub), as well as from hub to spoke 2 and from spoke 1 to spoke 2 (via hub).

5.3.3 Verifying crypto access lists

In Code Listings 5.4 to 5.6, we verify the crypto access lists that we have configured for the three routers by examining their respective crypto maps using the "show crypto map" command.

Code Listing 5.4:   Brussels-R1 crypto map

```
Brussels-R1#show crypto map
Crypto Map "chapter5" 10 ipsec-isakmp
    Peer = 172.20.20.3
    Extended IP access list 110
        access-list 110 permit ip 192.168.1.0 0.0.0.255 192.168.3.0 0.0.0.255
        access-list 110 permit ip 192.168.1.0 0.0.0.255 192.168.2.0 0.0.0.255
    Current peer: 172.20.20.3
    Security association lifetime: 4608000 kilobytes/3600 seconds
    PFS (Y/N): N
    Transform sets={ chapter5, }
    Interfaces using crypto map chapter5:
        FastEthernet0/0
```

From Code Listing 5.4, we can verify the crypto access list that we configured earlier for Brussels-R1 is encrypting IP traffic between subnets 192.168.1.0/24 and 192.168.3.0/24 (between spoke 1 and hub), as well as between subnets 192.168.1.0/24 and 192.168.2.0/24 (between spoke 1 and spoke 2).

From Code Listing 5.5, we can verify the crypto access list that we have configured earlier for Brussels-R2 is encrypting IP traffic between subnets 192.168.2.0/24 and 192.168.1.0/24 (between spoke 2 and spoke 1), as well as between subnets 192.168.2.0/24 and 192.168.3.0/24 (between spoke 2 and hub).

Code Listing 5.5:   Brussels-R2 crypto map

```
Brussels-R2#show crypto map
Crypto Map "chapter5" 10 ipsec-isakmp
    Peer = 172.20.20.3
    Extended IP access list 120
        access-list 120 permit ip 192.168.2.0 0.0.0.255 192.168.1.0 0.0.0.255
        access-list 120 permit ip 192.168.2.0 0.0.0.255 192.168.3.0 0.0.0.255
    Current peer: 172.20.20.3
    Security association lifetime: 4608000 kilobytes/3600 seconds
    PFS (Y/N): N
    Transform sets={ chapter5, }
    Interfaces using crypto map chapter5:
        FastEthernet0/0
```

From Code Listing 5.6, we can verify the two crypto access lists that we have configured earlier for Brussels-R3. In access list 110, we are encrypting IP traffic between subnets 192.168.3.0/24 and 192.168.1.0/24 (between hub and spoke 1), as well as between subnets 192.168.2.0/24 and 192.168.1.0/24 (between spoke 2 and spoke 1). In access list 120, we are encrypting IP traffic between subnets 192.168.3.0/24 and 192.168.2.0/24 (between hub and spoke 2), as well as between subnets 192.168.1.0/24 and 192.168.2.0/24 (between spoke 1 and spoke 2).

Code Listing 5.6:   Brussels-R3 Crypto Map

```
Brussels-R3#show crypto map
Crypto Map "chapter5" 10 ipsec-isakmp
    Peer = 172.20.20.1
    Extended IP access list 110
        access-list 110 permit ip 192.168.3.0 0.0.0.255 192.168.1.0 0.0.0.255
        access-list 110 permit ip 192.168.2.0 0.0.0.255 192.168.1.0 0.0.0.255
    Current peer: 172.20.20.1
    Security association lifetime: 4608000 kilobytes/3600 seconds
```

```
     PFS (Y/N): N
     Transform sets={ chapter5, }


Crypto Map "chapter5" 20 ipsec-isakmp
     Peer = 172.20.20.2
     Extended IP access list 120
         access-list 120 permit ip 192.168.3.0 0.0.0.255 192.168.2.0 0.0.0.255
         access-list 120 permit ip 192.168.1.0 0.0.0.255 192.168.2.0 0.0.0.255
     Current peer: 172.20.20.2
     Security association lifetime: 4608000 kilobytes/3600 seconds
     PFS (Y/N): N
     Transform sets={ chapter5, }
     Interfaces using crypto map chapter5:
        FastEthernet0/0
```

5.3.4 Monitoring and verifying hub-and-spoke IPSec operations

Code Listing 5.7 illustrates the ping test disseminated from spoke 1
(192.168.1.1) to spoke 2 (192.168.2.2). In this instance, the test is only 60
percent successful, that is, 3 of 5 ping packets went through to spoke 2.

Code Listing 5.7:   Ping test #1 from spoke 1 to spoke 2

```
Brussels-R1#ping
Protocol [ip]:
Target IP address: 192.168.2.2
Repeat count [5]:
Datagram size [100]:
Timeout in seconds [2]:
Extended commands [n]: y
Source address or interface: 192.168.1.1
Type of service [0]:
Set DF bit in IP header? [no]:
Validate reply data? [no]:
Data pattern [0xABCD]:
Loose, Strict, Record, Timestamp, Verbose[none]:
Sweep range of sizes [n]:
Type escape sequence to abort.
Sending 5, 100-byte ICMP Echos to 192.168.2.2, timeout is 2 seconds:
..!!!
Success rate is 60 percent (3/5), round-trip min/avg/max = 12/13/16 ms
```

Code Listing 5.8 illustrates the results for ping test #1 conducted in Code
Listing 5.7. In the command traces below, italic comments are inserted at

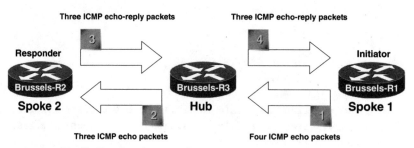

Figure 5.2: Traffic flow for ping test #1.

specific "checkpoints" for description purposes. We shall examine the traces from Brussels-R1, Brussels-R3, and Brussels-R2 in that order. The traffic flow for ping test #1 is illustrated in Figure 5.2.

Code Listing 5.8:   Results for ping test #1

We first examine the respective command outputs at Brussels-R1. The "show crypto isakmp sa" command at Brussels-R1 (Spoke 1) indicates that the IKE tunnel is created from spoke 1 (172.20.20.1) to hub (172.20.20.3) when the IP traffic flows from spoke 1 to spoke 2. Note that the connection ID assigned is 1.

```
Brussels-R1#show crypto isakmp sa
dst             src             state         conn-id   slot
172.20.20.3     172.20.20.1     QM_IDLE           1     · 0
```

From the "show crypto ipsec sa" command at Brussels-R1 (spoke 1), for the IPSec tunnel created between spoke 1 and hub, we can gather that four of five outgoing IP (ICMP echo) packets to spoke 2 have been encrypted (by DES) and digested (by MD5), and only three incoming IP (ICMP echo-reply) packets from spoke 2 have been decrypted (by DES) and verified (by MD5).

```
Brussels-R1#show crypto ipsec sa
interface: FastEthernet0/0
    Crypto map tag: chapter5, local addr. 172.20.20.1
```

Traffic flow is from local subnet 192.168.1.0/24 to remote subnet 192.168.2.0/24.
```
    local ident (addr/mask/prot/port): (192.168.1.0/255.255.255.0/0/0)
    remote ident (addr/mask/prot/port): (192.168.2.0/255.255.255.0/0/0)
```

```
current_peer: 172.20.20.3
  PERMIT, flags={origin_is_acl,}
```

Four outgoing (ICMP echo) packets from spoke 1 to spoke 2 are encrypted and digested (see Fig. 5.2).
```
#pkts encaps: 4, #pkts encrypt: 4, #pkts digest 4
```

Three incoming (ICMP echo-reply) packets from spoke 2 are decrypted and verified (see Fig. 5.2).
```
#pkts decaps: 3, #pkts decrypt: 3, #pkts verify 3
<Output Omitted>
```

At spoke 1, of the five packets sent, one is bad.
```
#send errors 1, #recv errors 0
```

The IPSec tunnel originates from spoke 1 (172.20.20.1) and terminates at hub (172.20.20.3).
```
local crypto endpt.: 172.20.20.1, remote crypto endpt.: 172.20.20.3
<Output Omitted>
```

For the inbound IPSec SA from hub to spoke 1, the SPI is 1605659124 and the connection ID is 2000.
```
inbound esp sas:
spi: 0x5FB469F4(1605659124)
  transform: esp-des esp-md5-hmac ,
  in use settings ={Tunnel, }
  slot: 0, conn id: 2000, flow_id: 1, crypto map: chapter5
  <Output Omitted>
```

For the outbound IPSec SA from spoke 1 to hub, the SPI is 1837161637 and the connection ID is 2001.
```
outbound esp sas:
spi: 0x6D80DCA5(1837161637)
  transform: esp-des esp-md5-hmac ,
  in use settings ={Tunnel, }
  slot: 0, conn id: 2001, flow_id: 2, crypto map: chapter5
  <Output Omitted>
```

The "show crypto engine connection active" command at Brussels-R1 concurs with the "show crypto ipsec sa" command reflecting four outbound (ICMP echo) packets being encrypted (and digested) and three inbound (ICMP echo-reply) packets being decrypted and verified. The connection IDs for the inbound and outbound IPSec SAs are 2000 and 2001, respectively, which correspond to the same connection ID values illustrated earlier in the "show crypto ipsec sa" command.

```
Brussels-R1#show crypto engine connection active

  ID  Interface      IP-Address     State  Algorithm              Encrypt  Decrypt
   1  <none>         <none>         set    HMAC_MD5+DES_56_CB        0        0
2000  FastEthernet0/0 172.20.20.1   set    HMAC_MD5+DES_56_CB        0        3
2001  FastEthernet0/0 172.20.20.1   set    HMAC_MD5+DES_56_CB        4        0
```

Next, we examine the respective command outputs at Brussels-R3. The "show crypto isakmp sa" command at Brussels-R3 (hub) indicates that two IKE tunnels are created when the IP traffic flows from spoke 1 to spoke 2, one from spoke 1 (172.20.20.1) to hub (172.20.20.3) and the other from hub (172.20.20.3) to spoke 2 (172.20.20.2). Note that the connection IDs assigned are 1 and 2, respectively.

```
Brussels-R3#show crypto isakmp sa
dst              src              state    conn-id    slot
172.20.20.2      172.20.20.3      QM_IDLE      2        0
172.20.20.3      172.20.20.1      QM_IDLE      1        0
```

From the "show crypto ipsec sa" command at Brussels-R3 (hub), for the IPSec tunnel created between hub and spoke 1, we can gather that three outgoing IP (ICMP echo-reply) packets from spoke 2 to spoke 1 have been encrypted, as well as digested, and four incoming IP (ICMP echo) packets from spoke 1 to spoke 2 have been decrypted and verified. For the IPSec tunnel created between hub and spoke 2, we can gather that three of four outgoing IP (ICMP echo) packets from spoke 1 to spoke 2 have been encrypted and digested, and three incoming IP (ICMP echo-reply) packets from spoke 2 to spoke 1 have been decrypted and verified.

```
Brussels-R3#show crypto ipsec sa

interface: FastEthernet0/0
    Crypto map tag: chapter5, local addr. 172.20.20.3
```

Traffic flow is from local subnet 192.168.2.0/24 to remote subnet 192.168.1.0/24.
```
    local ident (addr/mask/prot/port): (192.168.2.0/255.255.255.0/0/0)
    remote ident (addr/mask/prot/port): (192.168.1.0/255.255.255.0/0/0)
    current_peer: 172.20.20.1
      PERMIT, flags={origin_is_acl,}
```

Three outgoing (ICMP echo-reply) packets from spoke 2 to spoke 1 are encrypted and digested (see Fig. 5.2).
```
    #pkts encaps: 3, #pkts encrypt: 3, #pkts digest 3
```

Four incoming (ICMP echo) packets from spoke 1 to spoke 2 are decrypted and verified (see Fig. 5.2).

    #pkts decaps: 4, **#pkts decrypt: 4, #pkts verify 4**
    <Output Omitted>

This first IPSec tunnel originates from hub (172.20.20.3) and terminates at spoke 1 (172.20.20.1).

    local crypto endpt.: 172.20.20.3, remote crypto endpt.: 172.20.20.1
    <Output Omitted>

For the inbound IPSec SA from spoke 1 to hub, the SPI is 1837161637 and the connection ID is 2000.

    inbound esp sas:
    spi: 0x6D80DCA5(1837161637)
      transform: esp-des esp-md5-hmac ,
      in use settings ={Tunnel, }
      slot: 0, **conn id: 2000**, flow_id: 1, crypto map: chapter5
      <Output Omitted>

For the outbound IPSec SA from hub to spoke 1, the SPI is 1605659124 and the connection ID is 2001.

    outbound esp sas:
    spi: 0x5FB469F4(1605659124)
      transform: esp-des esp-md5-hmac ,
      in use settings ={Tunnel, }
      slot: 0, **conn id: 2001**, flow_id: 2, crypto map: chapter5
      <Output Omitted>

Traffic flow is from local subnet 192.168.1.0/24 to remote subnet 192.168.2.0/24.

    local ident (addr/mask/prot/port): **(192.168.1.0/255.255.255.0**/0/0)
    remote ident (addr/mask/prot/port): **(192.168.2.0/255.255.255.0**/0/0)
    current_peer: 172.20.20.2
      PERMIT, flags={origin_is_acl,}

Three outgoing (ICMP echo) packets from spoke 1 to spoke 2 are encrypted and digested (see Fig. 5.2).

    #pkts encaps: 3, **#pkts encrypt: 3, #pkts digest 3**

Three incoming (ICMP echo-reply) packets from spoke 2 to spoke 1 are decrypted and verified (see Fig. 5.2).

    #pkts decaps: 3, **#pkts decrypt: 3, #pkts verify 3**
    <Output Omitted>

At the hub, of the four traversing packets from spoke 1 to spoke 2, one is bad.

    #send errors 1, #recv errors 0

This other IPSec tunnel originates from hub (172.20.20.3) and terminates at spoke 2 (172.20.20.2).

    local crypto endpt.: 172.20.20.3, remote crypto endpt.: 172.20.20.2

    <Output Omitted>

For the inbound IPSec SA from spoke 2 to hub, the SPI is 2981567991 and the connection ID is 2002.

    inbound esp sas:

    spi: 0xʙ1B71DF7(2981567991)

      transform: esp-des esp-md5-hmac ,

      in use settings ={Tunnel, }

      slot: 0, **conn id: 2002**, flow_id: 3, crypto map: chapter5

      <Output Omitted>

For the outbound IPSec SA from hub to spoke 2, the SPI is 2142640922, and the connection ID is 2003.

    outbound esp sas:

    spi: 0x7FB61B1A(2142640922)

      transform: esp-des esp-md5-hmac ,

      in use settings ={Tunnel, }

      slot: 0, **conn id: 2003**, flow_id: 4, crypto map: chapter5

      <Output Omitted>

The "show crypto engine connection active" command at Brussels-R3 concurs with the "show crypto ipsec sa" command. For the IPSec tunnel between hub and spoke 1, the command reflects three outbound (ICMP echo-reply) packets being encrypted and digested together with four inbound (ICMP echo) packets being decrypted and verified, and the connections IDs for the inbound and outbound IPSec SAs between hub and spoke 1 are 2000 and 2001, respectively. For the IPSec tunnel between hub and spoke 2, the command reflects three outbound (ICMP echo) packets being encrypted and digested together with three inbound (ICMP echo-reply) packets being decrypted and verified, and the connection IDs for the inbound and outbound IPSec SAs between hub and spoke 2 correspond with 2002 and 2003, respectively.

Brussels-R3#show crypto engine connection active

ID	Interface	IP-Address	State	Algorithm	Encrypt	Decrypt
1	FastEthernet0/0	172.20.20.3	set	HMAC_MD5+DES_56_CB	0	0
2	<none>	<none>	set	HMAC_MD5+DES_56_CB	0	0
2000	FastEthernet0/0	172.20.20.3	set	HMAC_MD5+DES_56_CB	0	4
2001	FastEthernet0/0	172.20.20.3	set	HMAC_MD5+DES_56_CB	3	0
2002	FastEthernet0/0	172.20.20.3	set	HMAC_MD5+DES_56_CB	0	3
2003	FastEthernet0/0	172.20.20.3	set	HMAC_MD5+DES_56_CB	3	0

We now examine the respective command outputs at Brussels-R2. The "show crypto isakmp sa" command at Brussels-R2 (spoke 2) indicates that the IKE tunnel is created from hub (172.20.20.3) to spoke 2 (172.20.20.2) when the IP traffic flows from spoke 1 to spoke 2. Note the connection ID assigned is 1.

Brussels-R2#show crypto isakmp sa
```
dst            src            state         conn-id   slot
172.20.20.2    172.20.20.3    QM_IDLE       1         0
```

From the "show crypto ipsec sa" command at Brussels-R2 (spoke 2), for the IPSec tunnel created between spoke 2 and hub, we can gather that three outgoing IP (ICMP echo-reply) packets to spoke 1 have been encrypted and digested, and three incoming IP (ICMP echo) packets from spoke 1 have been decrypted and verified.

Brussels-R2#show crypto ipsec sa

```
interface: FastEthernet0/0
  Crypto map tag: chapter5, local addr. 172.20.20.2
```

Traffic flow is from local subnet 192.168.2.0/24 to remote subnet 192.168.1.0/24.
```
  local ident (addr/mask/prot/port): (192.168.2.0/255.255.255.0/0/0)
  remote ident (addr/mask/prot/port): (192.168.1.0/255.255.255.0/0/0)
  current_peer: 172.20.20.3
    PERMIT, flags={origin_is_acl,}
```

Three outgoing (ICMP echo-reply) packets from spoke 2 to spoke 1 are encrypted and digested (see Fig. 5.2).
```
  #pkts encaps: 3, #pkts encrypt: 3, #pkts digest 3
```

Three incoming (ICMP echo) packets from spoke 1 are decrypted and verified (see Fig. 5.2).
```
  #pkts decaps: 3, #pkts decrypt: 3, #pkts verify 3
  <Output Omitted>
```

The IPSec tunnel originates from spoke 2 (172.20.20.2) and terminates at hub (172.20.20.3).
```
    local crypto endpt.: 172.20.20.2, remote crypto endpt.: 172.20.20.3
    <Output Omitted>
```

For the inbound IPSec SA from hub to spoke 2, the SPI is 2142640922 and the connection ID is 2000.
```
  inbound esp sas:
  spi: 0x7FB61B1A(2142640922)
```

```
transform: esp-des esp-md5-hmac ,
in use settings ={Tunnel, }
slot: 0, conn id: 2000, flow_id: 1, crypto map: chapter5
sa timing: remaining key lifetime (k/sec): (4607999/3435)
<Output Omitted>
```

For the outbound IPSec SA from spoke 2 to hub, the SPI is 2981567991 and the connection ID is 2001.

```
outbound esp sas:
spi: 0xB1B71DF7(2981567991)
    transform: esp-des esp-md5-hmac ,
    in use settings ={Tunnel, }
    slot: 0, conn id: 2001, flow_id: 2, crypto map: chapter5
    <Output Omitted>
```

The "show crypto engine connection active" command at Brussels-R2 concurs with the "show crypto ipsec sa" command reflecting three outbound (ICMP echo-reply) packets being encrypted and digested, and three inbound (ICMP echo) packets being decrypted and verified. The connections IDs for the inbound and outbound IPSec SAs are 2000 and 2001, respectively, which correspond to the same connection ID values illustrated earlier in the "show crypto ipsec sa" command.

Brussels-R2#show crypto engine connection active

ID	Interface	IP-Address	State	Algorithm	Encrypt	Decrypt
1	FastEthernet0/0	172.20.20.3	set	HMAC_MD5+DES_56_CB	0	0
2000	FastEthernet0/0	172.20.20.3	set	HMAC_MD5+DES_56_CB	0	3
2001	FastEthernet0/0	172.20.20.3	set	HMAC_MD5+DES_56_CB	3	0

Code Listing 5.9 illustrates the ping test disseminated from spoke 2 (192.168.2.2) to spoke 1 (192.168.1.1). In this instance, the test is 100 percent successful, that is, 5 of 5 ping packets went through to spoke 1.

Code Listing 5.9:   Ping Test #2 from spoke 2 to spoke 1

```
Brussels-R2#ping
Protocol [ip]:
Target IP address: 192.168.1.1
Repeat count [5]:
Datagram size [100]:
Timeout in seconds [2]:
Extended commands [n]: y
```

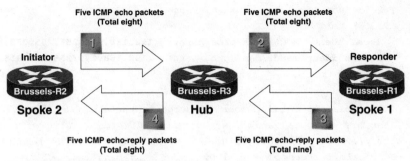

Figure 5.3: Traffic flow for ping test #2.

```
Source address or interface: 192.168.2.2
Type of service [0]:
Set DF bit in IP header? [no]:
Validate reply data? [no]:
Data pattern [0xABCD]:
Loose, Strict, Record, Timestamp, Verbose[none]:
Sweep range of sizes [n]:
Type escape sequence to abort.
Sending 5, 100-byte ICMP Echos to 192.168.1.1, timeout is 2 seconds:
!!!!!
Success rate is 100 percent (5/5), round-trip min/avg/max = 12/13/16 ms
```

Code Listing 5.10 illustrates the results for ping test #2 conducted in Code Listing 5.9. In the command traces below, italic comments are inserted at specific "checkpoints" for description purposes. We shall examine the traces from Brussels-R2, Brussels-R3, and Brussels-R1 in that order. The traffic flow for ping test #2 is illustrated in Figure 5.3.

Code Listing 5.10: Results for Ping Test #2

We first examine the respective command outputs at Brussels-R2. From the "show crypto ipsec sa" command at Brussels-R2 (spoke 2), for the IPSec tunnel created between spoke 2 and hub, we can gather that five outgoing IP (ICMP echo) packets to spoke 1 have been encrypted and digested, and five incoming IP (ICMP echo-reply) packets from Spoke 1 have been decrypted and verified.

Brussels-R2#show crypto ipsec sa

```
<Output Omitted>
```

Traffic flow is from local subnet 192.168.2.0/24 to remote subnet 192.168.1.0/24.
> **local ident** (addr/mask/prot/port): (**192.168.2.0/255.255.255.0**/0/0)
> **remote ident** (addr/mask/prot/port): (**192.168.1.0/255.255.255.0**/0/0)
> <Output Omitted>

The cumulative total number of outgoing packets from spoke 2 to spoke 1 that are encrypted and digested stands at eight. Three of them were from the previous ping test #1 (see Code Listing 5.8). The remaining five were ICMP echo packets generated by spoke 2 in ping test #2 (see Code Listing 5.9). Refer to Figure 5.3 for the traffic flow illustration.
> #pkts encaps: 8, **#pkts encrypt: 8, #pkts digest 8**

The cumulative total number of incoming packets from spoke 1 to spoke 2 that are decrypted and verified stands at eight. Three of them were from the previous ping test #1 (see Code Listing 5.8). The remaining five were the response (ICMP echo-reply) packets generated by spoke 1 in ping test #2 (see Code Listing 5.9). Refer to Figure 5.3 for the traffic flow illustration.
> #pkts decaps: 8, **#pkts decrypt: 8, #pkts verify 8**
> <Output Omitted>

The "show crypto engine connection active" command at Brussels-R2 concurs with the "show crypto ipsec sa" command reflecting a total of eight (3 + 5) outbound packets being encrypted and digested, and a total of eight (3 + 5) inbound packets being decrypted and verified.

Brussels-R2#show crypto engine connection active

ID	Interface	IP-Address	State	Algorithm	Encrypt	Decrypt
1	FastEthernet0/0	172.20.20.2	set	HMAC_MD5+DES_56_CB	0	0
2000	FastEthernet0/0	172.20.20.2	set	HMAC_MD5+DES_56_CB	0	8
2001	FastEthernet0/0	172.20.20.2	set	HMAC_MD5+DES_56_CB	8	0

Next, we examine the respective command outputs at Brussels-R3. From the "show crypto ipsec sa" command at Brussels-R3 (hub), for the IPSec tunnel created between hub and spoke 1, we can gather that five outgoing IP (ICMP echo) packets from spoke 2 to spoke 1 have been encrypted and digested, and five incoming IP (ICMP echo-reply) packets from spoke 1 to spoke 2 have been decrypted and verified. For the IPSec tunnel created between hub and spoke 2, we can gather that five outgoing IP (ICMP echo-reply) packets from spoke 1 to spoke 2 have been encrypted and digested, and five incoming IP (ICMP echo) packets from spoke 2 to spoke 1 have been decrypted and verified.

```
Brussels-R3#show crypto ipsec sa
```

`<Output Omitted>`

Traffic flow is from local subnet 192.168.2.0/24 to remote subnet 192.168.1.0/24 for the IPSec tunnel between hub and spoke 1.
   local ident (addr/mask/prot/port): (**192.168.2.0/255.255.255.0**/0/0)
   remote ident (addr/mask/prot/port): (**192.168.1.0/255.255.255.0**/0/0)
   `<Output Omitted>`

The cumulative total number of outgoing packets from spoke 2 to spoke 1 that are encrypted and digested stands at eight. Three of them were from the previous ping test #1 (see Code Listing 5.8). The remaining five were the ICMP echo packets generated by spoke 2 in ping test #2 (see Code Listing 5.9). Refer to Figure 5.3 for the traffic flow illustration.
   #pkts encaps: 8, **#pkts encrypt: 8, #pkts digest 8**

The cumulative total number of incoming packets from spoke 1 to spoke 2 that are decrypted and verified stands at nine. Four of them were from the previous ping test #1 (see Code Listing 5.8). The remaining five were the response (ICMP echo-reply) packets generated by spoke 1 in ping test #2 (see Code Listing 5.9). Refer to Figure 5.3 for the traffic flow illustration.
   #pkts decaps: 9, **#pkts decrypt: 9, #pkts verify 9**
   `<Output Omitted>`

Traffic flow is from local subnet 192.168.1.0/24 to remote subnet 192.168.2.0/24 for the IPSec tunnel between hub and spoke 2.
 local ident (addr/mask/prot/port): (**192.168.1.0/255.255.255.0**/0/0)
   remote ident (addr/mask/prot/port): (**192.168.2.0/255.255.255.0**/0/0)
   `<Output Omitted>`

The cumulative total number of outgoing packets from spoke 1 to spoke 2 that are encrypted and digested stands at eight. Three of them were from the previous ping test #1 (see Code Listing 5.8). The remaining five were the response (ICMP echo-reply) packets generated by spoke 1 in ping test #2 (see Code Listing 5.9). Refer to Figure 5.3 for the traffic flow illustration.
   #pkts encaps: 8, **#pkts encrypt: 8, #pkts digest 8**

The cumulative total number of incoming packets from spoke 2 to spoke 1 that are decrypted and verified stands at eight. Three of them were from the previous ping test #1 (see Code Listing 5.8). The remaining five were the ICMP echo packets generated by spoke 2 in ping test #2 (see Code Listing 5.9). Refer to Figure 5.3 for the traffic flow illustration.
   #pkts decaps: 8, **#pkts decrypt: 8, #pkts verify 8**
   `<Output Omitted>`

The "show crypto engine connection active" command at Brussels-R3 concurs with the "show crypto ipsec sa" command. For the IPSec tunnel between hub and spoke 1, the command reflects eight (3 + 5) outbound (ICMP echo) packets being encrypted and digested together with nine (4 + 5) inbound (ICMP echo-reply) packets being decrypted (and verified), and the connections IDs for the inbound and outbound IPSec SAs between hub and spoke 1 are 2000 and 2001, respectively.

For the IPSec tunnel between hub and spoke 2, the command reflects eight (3 + 5) outbound packets being encrypted and digested together with eight (3 + 5) inbound packets being decrypted and verified, and the connections IDs for the inbound and outbound IPSec SAs between hub and spoke 2 correspond with 2002 and 2003, respectively.

Brussels-R3#show crypto engine connection active

ID	Interface	IP-Address	State	Algorithm	Encrypt	Decrypt
1	FastEthernet0/0	172.20.20.3	set	HMAC_MD5+DES_56_CB	0	0
2	\<none>	\<none>	set	HMAC_MD5+DES_56_CB	0	0
2000	FastEthernet0/0	172.20.20.3	set	HMAC_MD5+DES_56_CB	0	9
2001	FastEthernet0/0	172.20.20.3	set	HMAC_MD5+DES_56_CB	8	0
2002	FastEthernet0/0	172.20.20.3	set	HMAC_MD5+DES_56_CB	0	8
2003	FastEthernet0/0	172.20.20.3	set	HMAC_MD5+DES_56_CB	8	0

We now examine the respective command outputs at Brussels-R1. From the "show crypto ipsec sa" command at Brussels-R1 (spoke 1), for the IPSec tunnel created between spoke 1 and hub, we can gather that five outgoing IP (ICMP echo-reply) packets to spoke 2 have been encrypted and digested, and five incoming IP (ICMP echo) packets from spoke 2 have been decrypted and verified.

Brussels-R1#show crypto ipsec sa

\<Output Omitted>

Traffic flow is from local subnet 192.168.1.0/24 to remote subnet 192.168.2.0/24.
   local ident (addr/mask/prot/port): (**192.168.1.0/255.255.255.0**/0/0)
   remote ident (addr/mask/prot/port): (**192.168.2.0/255.255.255.0**/0/0)
   \<Output Omitted>

The cumulative total number of outgoing packets from spoke 1 to spoke 2 that are encrypted and digested stands at nine. Four of them were from the previous ping test #1 (see Code Listing 5.8). The remaining five were the response (ICMP echo-reply) packets generated by spoke 1 in ping test #2 (see Code Listing 5.9). Refer to Figure 5.3 for the traffic flow illustration.

```
#pkts encaps: 9, #pkts encrypt: 9, #pkts digest 9
```

The cumulative total number of incoming packets from spoke 2 to spoke 1 that are decrypted and verified stands at eight. Three of them were from the previous ping test #1 (see Code Listing 5.8). The remaining five were the ICMP echo packets generated by spoke 2 in ping test #2 (see Code Listing 5.9). Refer to Figure 5.3 for the traffic flow illustration.

```
#pkts decaps: 8, #pkts decrypt: 8, #pkts verify 8
<Output Omitted>
```

The "show crypto engine connection active" command at Brussels-R2 concurs with the "show crypto ipsec sa" command reflecting a total of nine (4 + 5) outbound packets being encrypted and digested, and a total of eight (3 + 5) inbound packets being decrypted and verified.

Brussels-R1#show crypto engine connection active

ID	Interface	IP-Address	State	Algorithm	Encrypt	Decrypt
1	<none>	<none>	set	HMAC_MD5+DES_56_CB	0	0
2000	FastEthernet0/0	172.20.20.1	set	HMAC_MD5+DES_56_CB	0	8
2001	FastEthernet0/0	172.20.20.1	set	HMAC_MD5+DES_56_CB	9	0

Code Listing 5.11 illustrates the ping test disseminated from spoke 1 (192.168.1.1) to hub (192.168.3.3). In this instance, the test is only 80 percent successful, that is, 4 of 5 ping packets went through to Hub.

Code Listing 5.11:   Ping test #3 from spoke 1 to hub

```
Brussels-R1#ping
Protocol [ip]:
Target IP address: 192.168.3.3
Repeat count [5]:
Datagram size [100]:
Timeout in seconds [2]:
Extended commands [n]: y
Source address or interface: 192.168.1.1
Type of service [0]:
Set DF bit in IP header? [no]:
Validate reply data? [no]:
Data pattern [0xABCD]:
Loose, Strict, Record, Timestamp, Verbose[none]:
Sweep range of sizes [n]:
```

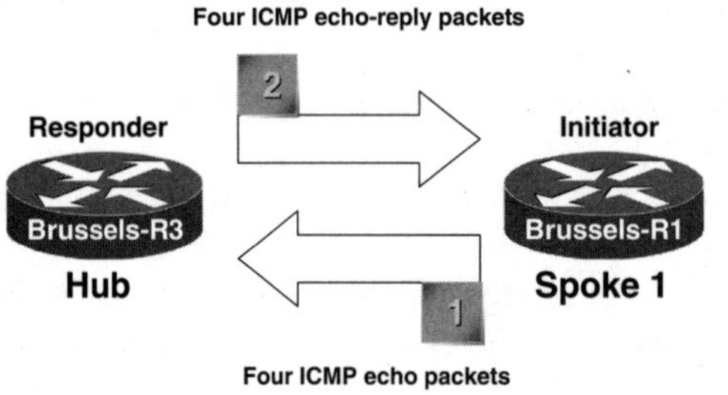

Four ICMP echo-reply packets

Responder   Initiator

Brussels-R3   Brussels-R1

Hub   **Spoke 1**

Figure 5.4: Traffic flow for ping test #3.

Four ICMP echo packets

```
Type escape sequence to abort.
Sending 5, 100-byte ICMP Echos to 192.168.3.3, timeout is 2 seconds:
.!!!!
Success rate is 80 percent (4/5), round-trip min/avg/max = 4/5/8 ms
```

Code Listing 5.12 illustrates the results for ping test #3 conducted in Code Listing 5.11. In the command traces below, italic comments are inserted at specific "checkpoints" for description purposes. We shall examine the traces for Brussels-R1 and Brussels-R3. The traffic flow for ping test #3 is illustrated in Figure 5.4.

Code Listing 5.12:   Results for Ping Test #3

We first examine the respective command outputs at Brussels-R1. From the "show crypto ipsec sa" command at Brussels-R1 (spoke 1), for the IPSec tunnel created between spoke 1 and hub, we can gather that four of five outgoing IP (ICMP echo) packets to hub have been encrypted and digested, and four incoming IP (ICMP echo-reply) packets from hub have been decrypted and verified.

Brussels-R1#show crypto ipsec sa

```
<Output Omitted>
```

Traffic flow is from local subnet 192.168.1.0/24 to remote subnet 192.168.3.0/24.
```
    local ident (addr/mask/prot/port): (192.168.1.0/255.255.255.0/0/0)
    remote ident (addr/mask/prot/port): (192.168.3.0/255.255.255.0/0/0)
    <Output Omitted>
```

Four outgoing (ICMP echo) packets from spoke 1 to hub are encrypted and digested (see Fig. 5.4).
```
    #pkts encaps: 4, #pkts encrypt: 4, #pkts digest 4
```

Four incoming (ICMP echo-reply) packets from hub are decrypted and verified (see Fig. 5.4).

  #pkts decaps: 4, **#pkts decrypt: 4, #pkts verify 4**

  <Output Omitted>

At spoke 1, of the five packets sent, one is bad.

  #send errors 1, #recv errors 0

The IPSec tunnel originates from spoke 1 (172.20.20.1) and terminates at hub (172.20.20.3).

  local crypto endpt.: 172.20.20.1, remote crypto endpt.: 172.20.20.3

  <Output Omitted>

For the inbound IPSec SA from hub to spoke 1, the SPI is 4206479133 and the connection ID is 2002.

  inbound esp sas:

  spi: 0xFAB9C71D(4206479133)

    transform: esp-des esp-md5-hmac ,

    in use settings ={Tunnel, }

    slot: 0, **conn id: 2002**, flow_id: 3, crypto map: chapter5

    <Output Omitted>

For the outbound IPSec SA from spoke 1 to hub, the SPI is 2929535312 and the connection ID is 2003.

  outbound esp sas:

  spi: 0xAE9D2950(2929535312)

    transform: esp-des esp-md5-hmac ,

    in use settings ={Tunnel, }

    slot: 0, **conn id: 2003**, flow_id: 4, crypto map: chapter5

    <Output Omitted>

The "show crypto engine connection active" command at Brussels-R1 concurs with the "show crypto ipsec sa" command reflecting four outbound (ICMP echo) packets to hub being encrypted and digested, and four inbound (ICMP echo-reply) packets from hub being decrypted and verified. The connections IDs for the inbound and outbound IPSec SAs are 2002 and 2003, respectively, which correspond to the same connection ID values illustrated earlier in the "show crypto ipsec sa" command.

Brussels-R1#show crypto engine connection active

ID	Interface	IP-Address	State	Algorithm	Encrypt	Decrypt
1	<none>	<none>	set	HMAC_MD5+DES_56_CB	0	0
2000	FastEthernet0/0	172.20.20.1	set	HMAC_MD5+DES_56_CB	0	8
2001	FastEthernet0/0	172.20.20.1	set	HMAC_MD5+DES_56_CB	9	0

2002	FastEthernet0/0	172.20.20.1	set	HMAC_MD5+DES_56_CB	**0**	**4**
2003	FastEthernet0/0	172.20.20.1	set	HMAC_MD5+DES_56_CB	**4**	**0**

We now examine the respective command outputs at Brussels-R3. From the "show crypto ipsec sa" command at Brussels-R3 (hub), for the IPSec tunnel created between hub and spoke 1, we can gather that four outgoing IP (ICMP echo-reply) packets to spoke 1 have been encrypted and digested, and four incoming IP (ICMP echo) packets from spoke 1 have been decrypted and verified.

Brussels-R3#show crypto ipsec sa

<Output Omitted>

Traffic flow is from local subnet 192.168.3.0/24 to remote subnet 192.168.1.0/24.
    local ident (addr/mask/prot/port): (**192.168.3.0/255.255.255.0**/0/0)
    remote ident (addr/mask/prot/port): (**192.168.1.0/255.255.255.0**/0/0)
    <Output Omitted>

Four outgoing (ICMP echo-reply) packets from hub to spoke 1 are encrypted and digested (see Fig. 5.4).
    #pkts encaps: 4, **#pkts encrypt: 4, #pkts digest 4**

Four incoming (ICMP echo) packets from spoke 1 are decrypted and verified (see Fig. 5.4).
    #pkts decaps: 4, **#pkts decrypt: 4, #pkts verify 4**
    <Output Omitted>

The IPSec tunnel originates from hub (172.20.20.3) and terminates at spoke 1 (172.20.20.1).
    local crypto endpt.: 172.20.20.3, remote crypto endpt.: 172.20.20.1
    <Output Omitted>

For the inbound IPSec SA from spoke 1 to hub, the SPI is 2929535312 and the connection ID is 2004.
    inbound esp sas:
    spi: 0xAE9D2950(2929535312)
      transform: esp-des esp-md5-hmac ,
      in use settings ={Tunnel, }
      slot: 0, **conn id: 2004**, flow_id: 5, crypto map: chapter5
      <Output Omitted>

For the outbound IPSec SA from Hub to Spoke 1, the SPI is 4206479133 and the connection ID is 2005.
    outbound esp sas:
    spi: 0xFAB9C71D(4206479133)

```
transform: esp-des esp-md5-hmac ,
in use settings ={Tunnel, }
slot: 0, conn id: 2005, flow_id: 6, crypto map: chapter5
<Output Omitted>
```

The "show crypto engine connection active" command at Brussels-R3 concurs with the "show crypto ipsec sa" command reflecting four outbound (ICMP echo-reply) packets to spoke 1 being encrypted and digested, and four inbound (ICMP echo) packets from spoke 1 being decrypted and verified. The connections IDs for the inbound and outbound IPSec SAs are 2004 and 2005, respectively, which correspond to the same connection ID values illustrated earlier in the "show crypto ipsec sa" command.

Brussels-R3#show crypto engine connection active

ID	Interface	IP-Address	State	Algorithm	Encrypt	Decrypt
1	FastEthernet0/0	172.20.20.3	set	HMAC_MD5+DES_56_CB	0	0
2	<none>	<none>	set	HMAC_MD5+DES_56_CB	0	0
2000	FastEthernet0/0	172.20.20.3	set	HMAC_MD5+DES_56_CB	0	9
2001	FastEthernet0/0	172.20.20.3	set	HMAC_MD5+DES_56_CB	8	0
2002	FastEthernet0/0	172.20.20.3	set	HMAC_MD5+DES_56_CB	0	8
2003	FastEthernet0/0	172.20.20.3	set	HMAC_MD5+DES_56_CB	8	0
2004	FastEthernet0/0	172.20.20.3	set	HMAC_MD5+DES_56_CB	**0**	**4**
2005	FastEthernet0/0	172.20.20.3	set	HMAC_MD5+DES_56_CB	**4**	**0**

Code Listing 5.13 illustrates the ping test disseminated from spoke 2 (192.168.2.2) to hub (192.168.3.3). In this instance, the test is only 80 percent successful, that is, 4 of 5 ping packets went through to hub.

Code Listing 5.13:   Ping test #4 from spoke 2 to hub

```
Brussels-R2#ping
Protocol [ip]:
Target IP address: 192.168.3.3
Repeat count [5]:
Datagram size [100]:
Timeout in seconds [2]:
Extended commands [n]: y
Source address or interface: 192.168.2.2
Type of service [0]:
Set DF bit in IP header? [no]:
Validate reply data? [no]:
Data pattern [0xABCD]:
Loose, Strict, Record, Timestamp, Verbose[none]:
```

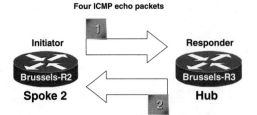

Figure 5.5: Traffic flow for ping test #4.

```
Sweep range of sizes [n]:
Type escape sequence to abort.
Sending 5, 100-byte ICMP Echos to 192.168.3.3, timeout is 2 seconds:
.!!!!
Success rate is 80 percent (4/5), round-trip min/avg/max = 4/5/8 ms
```

Code Listing 5.14 illustrates the results for ping test #4 conducted in Code Listing 5.13. In the command traces below, italic comments are inserted at specific "checkpoints" for description purposes. We shall examine the traces for Brussels-R2 and Brussels-R3. The traffic flow for ping test #4 is illustrated in Figure 5.5.

Code Listing 5.14:   Results for Ping Test #4

We first examine the respective command outputs at Brussels-R2. From the "show crypto ipsec sa" command at Brussels-R2 (spoke 2), for the IPSec tunnel created between spoke 2 and hub, we can gather that four of five outgoing IP (ICMP echo) packets to hub have been encrypted and digested, and four incoming IP (ICMP echo-reply) packets from hub have been decrypted and verified.

Brussels-R2#show crypto ipsec sa

<Output Omitted>

Traffic flow is from local subnet 192.168.2.0/24 to remote subnet 192.168.3.0/24.
   local ident (addr/mask/prot/port): (**192.168.2.0/255.255.255.0**/0/0)
   remote ident (addr/mask/prot/port): (**192.168.3.0/255.255.255.0**/0/0)
   <Output Omitted>

Four outgoing (ICMP echo) packets from spoke 2 to hub are encrypted and digested (see Fig. 5.5).
   #pkts encaps: 4, **#pkts encrypt: 4, #pkts digest 4**

*Four incoming (ICMP echo-reply) packets from hub are decrypted and verified
(see Fig. 5.5).*

   #pkts decaps: 4, **#pkts decrypt: 4, #pkts verify 4**

   <Output Omitted>

At Spoke 2, of the five packets sent, one is bad.

   #send errors 1, #recv errors 0

*The IPSec tunnel originates from spoke 2 (172.20.20.2) and terminates at
Hub (172.20.20.3).*

   local crypto endpt.: 172.20.20.2, remote crypto endpt.: 172.20.20.3

   <Output Omitted>

*For the inbound IPSec SA from hub to spoke 2, the SPI is 2266443490 and
the connection ID is 2002.*

   inbound esp sas:

   spi: 0x87172EE2(2266443490)

     transform: esp-des esp-md5-hmac ,

     in use settings ={Tunnel, }

     slot: 0, **conn id: 2002**, flow_id: 3, crypto map: chapter5

     <Output Omitted>

*For the outbound IPSec SA from spoke 2 to hub, the SPI is 618541355 and
the connection ID is 2003.*

   outbound esp sas:

   spi: 0x24DE312B(618541355)

     transform: esp-des esp-md5-hmac ,

     in use settings ={Tunnel, }

     slot: 0, **conn id: 2003**, flow_id: 4, crypto map: chapter5

     <Output Omitted>

*The "show crypto engine connection active" command at Brussels-R2 concurs
with the "show crypto ipsec sa" command reflecting four outbound (ICMP echo)
packets to hub being encrypted and digested, and four inbound (ICMP echo-
reply) packets from Hub being decrypted and verified. The connections IDs
for the inbound and outbound IPSec SAs are 2002 and 2003, respectively,
which correspond to the same connection ID values illustrated earlier in
the "show crypto ipsec sa" command.*

Brussels-R2#show crypto engine connection active

ID	Interface	IP-Address	State	Algorithm	Encrypt	Decrypt
1	FastEthernet0/0	172.20.20.2	set	HMAC_MD5+DES_56_CB	0	0
2000	FastEthernet0/0	172.20.20.2	set	HMAC_MD5+DES_56_CB	0	8
2001	FastEthernet0/0	172.20.20.2	set	HMAC_MD5+DES_56_CB	8	0

```
2002 FastEthernet0/0 172.20.20.2    set    HMAC_MD5+DES_56_CB       0       4
2003 FastEthernet0/0 172.20.20.2    set    HMAC_MD5+DES_56_CB       4       0
```

we now examine the respective command outputs at Brussels-R3. From the "show crypto ipsec sa" command at Brussels-R3 (hub), for the IPSec tunnel created between hub and spoke 2, we can gather that four outgoing IP (ICMP echo-reply) packets to spoke 2 have been encrypted and digested, and four incoming IP (ICMP echo) packets from spoke 2 have been decrypted and verified.

Brussels-R3#show crypto ipsec sa

```
<Output Omitted>
```

Traffic flow is from local subnet 192.168.3.0/24 to remote subnet 192.168.2.0/24.
    local ident (addr/mask/prot/port): (**192.168.3.0/255.255.255.0**/0/0)
    remote ident (addr/mask/prot/port): (**192.168.2.0/255.255.255.0**/0/0)
    <Output Omitted>

Four outgoing (ICMP echo-reply) packets from hub to spoke 2 are encrypted and digested (see Fig. 5.5).
    #pkts encaps: 4, **#pkts encrypt: 4, #pkts digest 4**

Four incoming (ICMP echo) packets from spoke 2 are decrypted and verified (see Fig. 5.5).
    #pkts decaps: 4, **#pkts decrypt: 4, #pkts verify 4**
    <Output Omitted>

The IPSec tunnel originates from hub (172.20.20.3) and terminates at spoke 2 (172.20.20.2).
    local crypto endpt.: 172.20.20.3, remote crypto endpt.: 172.20.20.2
    <Output Omitted>

For the inbound IPSec SA from spoke 2 to hub, the SPI is 618541355 and the connection ID is 2006.
    inbound esp sas:
    spi: 0x24DE312B(618541355)
      transform: esp-des esp-md5-hmac ,
      in use settings ={Tunnel, }
      slot: 0, **conn id: 2006**, flow_id: 7, crypto map: chapter5
      <Output Omitted>

For the outbound IPSec SA from hub to spoke 2, the SPI is 2266443490 and the connection ID is 2007.
    outbound esp sas:
    spi: 0x87172EE2(2266443490)

```
transform: esp-des esp-md5-hmac ,
in use settings ={Tunnel, }
slot: 0, conn id: 2007, flow_id: 8, crypto map: chapter5
<Output Omitted>
```

The "show crypto engine connection active" command at Brussels-R3 concurs with the "show crypto ipsec sa" command reflecting four outbound (ICMP echo-reply) packets to spoke 2 being encrypted and digested, and four inbound (ICMP echo) packets from spoke 2 being decrypted and verified. The connections IDs for the inbound and outbound IPSec SAs are 2006 and 2007, respectively, which correspond to the same connection ID values illustrated earlier in the "show crypto ipsec sa" command.

Brussels-R3#show crypto engine connection active

ID	Interface	IP-Address	State	Algorithm	Encrypt	Decrypt
1	FastEthernet0/0	172.20.20.3	set	HMAC_MD5+DES_56_CB	0	0
2	<none>	<none>	set	HMAC_MD5+DES_56_CB	0	0
2000	FastEthernet0/0	172.20.20.3	set	HMAC_MD5+DES_56_CB	0	9
2001	FastEthernet0/0	172.20.20.3	set	HMAC_MD5+DES_56_CB	8	0
2002	FastEthernet0/0	172.20.20.3	set	HMAC_MD5+DES_56_CB	0	8
2003	FastEthernet0/0	172.20.20.3	set	HMAC_MD5+DES_56_CB	8	0
2004	FastEthernet0/0	172.20.20.3	set	HMAC_MD5+DES_56_CB	0	4
2005	FastEthernet0/0	172.20.20.3	set	HMAC_MD5+DES_56_CB	4	0
2006	FastEthernet0/0	172.20.20.3	set	HMAC_MD5+DES_56_CB	**0**	**4**
2007	FastEthernet0/0	172.20.20.3	set	HMAC_MD5+DES_56_CB	**4**	**0**

5.4   Scaling IPSec VPNs

Typically, there are two ways to scale IPSec in large networks. Tunnel endpoint discovery (TED) can be used to scale partially or fully meshed networks, and the multihop crypto (or encryption) model can be used to scale existing hub-and-spoke networks. In the following sections, we briefly discuss these two IPSec scaling techniques.

5.4.1   Meshed networks and tunnel endpoint discovery

Tunnel endpoint discovery (see Chapter 3, Section 3.7 for details) allows IPSec to scale to large networks by reducing multiple encryptions, decreasing the setup time, and allowing for simpler configurations on participating peer routers. Each node requires only a straightforward configuration that defines the local network that the router protects and the required IPSec transforms. The limitation of the dynamic crypto map is that the initiating router cannot dynamically determine an IPSec peer; instead, only the receiving router has this ability. However, by defining a dynamic crypto map together with TED, the

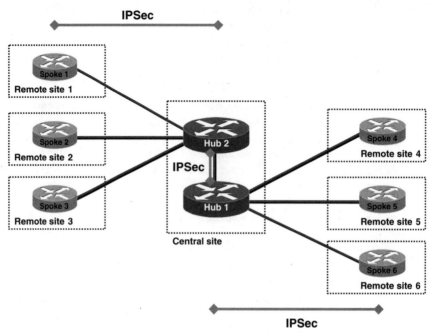

Figure 5.6:   The Multi-hop crypto model.

initiating router can dynamically determine an IPSec peer for secure IPSec communications. In other words, TED provides automated peer discovery, which eases scalability and configuration, as well as enhances availability by automating the failover IPSec peer switch. TED mechanisms work best in partially or fully meshed networks that require spoke-to-spoke connectivity on an occasional basis.

5.4.2   Hub-and-spoke networks and multihop crypto

The multihop crypto (encryption) model is another alternative that can be deployed to scale IPSec VPNs. The multihop model is implemented by combining multiple hub-and-spoke IPSec networks together as illustrated in Figure 5.6. Multihop crypto allows the distribution of IPSec traffic loads across multiple central site (hub) routers and requires separate subnets and crypto maps for each interface on the central site routers.

With multihop crypto, instead of the any-to-any IPSec communication with every router, IPSec communication is limited to between the hub-and-spoke routers, thus improving the network scalability significantly. For any-to-any (fully meshed) IPSec implementation, the configurations at every router can become lengthy and complex. In the multihop crypto model, configurations are fairly simple for the (spoke) routers at the remote sites; however, the configurations for the (hub) routers at the central site can become rather complex and difficult to maintain. In other words, the multihop crypto model inherited all the pros and cons of a hub-and-spoke implementation as discussed in Section 5.2.

5.5 Summary

Chapter 5 gives the reader hands-on skills for implementing IPSec VPN in large enterprise networks using a hub-and-spoke topology. The chapter first gives an overview of meshed versus hub-and-spoke networks. Case Study 5.1 further substantiates the scalability of a hub-and-spoke topology with an IPSec hub-and-spoke VPN design and implementation. The concluding section of the chapter discusses how the tunnel endpoint discovery (TED) mechanism and the multihop crypto model can be used to scale existing meshed and hub-and-spoke networks, respectively.

6

Advanced IPSec VPN Deployment

6.1 Chapter Overview

This chapter is an extension to Chapter 5, in which we tackle the more advanced issues that affect interoperability, reliability, and performance of the network when deploying IPSec VPNs in large enterprises. Some of these issues include the incorporation of IP addressing services, such as NAT (Network Address Translation); network resiliency techniques, such as HSRP (hot standby router protocol); and performance optimization parameters, such as fragmentation, IKE SA lifetimes, and IKE keepalives. We address all these advanced topics in the following sections with supporting case studies.

6.2 Network Address Translation Overview

6.2.1 NAT background

Network address translation (NAT), first described in IETF document RFC 1631, is a method by which the source and/or destination IP addresses in the IP header are mapped from one address space to another in an attempt to provide transparent routing to hosts. Traditionally, NAT devices are used to connect an isolated address space with private unregistered addresses to an external address space with globally unique registered addresses. The need for NAT arises when a network's internal IP addresses cannot be used outside the network either because they are invalid for outside use or because the internal addressing must be kept private from the external network. Moreover, public IP addresses are limited and it is difficult to obtain a large block of these addresses. So most of the time private IP addresses (RFC 1918) are used for internal private networks. NAT allows hosts in a private network to communicate transparently with destinations on an external network and vice versa.

6.2.2 NAT terminologies

During NAT implementation on a router, an interface on the router can be defined as either inside (connecting to the inside network) or outside (connecting to the outside network); address translations occur only from inside to outside interfaces or vice versa. From the perspective of NAT, an IP address is either local or global. Local IP addresses are seen on the inside network, and global IP addresses are seen on the outside network. Some commonly used NAT terms are as follows:

- *Inside local (IL)*: This refers to the IP address assigned to a host on the inside network. This address may be globally unique or it may be allocated out of the private address space defined in RFC 1918.

- *Inside global (IG)*: This refers to the translated IP address of an inside host as it appears to the outside world. These addresses can be allocated from a globally unique address space, typically provided by an Internet service provider (ISP).

- *Outside local (OL)*: This refers to the IP address of an outside host as it appears to the inside network. These addresses can be allocated from the private address space defined in RFC 1918.

- *Outside global (OG)*: This refers to the IP address assigned to a host on the outside network.

6.2.3 NAT concepts

6.2.3.1 Static versus dynamic translations.   NAT translations are either static or dynamic:

- *Static translation*: Statically configured one-to-one address mapping between inside local and global addresses. The translation is entered directly into the configuration and is always in the translation table.

- *Dynamic translation*: Dynamic mapping between the inside local and global addresses. The translation is created when needed, and it uses access lists to distinguish the IP addresses for the NAT operation.

6.2.3.2 NAT versus PAT.   The two types of NAT techniques are simple network address translation (NAT) and port address translation (PAT). NAT operates bidirectionally and maps one IP address to another using a one-to-one translation. PAT (also known as extended network address translation or network address port translation) operates unidirectionally and maps one IP address and port pair to another using a one-to-many translation. In PAT, unique port numbers are used to identify translations on a single IP address.

NAT is used when there are sufficient IP addresses for one-to-one translations, and PAT is applied when there are an insufficient number of IP address to translate all the inside addresses. Therefore, PAT helps to conserve registered IP addresses.

6.2.3.3 Inside versus outside source translations. Inside source translation is used in cases when hosts on the inside network are required to be concealed from the outside network. Inside source translation translates the source IP address for packets going from the inside network to the outside network and the destination IP address for packets going from the outside network to the inside network.

Outside source translation is used in cases when identical IP addresses (overlapping address spaces) are being used on both the inside and outside network. Outside source address translation translates the source IP address for packets going from the outside network to the inside network and the destination IP address for packets going from the inside network to the outside network.

6.2.4 NAT and IPSec

In a way, IPSec and NAT contradict each other. IPSec AH and ESP protect the contents of the IP headers (including the source and destination addresses) from modification, while the fundamental role of NAT is to change the addresses in the IP header of a packet.

In IPSec transport mode, both AH and ESP have an integrity check covering the entire payload. When the payload is TCP or UDP, the TCP/UDP checksum is covered by the integrity check. When a NAT device modifies an address the checksum is no longer valid with respect to the new address. Typically NAT also updates the checksum, but this becomes ineffective when AH and ESP are used. Consequently, receivers will discard a packet either because it fails the IPSec integrity check (if the NAT device updates the checksum) or because the checksum is invalid (if the NAT device leaves the checksum unchanged).

Nevertheless, the IPSec tunnel mode ESP is permissible so long as the embedded packet contents are unaffected by the outer IP header translation. End-to-end ESP-based transport mode authentication and confidentiality are also permissible for packets such as (Internet Control Message Protocol ICMP), whose IP payload content is unaffected by the outer IP header translation.

It makes more sense to bypass the application of NAT to IPSec VPN traffic, because NAT can have an undesirable effect on the VPN traffic flow. Moreover, the actual IP addresses of devices utilizing the IPSec tunnel for transport are concealed through encryption, and only public IP addresses of the IPSec peers are visible. Hence, there is no real benefit for applying NAT to the IPSec VPN traffic. In the next section, we discuss how we can apply NAT only to Internet-bound traffic and at the same time bypass NAT for VPN-bound traffic.

6.3 Case Study 6.1: IPSec with NAT

6.3.1 Case overview

Figure 6.1 illustrates the network design and topology for Case Study 6.1. The case study uses six Cisco routers distributed across four different locations: Site 1, Site 2, Site 3, and Internet. Site 1, Site 2, and Site 3 are interconnected by

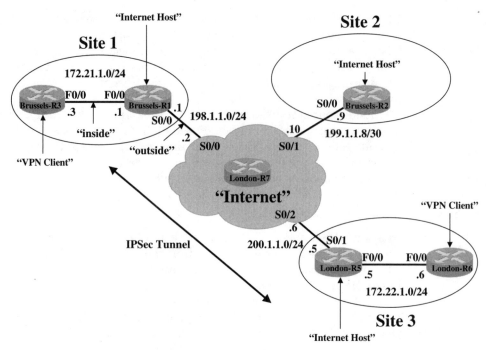

Figure 6.1: Network design and topology for Case Study 6.1

London-R7, which emulates the Internet by posing as an ISP gateway for all these sites.

Brussels-R1, Brussels-R3, London-R5, and London-R6 are routers belonging to the same VPN customer. Brussels-R1 and Brussels-R3 are interconnected in Site 1; London-R5 and London-R6 are interconnected together in Site 3. An IPSec tunnel is established between Brussels-R1 and London-R5, securing all IP traffic between private subnet 172.21.1.0/24 where Brussels-R3 is connected and private subnet 172.22.1.0/24 where London-R6 is connected. In other words, Brussels-R3 and London-R6 are VPN clients belonging to the same VPN.

Nevertheless, the subnet 172.21.1.0/24 (inside network), in which Brussels-R3 resides, is hidden from the "outside world" (in this case, subnets 198.1.1.0/24, 199.1.1.8/30, and 200.1.1.0/24) using NAT. With Brussels-R1 configured as a NAT router, any host traffic originating from subnet 172.21.1.0/24 will appear to the "outside world" as though they were originated from subnet 198.1.1.0/24. Furthermore, Site 2 with subnet 199.1.1.8/30, where Brussels-R2 is connected, is a public site that is visible and accessible to everyone including the VPN clients.

With all these preliminaries in place, we can move on to discuss the incorporation of IPSec and NAT with two different scenarios:

- *Scenario 1—NAT*: In this scenario, NAT is configured on Brussels-R1 only for simplicity reasons. In this case, NAT is not required when the VPN clients—Brussels-R3 and London-R6—communicate with each other between private subnets 172.21.1.0/24 and 172.22.1.0/24 over the IPSec tunnel. However, NAT has to be enabled for IP packets originating from private subnet 172.21.1.0/24 when the VPN clients on this subnet access Internet (public) destinations (subnets 198.1.1.0/24, 199.1.1.8/30, and 200.1.1.0/24). As documented in IETF document RFC 1918, private addresses cannot be used to access the Internet. To do so, private unregistered addresses must be translated to valid globally unique registered addresses using NAT (RFC 2663).

- *Scenario 2—NAT with Server*: Scenario 2 has the same requirements as scenario 1. On top of these, scenario 2 also requires an internal server that is connected to private subnet 172.21.1.0/24 to be accessible to Internet hosts and VPN clients. Since this internal server belongs to the same VPN as the VPN clients, no NAT is required when the VPN clients access the server and vice versa. Even so, NAT is still necessary when allowing the server access to Internet hosts (from subnets 198.1.1.0/24, 199.1.1.8/30, and 200.1.1.0/24) and concealing the server's real identity (IP address) from these hosts.

We illustrate these two scenarios in more detail in the subsequent sections.

6.3.2 Scenario 1—IPSec with plain NAT

6.3.2.1 Configurations. Code Listings 6.1 to 6.6 illustrate the configurations for the six different routers portrayed in Figure 6.1. In the configurations, comments (in italics) precede certain configuration lines to explain them.

Code Listing 6.1: Brussels-R1 configuration

```
! --- Brussels-R1 is the main focal point of the two
! --- illustrated scenarios because it is both an
! --- IPSec and a NAT router. It establishes an IPSec tunnel
! --- with London-R5 and performs NAT operation on packets
! --- originating from private subnet 172.21.1.0/24.
hostname Brussels-R1
!
! --- Supersede the default policy and use pre-shared keys
! --- for peer authentication.
crypto isakmp policy 110
 authentication pre-share
!
! --- Specify the pre-shared key "key15" to be used for
! --- the IPSec tunnel between Brussels-R1 and London-R5.
crypto isakmp key key15 address 200.1.1.5
```

```
!
! --- Specify the ESP transform settings for IPSec,
! --- which is later applied to the crypto map.
crypto ipsec transform-set chapter6 esp-des esp-sha-hmac
!
! --- Specify the crypto map chapter6 where we define
! --- our remote peer London-R5 (200.1.1.5), transform set chapter6,
! --- and our crypto access list 110.
 crypto map chapter6 10 ipsec-isakmp
 set peer 200.1.1.5
 set transform-set chapter6
 match address 110
!
interface FastEthernet0/0
 ip address 172.21.1.1 255.255.255.0
! --- Define FastEthernet0/0 as the interface
! --- to the inside network during NAT operation.
 ip nat inside
!
interface Serial0/0
 ip address 198.1.1.1 255.255.255.0
! --- Define Serial0/0 as the interface
! --- to the outside network during NAT operation.
 ip nat outside
! --- Apply crypto map chapter6 to Serial0/0
! --- to activate crypto engine.
 crypto map chapter6
!
! --- Globally registered addresses ranging from
! --- 198.1.1.20 to 198.1.1.29 are defined in
! --- address pool chapter6 to be used during
! --- the NAT operation for translating the
! --- private unregistered addresses to
! --- globally unique registered ones.
ip nat pool chapter6 198.1.1.20 198.1.1.29 netmask 255.255.255.0
! --- Perform dynamic NAT using address pool chapter6
! --- on the source addresses of IP packets originating
! --- from the range of subnets defined in access list 1
! --- and traversing FastEthernet0/0.
ip nat inside source list 1 pool chapter6
!
! --- Routing from Brussels-R1 to Internet destinations
! --- is via a default route to Internet gateway London-R7
! --- with a next-hop address of 198.1.1.2.
ip route 0.0.0.0 0.0.0.0 198.1.1.2
```

```
!
! --- Access list 1 defines 172.21.1.0/24
! --- as the subnet for the NAT operation.
access-list 1 permit 172.21.1.0 0.0.0.255
!
! --- This is the crypto access list that we have
! --- referenced in crypto map chapter6.
! --- We are encrypting IP traffic between
! --- private subnets 172.21.1.0/24 and 172.22.1.0/24.
access-list 110 permit ip 172.21.1.0 0.0.0.255 172.22.1.0 0.0.0.255
```

Code Listing 6.2: Brussels-R2 configuration

```
! --- Brussels-R2 assumes the role of an Internet host
! --- in both of the illustrated scenarios.
hostname Brussels-R2
!
interface Serial0/0
 ip address 199.1.1.9 255.255.255.252
!
! --- Routing from Brussels-R2 to Internet destinations
! --- is via a default route to Internet gateway London-R7
! --- with a next-hop address of 199.1.1.10.
ip route 0.0.0.0 0.0.0.0 199.1.1.10
```

Code Listing 6.3: Brussels-R3 configuration

```
! --- Brussels-R3 assumes the role of a VPN client in scenario 1
! --- and an internal server in scenario 2.
hostname Brussels-R3
!
 interface FastEthernet0/0
 ip address 172.21.1.3 255.255.255.0
!
! --- Routing from Brussels-R3 to Internet destinations
! --- is via a default route to Brussels-R1
! --- with a next-hop address of 172.21.1.1.
ip route 0.0.0.0 0.0.0.0 172.21.1.1
```

Code Listing 6.4:   London-R5 configuration

```
! --- London-R5 is the next most strategic router
! --- after Brussels-R1 since it is also an IPSec router
! --- and establishes an IPSec tunnel with Brussels-R1.
hostname London-R5
!
! --- Supersede the default policy and use
! --- pre-shared keys for peer authentication.
crypto isakmp policy 110
 authentication pre-share
!
! --- Specify the pre-shared key "key15" to be used
! --- for the IPSec tunnel between Brussels-R1 and London-R5.
crypto isakmp key key15 address 198.1.1.1
!
! --- Specify the ESP transform settings for IPSec,
! --- which is later applied to the crypto map.
crypto ipsec transform-set chapter6 esp-des esp-sha-hmac
!
! --- Specify the crypto map chapter6 where we define
! --- our remote peer Brussels-R1 (198.1.1.1),
! --- transform set chapter6, and our crypto access list 110.
crypto map chapter6 10 ipsec-isakmp
 set peer 198.1.1.1
 set transform-set chapter6
 match address 110
!
interface FastEthernet0/0
 ip address 172.22.1.5 255.255.255.0
!
interface Serial0/1
 ip address 200.1.1.5 255.255.255.0
! --- Apply crypto map chapter6 to Serial0/1
! --- to activate crypto engine.
 crypto map chapter6
!
! --- Routing from London-R5 to Internet destinations
! --- is via a default route to Internet gateway London-R7
! --- with a next-hop address of 200.1.1.6.
ip route 0.0.0.0 0.0.0.0 200.1.1.6
!
! --- This is the crypto access list that we have
! --- referenced in crypto map chapter6.
```

```
! --- We are encrypting IP traffic between
! --- private subnets 172.22.1.0/24 and 172.21.1.0/24.
access-list 110 permit ip 172.22.1.0 0.0.0.255 172.21.1.0 0.0.0.255
```

Code Listing 6.5: London-R6 configuration

```
! --- London-R6 assumes the role of a VPN client
! --- in both of the illustrated scenarios.
hostname London-R6
!
interface FastEthernet0/0
 ip address 172.22.1.6 255.255.255.0
!
! --- Routing from London-R6 to Internet destinations
! --- is via a default route to London-R5
! --- with a next-hop address of 172.22.1.5.
ip route 0.0.0.0 0.0.0.0 172.22.1.5
```

Code Listing 6.6: London-R7 configuration

```
! --- London-R7 is emulating the Internet
! --- by assuming the role of an ISP gateway.
hostname London-R7
!
interface Serial0/0
! --- 198.1.1.2 is the default next-hop for Brussels-R1.
 ip address 198.1.1.2 255.255.255.0
 clockrate 64000
!
interface Serial0/1
! --- 199.1.1.10 is the default next-hop for Brussels-R2.
 ip address 199.1.1.10 255.255.255.252
 clockrate 64000
!
interface Serial0/2
! --- 200.1.1.6 is the default next-hop for London-R5.
 ip address 200.1.1.6 255.255.255.0
 clockrate 64000
!
! --- For the sake of simplicity, we did not implement
! --- NAT on London-R5 for London-R6.
! --- Therefore, a static route is required to direct
! --- traffic to private subnet 172.22.1.0/24.
ip route 172.22.1.0 255.255.255.0 200.1.1.5
```

6.3.2.2 Verification test results. Code Listings 6.7 to 6.11 illustrate the verification test results for the IPSec and NAT configurations implemented in the previous section.

Code Listing 6.7 is used to verify the IP connectivity between the VPN clients residing in private subnets 172.21.1.0/24 and 172.22.1.0/24. From the listing, we can see that the ping test conducted from Brussels-R3 (a VPN client at private subnet 172.21.1.0/24 with host address 172.21.1.3) to London-R6 (a VPN client at private subnet 172.22.1.0/24 with host address 172.22.1.6) is not successful.

Code Listing 6.7:   Ping test from Brussels-R3 (172.21.1.3) to London-R6 (172.22.1.6)

```
Brussels-R3#ping 172.22.1.6
Type escape sequence to abort.
Sending 5, 100-byte ICMP Echos to 172.22.1.6, timeout is 2 seconds:
.....
Success rate is 0 percent (0/5)
```

The "show ip nat translation" command in Code Listing 6.8 displays the NAT translation table of Brussels-R1. We can see that a successful address translation has taken place for which the private host address 172.21.1.3 has been translated to global address 198.1.1.20 (first address in address pool chapter6 defined in Code Listing 6.1). However, the NAT operation just stop short after the outbound translation, as there is no further returning inbound traffic.

Code Listing 6.8:   Brussels-R1 NAT translation table

```
Brussels-R1#show ip nat translation
Pro Inside global     Inside local      Outside local      Outside global
--- 198.1.1.20        172.21.1.3        ---                ---
```

Even though the NAT operation can be considered as successful here, it is not desirable in this scenario since one of the requirements is to bypass NAT for IP packets that are exchanged between the two private subnets 172.21.1.0/24 and 172.22.1.0/24 through the IPSec tunnel.

From the crypto engine connection table of Brussels-R1 illustrated in Code Listing 6.9, we can see that no IPSec tunnels are built. This is because the NAT operation occurs before IPSec (see Code Listing 6.8). Therefore, the source addresses of IP packets originating from private subnet 172.21.1.0/24 will no longer match the crypto access list (access list 110) defined in Code Listing 6.1. As such, the IPSec process cannot be initiated, and IP packets originating from private subnet 172.21.1.0/24 to destinations in private subnet 172.22.1.0/24 are ignored and dropped.

Code Listing 6.9:   Brussels-R1 crypto engine connection table

---

```
Brussels-R1#show crypto engine connection active

 ID Interface     IP-Address     State Algorithm          Encrypt  Decrypt
```

Code Listing 6.10 is used to verify the IP connectivity between a VPN client (Brussels-R3) residing in private subnet 172.21.1.0/24 and an Internet host (Brussels-R2) residing in public subnet 199.1.1.8/30. From the listing, we can see that the ping test is conducted successfully from Brussels-R3 (172.21.1.3) to Brussels-R2 (199.1.1.9), thus fulfilling the requirement that gives VPN clients (on private subnet 172.21.1.0/24) access to the Internet. In this case, Brussels-R3 will appear to Brussels-R2 as a host in public subnet 198.1.1.0/24 with an IP address of 198.1.1.20 (see Code Listing 6.8) after the NAT operation.

Code Listing 6.10:   Ping test from Brussels-R3 (172.21.1.3) to Brussels-R2 (199.1.1.9)

```
Brussels-R3#ping 199.1.1.9

Type escape sequence to abort.
Sending 5, 100-byte ICMP Echos to 199.1.1.9, timeout is 2 seconds:
!!!!!
Success rate is 100 percent (5/5), round-trip min/avg/max = 56/56/60 ms
```

Code Listing 6.11 is used to verify the IP connectivity between a VPN client (Brussels-R3) residing in private subnet 172.21.1.0/24 and an Internet host (London-R5) residing in public subnet 200.1.1.0/24. From the listing, we can see that the ping test is conducted successfully from Brussels-R3 (172.21.1.3) to London-R5 (200.1.1.5), thus fulfilling the requirement that gives VPN clients (on private subnet 172.21.1.0/24) access to the Internet. In this case, Brussels-R3 will appear to London-R5 as a host in public subnet 198.1.1.0/24 with an IP address of 198.1.1.20 (see Code Listing 6.8) after the NAT operation.

Code Listing 6.11:   Ping test from Brussels-R3 (172.21.1.3) to London-R5 (200.1.1.5)

```
Brussels-R3#ping 200.1.1.5

Type escape sequence to abort.
Sending 5, 100-byte ICMP Echos to 200.1.1.5, timeout is 2 seconds:
!!!!!
Success rate is 100 percent (5/5), round-trip min/avg/max = 56/58/60 ms
```

6.3.2.3 Proposed solution. From the derived test results in section 6.3.2.2, we can conclude that the NAT configuration listed in Code Listing 6.1 for Brussels-R1 complies with only half the prerequisites defined in section 6.3.1 for scenario 1. To fulfill all the requirements, we have to modify the NAT configuration in Brussels-R1 so that it will do NAT only for Internet-bound traffic (from private subnet 172.21.1.0/24 to subnets 198.1.1.0/24, 199.1.1.8/30, and 200.1.1.0/24) and not for VPN-bound traffic (from private subnet 172.21.1.0/24 to private subnet 172.22.1.0/24).

Code Listing 6.12 illustrates the modified NAT configuration for Brussels-R1. In this case, a route-map (route-map nonat) is used to enable NAT only for Internet bound traffic.

Code Listing 6.12:   Brussels-R1 modified NAT configuration

```
hostname Brussels-R1
!
<Output Omitted>
ip nat pool chapter6 198.1.1.20 198.1.1.29 netmask 255.255.255.0


! --- Perform NAT based on the matching conditions
! --- defined in route-map nonat.
ip nat inside source route-map nonat pool chapter6
!
<Output Omitted>
! --- Deny VPN-bound traffic.
access-list 120 deny    ip 172.21.1.0 0.0.0.255 172.22.1.0 0.0.0.255
! --- Permit Internet-bound traffic.
access-list 120 permit ip 172.21.1.0 0.0.0.255 any
!
! --- Create route-map nonat.
route-map nonat permit 10
 match ip address 120 ! --- Match access-list 120.
```

Code Listing 6.13 is used to verify the IP connectivity between the VPN clients residing in private subnets 172.21.1.0/24 and 172.22.1.0/24. From the listing, we can see that the ping test conducted from Brussels-R3 (172.21.1.3) to London-R6 (172.22.1.6) is now successful (80%).

Code Listing 6.13:   Ping test from Brussels-R3 (172.21.1.3) to London-R6 (172.22.1.6)

```
Brussels-R3#ping 172.22.1.6

Type escape sequence to abort.
Sending 5, 100-byte ICMP Echos to 172.22.1.6, timeout is 2 seconds:
.!!!!
Success rate is 80 percent (4/5), round-trip min/avg/max = 92/92/92 ms
```

The "show ip nat translation" command in Code Listing 6.14 displays the NAT translation table of Brussels-R1. We can see that no address translation has taken place, indicating that NAT has been bypassed for IP packets that are exchanged between the two private subnets 172.21.1.0/24 and 172.22.1.0/24 through the IPSec tunnel. This is the desirable outcome that fulfills the first part of the requirement for scenario 1 (see section 6.3.1).

Code Listing 6.14:   Brussels-R1 NAT translation table

```
Brussels-R1#show ip nat translation

Brussels-R1#
```

From the crypto engine connection table of Brussels-R1 illustrated in Code Listing 6.15, we can see that two unidirectional IPSec tunnels (see Chapter 2, section 2.9.3.2 for more details) have been built. In addition, the four outbound ICMP echo packets originating from Brussels-R3 to London-R6 (generated previously in Code Listing 6.13 during the ping test) are encrypted, and the four returning inbound ICMP echo-reply packets from London-R6 to Brussels-R3 are decrypted accordingly. This further attests that NAT has been appropriately bypassed for VPN-bound traffic; IPSec has been properly initiated and is fully functional.

Code Listing 6.15:   Brussels-R1 crypto engine connection table

---

```
Brussels-R1#show crypto engine connection active

  ID Interface   IP-Address   State  Algorithm           Encrypt   Decrypt
  52 no idb      no address   set    DES_56_CBC               0         0
  53 Serial0/0   198.1.1.1    set    HMAC_SHA+DES_56_CB       0         4
  54 Serial0/0   198.1.1.1    set    HMAC_SHA+DES_56_CB       4         0
```

The ping test conducted successfully (100%) in Code Listing 6.16 from London-R6 (172.22.1.6) to Brussels-R3 (172.21.1.3) is used to ensure that IP connectivity is also available in the reverse direction between the two VPN clients (London-R6 and Brussels-R3).

Code Listing 6.16:   Ping test from London-R6 (172.22.1.6) to Brussels-R3 (172.21.1.3)

```
London-R6#ping 172.21.1.3

Type escape sequence to abort.
Sending 5, 100-byte ICMP Echos to 172.21.1.3, timeout is 2 seconds:
!!!!!
Success rate is 100 percent (5/5), round-trip min/avg/max = 88/91/92 ms
```

The crypto engine connection table of London-R5 in Code Listing 6.17 further consolidates the findings in Code Listing 6.16. In this instance, the five outbound ICMP echo packets originating from London-R6 to Brussels-R3 (generated previously in Code Listing 6.16 during the ping test) are encrypted, and the five returning inbound ICMP echo-reply packets from Brussels-R3 to London-R6 are decrypted accordingly. This also indicates that IPSec is working fine in the reverse direction. Note that the five encrypted outbound and five decrypted inbound packets are aggregated with those four listed in Code Listing 6.15 to give a cumulative total of nine.

Code Listing 6.17:   London-R5 crypto engine connection table

---

```
London-R5#show crypto engine connection active

  ID Interface   IP-Address   State  Algorithm              Encrypt   Decrypt
   1 <none>      <none>       set    HMAC_SHA+DES_56_CB         0         0
2000 Serial0/1   200.1.1.5    set    HMAC_SHA+DES_56_CB         0         9
2001 Serial0/1   200.1.1.5    set    HMAC_SHA+DES_56_CB         9         0
```

Code Listing 6.18 is used to verify the IP connectivity between a VPN client (Brussels-R3) residing in private subnet 172.21.1.0/24 and an Internet host (London-R5) residing in public subnet 200.1.1.0/24. From the listing, we can see that the ping test is conducted successfully (100%) from Brussels-R3 (172.21.1.3) to London-R5 (200.1.1.5), thus fulfilling the second part of the requirement for scenario 1 (see section 6.3.1), which gives VPN clients (on private subnet 172.21.1.0/24) access to the Internet. In this case, Brussels-R3 will appear to London-R5 as a host in public subnet 198.1.1.0/24 with an IP address of 198.1.1.20 (see Code Listing 6.19) after the NAT operation.

Code Listing 6.18:   Ping test from Brussels-R3 (172.21.1.3) to London-R5 (200.1.1.5)

```
Brussels-R3#ping 200.1.1.5

Type escape sequence to abort.
Sending 5, 100-byte ICMP Echos to 200.1.1.5, timeout is 2 seconds:
!!!!!
Success rate is 100 percent (5/5), round-trip min/avg/max = 56/59/60 ms
```

The NAT translation table of Brussels-R1 in Code Listing 6.19 reflects that address translation has taken place during the ping test conducted in Code Listing 6.18 for the five outgoing and incoming ICMP packets. This further attests that NAT has been appropriately enabled for Internet bound traffic.

Code Listing 6.19:   Brussels-R1 NAT translation table

```
Brussels-R1#show ip nat translation
Pro Inside global       Inside local        Outside local       Outside global
icmp 198.1.1.20:4734    172.21.1.3:4734     200.1.1.5:4734      200.1.1.5:4734
icmp 198.1.1.20:4733    172.21.1.3:4733     200.1.1.5:4733      200.1.1.5:4733
icmp 198.1.1.20:4732    172.21.1.3:4732     200.1.1.5:4732      200.1.1.5:4732
icmp 198.1.1.20:4731    172.21.1.3:4731     200.1.1.5:4731      200.1.1.5:4731
icmp 198.1.1.20:4730    172.21.1.3:4730     200.1.1.5:4730      200.1.1.5:4730
```

Code Listing 6.20 is used to verify the IP connectivity between a VPN client (Brussels-R3) residing in private subnet 172.21.1.0/24 and an Internet host (Brussels-R2) residing in public subnet 199.1.1.8/30. From the listing, we can see that the ping test is conducted successfully from Brussels-R3 (172.21.1.3) to Brussels-R2 (199.1.1.9), thus fulfilling the second part of the requirement for scenario 1 (see Section 6.3.1), which gives VPN clients (on private subnet 172.21.1.0/24) access to the Internet. In this case, Brussels-R3 will appear to Brussels-R2 as a host in public subnet 198.1.1.0/24 with an IP address of 198.1.1.20 (see Code Listing 6.21) after the NAT operation.

Code Listing 6.20:   Ping test from Brussels-R3 (172.21.1.3) to Brussels-R2 (199.1.1.9)

```
Brussels-R3#ping 199.1.1.9

Type escape sequence to abort.
Sending 5, 100-byte ICMP Echos to 199.1.1.9, timeout is 2 seconds:
!!!!!
Success rate is 100 percent (5/5), round-trip min/avg/max = 56/58/60 ms
```

The NAT translation table of Brussels-R1 in Code Listing 6.21 is built on Code Listing 6.19, and it reflects that address translation has taken place during the ping test conducted in Code Listing 6.20 for the five outgoing and five incoming ICMP packets. This further attests that NAT has been appropriately enabled for Internet bound traffic.

Code Listing 6.21:   Brussels-R1 NAT translation table

```
Brussels-R1#show ip nat translation
Pro Inside global       Inside local        Outside local       Outside global
icmp 198.1.1.20:2569    172.21.1.3:2569     199.1.1.9:2569      199.1.1.9:2569
icmp 198.1.1.20:2568    172.21.1.3:2568     199.1.1.9:2568      199.1.1.9:2568
```

```
icmp 198.1.1.20:2567    172.21.1.3:2567    199.1.1.9:2567    199.1.1.9:2567
icmp 198.1.1.20:2566    172.21.1.3:2566    199.1.1.9:2566    199.1.1.9:2566
icmp 198.1.1.20:2565    172.21.1.3:2565    199.1.1.9:2565    199.1.1.9:2565
icmp 198.1.1.20:4734    172.21.1.3:4734    200.1.1.5:4734    200.1.1.5:4734
icmp 198.1.1.20:4733    172.21.1.3:4733    200.1.1.5:4733    200.1.1.5:4733
icmp 198.1.1.20:4732    172.21.1.3:4732    200.1.1.5:4732    200.1.1.5:4732
icmp 198.1.1.20:4731    172.21.1.3:4731    200.1.1.5:4731    200.1.1.5:4731
icmp 198.1.1.20:4730    172.21.1.3:4730    200.1.1.5:4730    200.1.1.5:4730
```

6.3.3   Scenario 2—IPSec and NAT with internal server

Scenario 2 is an extension to scenario 1. Besides inheriting all the previous criteria for scenario 1, scenario 2 also requires Internet hosts as well as VPN clients to access an internal server that is connected to private subnet 172.21.1.0/24. Likewise, the internal server must be able to access the Internet hosts (from subnets 198.1.1.0/24, 199.1.1.8/30, and 200.1.1.0/24) and the VPN clients (from private subnets 172.21.1.0/24 and 172.22.1.0/24) without fail. Brussels-R3 (172.21.1.3) will be the appointed internal server in this scenario.

Before making any modifications to the NAT configuration on Brussels-R1, let's do some ping tests from the Internet hosts (Brussels-R2 and London-R5) to the internal server (Brussels-R3), as illustrated in Code Listings 6.22 and 6.23, to verify the current IP connectivity from the Internet hosts to the internal server.

The ping test conducted in Code Listing 6.22 from Internet host Brussels-R2 (199.1.1.9) to internal server Brussels-R3 (172.21.1.3) gives us a negative result, indicating that there is currently no IP connectivity from Brussels-R2 to Brussels-R3.

Code Listing 6.22:   Ping test from Brussels-R2 (199.1.1.9) to Brussels-R3 (172.21.1.3)

```
Brussels-R2#ping 172.21.1.3

Type escape sequence to abort.
Sending 5, 100-byte ICMP Echos to 172.21.1.3, timeout is 2 seconds:
.....
Success rate is 0 percent (0/5)
```

The ping test conducted in Code Listing 6.23 from Internet host London-R5 (200.1.1.5) to internal server Brussels-R3 (172.21.1.3) gives us a negative result, indicating that there is currently no IP connectivity from London-R5 to Brussels-R3.

Code Listing 6.23:   Ping test from London-R5 (200.1.1.5) to Brussels-R3 (172.21.1.3)

```
London-R5#ping 172.21.1.3

Type escape sequence to abort.
Sending 5, 100-byte ICMP Echos to 172.21.1.3, timeout is 2 seconds:
.....
Success rate is 0 percent (0/5)
```

6.3.3.1 Configurations. Code Listing 6.24 illustrates the modified NAT config-uration for Brussels-R1, which enables the Internet hosts to access the inter-nal server via global address 198.1.1.10. This is achieved with a static NAT statement that translates local address 172.21.1.3 of the internal server to the global address 198.1.1.10 permanently.

Code Listing 6.24:   Brussels-R1 modified NAT configuration for internal server (172.21.1.3)

```
hostname Brussels-R1
!
<Output Omitted>
!
ip nat pool chapter6 198.1.1.20 198.1.1.29 netmask 255.255.255.0
ip nat inside source route-map nonat pool chapter6
! --- Add static NAT statement for access
! --- to internal server 172.21.1.3.
! --- The Internet hosts now know the
! --- internal server as 198.1.1.10.
ip nat inside source static 172.21.1.3 198.1.1.10
!
<Output Omitted>
!
access-list 120 deny   ip 172.21.1.0 0.0.0.255 172.22.1.0 0.0.0.255
access-list 120 permit ip 172.21.1.0 0.0.0.255 any
route-map nonat permit 10
 match ip address 120
```

6.3.3.2 Verification test results. Code Listings 6.25 to 6.31 illustrates the veri-fication test results for the configurations implemented in the previous section.

The NAT translation table of Brussels-R1 in Code Listing 6.25 attests that the static NAT translation defined in Code Listing 6.24 is in place. Brussels-R3 (172.21.1.3) is now accessible to the Internet hosts via global address 198.1.1.10.

Code Listing 6.25:   Brussels-R1 NAT translation table

```
Brussels-R1#show ip nat translation
Pro Inside global      Inside local      Outside local      Outside global
--- 198.1.1.10         172.21.1.3        ---                ---
```

From the positive ping test results illustrated in Code Listing 6.26, we can conclude that Brussels-R2 now has the IP connectivity to Brussels-R3 via its global address 198.1.1.10.

Code Listing 6.26:   Ping test from Brussels-R2 (199.1.1.9) to Brussels-R3 (198.1.1.10)

```
Brussels-R2#ping 198.1.1.10

Type escape sequence to abort.
Sending 5, 100-byte ICMP Echos to 198.1.1.10, timeout is 2 seconds:
!!!!!
Success rate is 100 percent (5/5), round-trip min/avg/max = 56/58/60 ms
```

From the positive ping test results illustrated in Code Listing 6.27, we can conclude that London-R5 now has the IP connectivity to Brussels-R3 via its global address 198.1.1.10.

Code Listing 6.27:   Ping test from London-R5 (200.1.1.5) to Brussels-R3 (198.1.1.10)

```
London-R5#ping 198.1.1.10

Type escape sequence to abort.
Sending 5, 100-byte ICMP Echos to 198.1.1.10, timeout is 2 seconds:
!!!!!
Success rate is 100 percent (5/5), round-trip min/avg/max = 56/58/60 ms
```

However, the ping test conducted from Brussels-R3 (172.21.1.3) to London-R6 (172.22.1.6) as illustrated in Code Listing 6.28 still gives us a negative result. This is because the static NAT operation will always be performed on the source addresses of outbound IP packets originating from Brussels-R3 (internal server). So, once again, this traffic will not match the crypto access list in Brussels-R1 and initiate the IPSec tunnel.

Code Listing 6.28:   Ping test from Brussels-R3 (172.21.1.3) to London-R6 (172.22.1.6)

```
Brussels-R3#ping 172.22.1.6

Type escape sequence to abort.
Sending 5, 100-byte ICMP Echos to 172.22.1.6, timeout is 2 seconds:
.....
Success rate is 0 percent (0/5)
```

The blank crypto engine connection table for Brussels-R1 in Code Listing 6.29 further consolidates our findings in Code Listing 6.28. The IPSec tunnels did not start at all.

Code Listing 6.29:   Brussels-R1 crypto engine connection table

---

```
Brussels-R1#show crypto engine connection active

  ID Interface     IP-Address     State  Algorithm    Encrypt  Decrypt
```

How about conducting the ping test in the reverse direction from London-R6 (172.22.1.6) to Brussels-R3 (172.21.1.3)? As illustrated in Code Listing 6.30, the outcome is still negative.

Code Listing 6.30:   Ping test from London-R6 (172.22.1.6) to Brussels-R3 (172.21.1.3)

```
London-R6#ping 172.21.1.3

Type escape sequence to abort.
Sending 5, 100-byte ICMP Echos to 172.21.1.3, timeout is 2 seconds:
.....
Success rate is 0 percent (0/5)
```

However, the crypto engine connection table of Brussels-R1 in Code Listing 6.31 now reveals something interesting—four inbound encrypted ICMP echo packets from London-R6 were decrypted by Brussels-R1. However, there are no returning outbound ICMP echo-reply packets from Brussels-R3 to encrypt at all.

Code Listing 6.31:   Brussels-R1 crypto engine connection table

---

```
Brussels-R1#show crypto engine connection active

ID Interface    IP-Address    State  Algorithm            Encrypt   Decrypt
55 no idb       no address    set    DES_56_CBC              0         0
56 Serial0/0    198.1.1.1     set    HMAC_SHA+DES_56_CB      0         4
57 Serial0/0    198.1.1.1     set    HMAC_SHA+DES_56_CB      0         0
```

When the ping test is performed from London-R6 to Brussels-R3 in Code Listing 6.30, the ICMP echo packets come into the IPSec tunnel with a source address of 172.22.1.6 and a destination address of 172.21.1.3. In this instance, NAT is not applicable since the destination address is not 198.1.1.10. Therefore, the encrypted ICMP echo packets are decrypted by Brussels-R1 and delivered to Brussels-R3. Brussels-R3 in turn responded with ICMP echo-reply packets. Static NAT is performed on the source addresses of these return packets before IPSec. As such, each return packet will now have a source address of 198.1.1.10 and a destination address of 172.22.1.6. This will not match the crypto access list in Brussels-R1, so the return traffic will not go back into the IPSec tunnel, and thus the ping test conducted earlier in Code Listing 6.30 will register a negative result.

6.3.3.3 Proposed solution. From the derived test results in section 6.3.3.2, Brussels-R3 (internal server) and London-R6 (VPN client) still do not have the required IP connectivity between them. The static NAT configuration in Brussels-R1 only caters for Brussels-R3 access to the Internet hosts and vice versa. The configuration does not take into consideration the access between Brussels-R3 and London-R6 via the IPSec tunnel.

To fulfill the complete requirement for scenario 2, we have to again modify the NAT configuration in Brussels-R1. This time we make use of policy routing to help us bypass the static NAT statement defined earlier in Code Listing 6.24.

Code Listing 6.32 illustrates the modified NAT configuration with policy routing for Brussels-R1, which allows IP packets exchanged between Brussels-R3 (internal server) and private subnet 172.22.1.0/24 to bypass the NAT operation. This is achieved with a policy route-map that is applied on the NAT inside interface (FastEthernet0/0), matching traffic from Brussels-R3 (172.21.1.3) to private subnet 172.22.1.0/24. The matched IP packet is forwarded to a loopback network through Loopback0 (not marked as a NAT interface) back to the router and then routed out of Serial0/0, thus bypassing the NAT operation.

Code Listing 6.32: Brussels-R1 modified NAT configuration with policy routing

---

```
hostname Brussels-R1
!
<Output Omitted>
!
! --- Create loopback interface with address 1.1.1.1.
interface Loopback0
 ip address 1.1.1.1 255.255.255.252
!
interface FastEthernet0/0
 ip address 172.21.1.1 255.255.255.0
 ip nat inside
! --- Enable Cisco fast switching (cache switching) for policy routing.
 ip route-cache policy
! --- Apply policy route-map rtmap on FastEthernet0/0.
 ip policy route-map rtmap
!
interface Serial0/0
 ip address 198.1.1.1 255.255.255.0
 ip nat outside
 crypto map chapter6
!
ip nat pool chapter6 198.1.1.20 198.1.1.29 netmask 255.255.255.0
ip nat inside source route-map nonat pool chapter6
ip nat inside source static 172.21.1.3 198.1.1.10
!
<Output Omitted>
!
access-list 120 deny   ip 172.21.1.0 0.0.0.255 172.22.1.0 0.0.0.255
access-list 120 permit ip 172.21.1.0 0.0.0.255 any

! --- Create access list 130 that matches
! --- any IP traffic from Brussels-R3 (172.21.1.3)
! --- to private subnet 172.22.1.0/24.
access-list 130 permit ip host 172.21.1.3 172.22.1.0 0.0.0.255
route-map nonat permit 10
 match ip address 120
!
! --- Define route-map rtmap.
route-map rtmap permit 10
 match ip address 130 ! --- Match access-list 130 and then forward packet to
 set ip next-hop 1.1.1.2 ! --- loopback network 1.1.1.0/30 at next-hop 1.1.1.2.
```

From Code Listing 6.33, we can see that the ping test conducted from Brussels-R3 (172.21.1.3) to London-R6 (172.22.1.6) is now successful (80%).

Code Listing 6.33:   Ping test from Brussels-R3 (172.21.1.3) to London-R6 (172.22.1.6)

```
Brussels-R3#ping 172.22.1.6

Type escape sequence to abort.
Sending 5, 100-byte ICMP Echos to 172.22.1.6, timeout is 2 seconds:
.!!!!
Success rate is 80 percent (4/5), round-trip min/avg/max = 92/93/96 ms
```

From the crypto engine connection table of Brussels-R1 illustrated in Code Listing 6.34, we can see that two unidirectional IPSec tunnels have been built. In addition, the four outbound ICMP echo packets originating from Brussels-R3 to London-R6 (generated previously in Code Listing 6.33 during the ping test) are encrypted, and the four returning inbound ICMP echo-reply packets from London-R6 to Brussels-R3 are decrypted accordingly. This further attests that NAT has been appropriately bypassed for VPN bound traffic; IPSec has been properly initiated and is fully functional.

Code Listing 6.34:   Brussels-R1 crypto engine connection table

---

```
Brussels-R1#show crypto engine connection active

ID Interface   IP-Address   State  Algorithm          Encrypt  Decrypt
63 no idb      no address   set    DES_56_CBC               0        0
64 Serial0/0   198.1.1.1    set    HMAC_SHA+DES_56_CB       0        4
65 Serial0/0   198.1.1.1    set    HMAC_SHA+DES_56_CB       4        0
```

From Code Listing 6.35, we can see that the ping test conducted in the reverse direction from London-R6 (172.22.1.6) to Brussels-R3 (172.21.1.3) is now successful (100%). The ICMP echo packet flow is illustrated in Figure 6.2, and the return ICMP echo-reply packet is shown in Figure 6.3.

Code Listing 6.35:   Ping test from London-R6 (172.22.1.6) to Brussels-R3 (172.21.1.3)

```
London-R6#ping 172.21.1.3

Type escape sequence to abort.
Sending 5, 100-byte ICMP Echos to 172.21.1.3, timeout is 2 seconds:
!!!!!
Success rate is 100 percent (5/5), round-trip min/avg/max = 92/92/92 ms
```

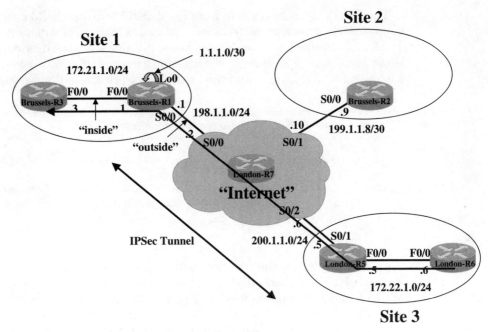

Figure 6.2: ICMP echo packet flow from London-R6 to Brussels-R3.

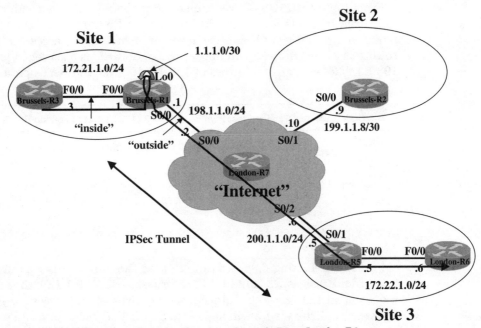

Figure 6.3: ICMP echo-reply packet flow from Brussels-R3 to London-R6.

The five ICMP echo packets from London-R6 to Brussels-R3 enter the IPSec tunnel with a source address of 172.22.1.6 and a destination address of 172.21.1.3. In this instance, NAT is not applicable since the destination address is not 198.1.1.10. The encrypted packets are decrypted at Brussels-R1 (see Code Listing 6.36) accordingly and then delivered to Brussels-R3 (see Fig. 6.2).

Code Listing 6.36: Brussels-R1 crypto engine connection table

---

```
Brussels-R1#show crypto engine connection active

  ID Interface    IP-Address    State  Algorithm              Encrypt   Decrypt
  63 no idb       no address    set    DES_56_CBC                 0         0
  64 Serial0/0    198.1.1.1     set    HMAC_SHA+DES_56_CB         0         9
  65 Serial0/0    198.1.1.1     set    HMAC_SHA+DES_56_CB         9         0
```

With policy routing, the five returning ICMP echo-reply packets (received at Brussels-R1 FastEthernet0/0) are first routed through Loopback0 to next-hop 1.1.1.2; the packets then return to Brussels-R1 again on Loopback0 (see Figure 6.3). The packets are then encrypted (see Code Listing 6.36) and routed out Brussels-R1 Serial0/0 normally. NAT is not invoked here because Loopback0 has not been defined as a NAT interface. Note that the five encrypted outbound and five decrypted inbound packets are aggregated with those four listed in Code Listing 6.34 to reflect a cumulative total of nine in Code Listing 6.36.

From Code Listing 6.37, we can see that the ping test is conducted successfully from Brussels-R3 (172.21.1.3) to London-R5 (200.1.1.5), thus fulfilling the requirement that gives the internal server (Brussels-R3) access to the Internet. In this case, Brussels-R3 will appear to London-R5 as a server in public subnet 198.1.1.0/24 with an IP address of 198.1.1.10 after the NAT operation.

Code Listing 6.37: Ping test from Brussels-R3 (172.21.1.3) to London-R5 (200.1.1.5)

```
Brussels-R3#ping 200.1.1.5

Type escape sequence to abort.
Sending 5, 100-byte ICMP Echos to 200.1.1.5, timeout is 2 seconds:
!!!!!
Success rate is 100 percent (5/5), round-trip min/avg/max = 56/56/60 ms
```

From Code Listing 6.38, we can see that the ping test is conducted successfully from Brussels-R3 (172.21.1.3) to Brussels-R2 (199.1.1.9), thus fulfilling the requirement that gives the internal server (Brussels-R3) access to the Internet. In this case, Brussels-R3 will appear to Brussels-R2 as a server in public subnet 198.1.1.0/24 with an IP address of 198.1.1.10 after the NAT operation.

Code Listing 6.38:   Ping test from Brussels-R3 (172.21.1.3) to Brussels-R2 (199.1.1.9)

```
Brussels-R3#ping 199.1.1.9

Type escape sequence to abort.
Sending 5, 100-byte ICMP Echos to 199.1.1.9, timeout is 2 seconds:
!!!!!
Success rate is 100 percent (5/5), round-trip min/avg/max = 56/56/60 ms
```

With the last code listing for this section, we have met all the requirements for scenario 2.

6.4   HSRP Overview

Network resiliency enables remote sites to locate another tunneling peer if the primary peer becomes unreachable or if there is a permanent loss of IP connectivity between peers. IPSec VPN resiliency can be achieved with Cisco hot standby router protocol (HSRP). In this section, we briefly discuss HSRP in general to give the reader a grasp on what HSRP is before implementing it with IPSec in the next section.

HSRP (RFC 2281) provides a mechanism that is designed to support nondisruptive failover of IP traffic in certain circumstances. In particular, the protocol protects against the failure of the first hop router when the source host cannot learn the IP address of the first hop router dynamically. The protocol is designed for use over multiaccess, multicast, or broadcast capable media (for example, Ethernet). HSRP is intended for legacy host implementations that do not support dynamic router discovery but are capable of configuring a default router. In other words, the objective of the protocol is to allow hosts to appear to use a single router and to maintain connectivity even if the actual first hop router they are using fails.

With HSRP, a set of routers works in concert to present the illusion of a single virtual router to the hosts on the LAN (local area network). This set is known as an HSRP group or a standby group. A single well-known virtual media access control (MAC) address is allocated to the group, as well as a virtual IP address. The virtual IP address should belong to the primary subnet in use on the LAN, but must differ from the addresses allocated as interface addresses on all routers and hosts on the LAN. A single router elected from the group is responsible for forwarding the packets that hosts send to the virtual router. This router is known as the active router. Another router is elected as the standby router. In the event that the active router fails, the standby assumes the packet-forwarding duties of the active router without a major interruption in the host's connectivity. Although an arbitrary number of routers may run HSRP, only the active router forwards the packets sent to the virtual router.

To minimize network traffic, only the active and the standby routers send periodic HSRP messages once the protocol has completed the election process. If the active router fails, the standby router takes over as the active router. If the standby router fails or becomes the active router, another router is elected as the standby router.

Each HSRP router in the group participates in the protocol by implementing a simple state machine as follows:

- *Initial state*: This is the starting state and indicates that HSRP is not running. This state is entered through a configuration change or when an interface first comes up.

- *Learn state*: The router has not determined the virtual IP address and has not yet seen an authenticated Hello message from the active router. In this state the router is still waiting to hear from the active router.

- *Listen state*: The router knows the virtual IP address but is neither the active router nor the standby router. It listens for Hello messages from those routers.

- *Speak state*: The router sends periodic Hello messages and actively participates in the election of the active and/or standby router. A router cannot enter Speak state unless it has the virtual IP address.

- *Standby state*: The router is a candidate to become the next active router and sends periodic Hello messages. Excluding transient conditions, there can be at most one router in the group in Standby state.

- *Active state*: The router is currently forwarding packets that are sent to the group's virtual MAC address. The router sends periodic Hello messages. Excluding transient conditions, there can be at most one router in Active state in the group.

6.5 Case Study 6.2: IPSec with HSRP

6.5.1 Case overview

Figure 6.4 illustrates the network design and topology for Case Study 6.2. The case study uses six Cisco routers distributed across three different locations: Site 1, Site 3, and Internet. Site 1 and Site 3 are interconnected by London-R7, which is emulating the Internet by posing as an Internet service provider (ISP) gateway for all these sites. Brussels-R1, Brussels-R2, and Brussels-R3 are interconnected in Site 1; London-R5 and London-R6 are interconnected in Site 3. Site 1 and Site 3 belong to the same VPN and communicate with each other securely over the Internet via an IPSec tunnel.

Network resiliency is implemented with the help of HSRP to achieve transparency to the VPN clients (including Brussels-R3) on subnet 172.21.1.0/24 when the network links of either Brussels-R1 or Brussels-R2 become unavailable. In other words, the VPN clients on subnet 172.21.1.0/24 point to standby

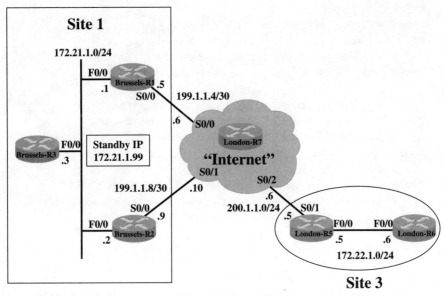

Figure 6.4: Network design and topology for Case Study 6.2.

address 172.21.1.99 (a virtual router representing both Brussels-R1 and Brussels-R2) as the default gateway address. With all these preliminaries in place, we can move on to discuss the incorporation of IPSec and HSRP with three different scenarios in the subsequent sections.

6.5.2 Scenario 1—IPSec dual peers with HSRP

6.5.2.1 Configurations. Code Listings 6.39 to 6.44 illustrate the configurations for the six different routers portrayed in Figure 6.4. The IPSec dual peering is implemented on London-R5 in Code Listing 6.42, and HSRP is configured on both Brussels-R1 in Code Listing 6.39 and Brussels-R2 in Code Listing 6.40. In the configurations, comments (in *italics*) precede certain configuration lines to explain them.

Code Listing 6.39:   Brussels-R1 configuration

```
hostname Brussels-R1
!
! --- Supersede the default policy and use
! --- pre-shared keys for peer authentication.
crypto isakmp policy 110
  authentication pre-share
```

```
!
! --- Specify the pre-shared key "key15"
! --- to be used for the IPSec tunnel
! --- between Brussels-R1 and London-R5 (200.1.1.5).
crypto isakmp key key15 address 200.1.1.5
!
! --- Specify the ESP transform settings for IPSec,
! --- which is later applied to the crypto map.
crypto ipsec transform-set chapter6 esp-des esp-sha-hmac
!
! --- Specify the crypto map chapter6 where
! --- we define our remote peer London-R5 (200.1.1.5),
! --- transform set chapter6, and our crypto access list 110.
 crypto map chapter6 10 ipsec-isakmp
 set peer 200.1.1.5
 set transform-set chapter6
 match address 110
!
! --- HSRP is configured on FastEthernet0/0 of Brussels-R1.
interface FastEthernet0/0
 ip address 172.21.1.1 255.255.255.0

! --- The standby priority determines which router
! --- assumes the active router role.
! --- In this case, Brussels-R1 is given a HSRP priority of 110,
! --- higher than priority 100 of Brussels-R2. Therefore,
! --- Brussels-R1 will assume the role of the active router.
! --- However, once Brussels-R1 is disabled, the standby router
! --- Brussels-R2 will automatically assume the role of
! --- the active router. When Brussels-R1 is re-enabled,
! --- Brussels-R2 will still retain the active router role
! --- even though it has a lower priority than Brussels-R1.
! --- Therefore, to ensure that Brussels-R1 resumes
! --- the active router role, the "preempt" keyword is used.
 standby priority 110 preempt

! --- Define virtual router IP address to be 172.21.1.99.
! --- For simplicity reasons, no standby
! --- group number (Group Number 0) is specified here.
 standby ip 172.21.1.99

! --- The HSRP tracking feature is enabled to reduce
! --- the possibility of Brussels-R1 assuming the role of
! --- the active router when Serial0/0 becomes unavailable.
! --- In this case, the HSRP priority of Brussels-R1 is
```

```
! --- decremented by an interface priority of 30 when Serial0/0
! --- becomes unavailable. The priority of Brussels-R1 is
! --- incremented back by this same amount when Serial0/0 becomes
! --- available again. Note that for HSRP tracking,
! --- it is important to enable "preempt" on both routers
! --- so that when the HSRP priority of the higher-priority router
! --- is decreased to a value lower than that of the standby router,
! --- the standby router will "preempt" and assume the role of
! --- the active router. Likewise, when the priority of the previous
! --- active router has been restored to its original value,
! --- this router should be able to reinstate its active
! --- router role by doing a preempt.
standby track Serial0/0 30
!
interface Serial0/0
 ip address 199.1.1.5 255.255.255.252
! --- Apply crypto map chapter6 to Serial0/0
! --- to activate crypto engine.
 crypto map chapter6
!
! --- Routing from Brussels-R1 to Internet destinations
! --- is via a default route to Internet gateway London-R7
! --- with a next-hop address of 199.1.1.6.
ip route 0.0.0.0 0.0.0.0 199.1.1.6
!
! --- This is the crypto access list that we have
! --- referenced in crypto map chapter6.
! --- We are encrypting IP traffic between
! --- private subnets 172.21.1.0/24 and 172.22.1.0/24.
access-list 110 permit ip 172.21.1.0 0.0.0.255 172.22.1.0 0.0.0.255
```

Code Listing 6.40:   Brussels-R2 configuration

```
hostname Brussels-R2
!
! --- Supersede the default policy and use
! --- pre-shared keys for peer authentication.
crypto isakmp policy 110
 authentication pre-share
!
! --- Specify the pre-shared key "key25"
! --- to be used for the IPSec tunnel
! --- between Brussels-R1 and London-R5 (200.1.1.5).
crypto isakmp key key25 address 200.1.1.5
```

```
!
! --- Specify the ESP transform settings for IPSec,
! --- which is later applied to the crypto map.
crypto ipsec transform-set chapter6 esp-des esp-sha-hmac
!
! --- Specify the crypto map chapter6 where we define
! --- our remote peer London-R5 (200.1.1.5),
! --- transform set chapter6, and our crypto access list 110.
 crypto map chapter6 10 ipsec-isakmp
 set peer 200.1.1.5
 set transform-set chapter6
 match address 110
!
! --- HSRP is configured on FastEthernet0/0 of Brussels-R2.
interface FastEthernet0/0
 ip address 172.21.1.2 255.255.255.0

! --- Brussels-R2 is given an HSRP priority of 100,
! --- lower than priority 110 of Brussels-R1.
! --- Therefore, Brussels-R2 will assume the role of
! --- the standby router. The HSRP preempt mechanism
! --- is also enabled on Brussels-R2.
 standby priority 100 preempt
! --- Define virtual router IP address to be 172.21.1.99.
! --- For simplicity reasons, no standby
! --- group number (group number 0) is specified here.
 standby ip 172.21.1.99
! --- The HSRP tracking feature is enabled
! --- on Brussels-R2 for Serial0/0.
 standby track Serial0/0 30
!
interface Serial0/0
 ip address 199.1.1.9 255.255.255.252
! --- Apply crypto map chapter6 to Serial0/0
! --- to activate crypto engine.
 crypto map chapter6
!
! --- Routing from Brussels-R2 to Internet destinations
! --- is via a default route to Internet gateway London-R7
! --- with a next-hop address of 199.1.1.10.
ip route 0.0.0.0 0.0.0.0 199.1.1.10
!
! --- This is the crypto access list that we have
! --- referenced in crypto map chapter6.
```

```
! --- We are encrypting IP traffic between
! --- private subnets 172.21.1.0/24 and 172.22.1.0/24.
access-list 110 permit ip 172.21.1.0 0.0.0.255 172.22.1.0 0.0.0.255
```

Code Listing 6.41:   Brussels-R3 configuration

```
! --- Brussels-R3 assumes the role of a VPN client
! --- throughout the case study.
hostname Brussels-R3
!
interface FastEthernet0/0
 ip address 172.21.1.3 255.255.255.0
!
! --- Routing from Brussels-R3 to Internet destinations
! --- is via a default route to the HSRP virtual router
! --- with a next-hop address of 172.21.1.99.
ip route 0.0.0.0 0.0.0.0 172.21.1.99
```

Code Listing 6.42:   London-R5 configuration

```
hostname London-R5
!
! --- Supersede the default policy and use
! --- pre-shared keys for peer authentication.
crypto isakmp policy 110
 authentication pre-share
!
! --- Specify the pre-shared key "key25"
! --- to be used for the IPSec tunnel
! --- between London-R5 and Brussels-R2 (199.1.1.9).
crypto isakmp key key25 address 199.1.1.9

! --- Specify the pre-shared key "key15"
! --- to be used for the IPSec tunnel
! --- between London-R5 and Brussels-R1 (199.1.1.5).
crypto isakmp key key15 address 199.1.1.5
!

! --- Specify the ESP transform settings for IPSec,
! --- which is later applied to the crypto map.
crypto ipsec transform-set chapter6 esp-des esp-sha-hmac
!
```

```
! --- Specify the crypto map chapter6 sequence number 10,
! --- where we define our first remote peer
! --- Brussels-R1 (199.1.1.5), transform set chapter6,
! --- and the same crypto access list 110.
crypto map chapter6 10 ipsec-isakmp
 set peer 199.1.1.5
 set transform-set chapter6
 match address 110

! --- Specify the crypto map chapter6 sequence number 20,
! --- where we define our second remote peer
! --- Brussels-R2 (199.1.1.9), transform set chapter6,
! --- and the same crypto access list 110.
crypto map chapter6 20 ipsec-isakmp
 set peer 199.1.1.9
 set transform-set chapter6
 match address 110
!
interface FastEthernet0/0
 ip address 172.22.1.5 255.255.255.0
!
interface Serial0/1
 ip address 200.1.1.5 255.255.255.252
! --- Apply crypto map chapter6 to Serial0/1
! --- to activate crypto engine.
 crypto map chapter6
!
! --- Routing from London-R5 to Internet destinations
! --- is via a default route to Internet gateway London-R7
! --- with a next-hop address of 200.1.1.6.
ip route 0.0.0.0 0.0.0.0 200.1.1.6
!
! --- This is the crypto access list that we have
! --- referenced in crypto map chapter6.
! --- We are encrypting IP traffic between
! --- private subnets 172.22.1.0/24 and 172.21.1.0/24.
access-list 110 permit ip 172.22.1.0 0.0.0.255 172.21.1.0 0.0.0.255
```

Code Listing 6.43:   London-R6 configuration

```
! --- London-R6 assumes the role of a VPN client
! --- throughout the case study.
hostname London-R6
!
```

```
interface FastEthernet0/0
 ip address 172.22.1.6 255.255.255.0
!
! --- Routing from London-R6 to Internet destinations
! --- is via a default route to London-R5
! --- with a next-hop address of 172.22.1.5.
ip route 0.0.0.0 0.0.0.0 172.22.1.5
```

Code Listing 6.44: London-R7 configuration

```
! --- London-R7 is emulating the Internet
! --- by assuming the role of an ISP gateway.
hostname London-R7
!
interface Serial0/0
! --- 199.1.1.6 is the default next-hop for Brussels-R1.
 ip address 199.1.1.6 255.255.255.252
 clockrate 64000
!
interface Serial0/1
! --- 199.1.1.10 is the default next-hop for Brussels-R2.
 ip address 199.1.1.10 255.255.255.252
 clockrate 64000
!
interface Serial0/2
! --- 200.1.1.6 is the default next-hop for London-R5.
 ip address 200.1.1.6 255.255.255.252
 clockrate 64000
!
! --- For simplicity reasons, three static routes
! --- are specified to direct traffic to
! --- subnets 172.21.1.0/24 and 172.22.1.0/24.
ip route 172.21.1.0 255.255.255.0 199.1.1.5
ip route 172.21.1.0 255.255.255.0 199.1.1.9
ip route 172.22.1.0 255.255.255.0 200.1.1.5
```

6.5.2.2 Verification before HSRP failover. Code Listings 6.45 to 6.50 illustrate some of the verification test results before the HSRP failover (making FastEthernet0/0 on Brussels-R1 unavailable) for the configurations implemented in the previous section.

The "show standby" command in Code Listing 6.45 illustrates the HSRP status (before failover) of Brussels-R1 as the current active router.

Code Listing 6.45:   HSRP status of Brussels-R1 before failover

---

```
Brussels-R1#show standby
FastEthernet0/0 - Group 0
  Local state is Active, priority 110, may preempt
  Hellotime 3 holdtime 10
  Next hello sent in 00:00:02.522
  Hot standby IP address is 172.21.1.99 configured
  Active router is local
  Standby router is 172.21.1.2 expires in 00:00:09
  Standby virtual mac address is 0000.0c07.ac00
```

The "show standby" command in Code Listing 6.46 illustrates the HSRP status (before failover) of Brussels-R2 as the current standby router.

Code Listing 6.46:   HSRP status of Brussels-R2 before failover

---

```
Brussels-R2#show standby
FastEthernet0/0 - Group 0
  Local state is Standby, priority 100, may preempt
  Hellotime 3 holdtime 10
  Next hello sent in 00:00:00.652
  Hot standby IP address is 172.21.1.99 configured
  Active router is 172.21.1.1 expires in 00:00:07
  Standby router is local
  Standby virtual mac address is 0000.0c07.ac00
```

As illustrated in Code Listing 6.47, the ping test is conducted successfully (80%) from Brussels-R3 (172.21.1.3) to London-R6 (172.22.1.6).

Code Listing 6.47:   Ping test from Brussels-R3 (172.21.1.3) to London-R6 (172.22.1.6)

```
Brussels-R3#ping
Protocol [ip]:
Target IP address: 172.22.1.6
Repeat count [5]:
Datagram size [100]:
Timeout in seconds [2]:
Extended commands [n]: y
Source address or interface: 172.21.1.3
Type of service [0]:
```

```
Set DF bit in IP header? [no]:
Validate reply data? [no]:
Data pattern [0xABCD]:
Loose, Strict, Record, Timestamp, Verbose[none]:
Sweep range of sizes [n]:
Type escape sequence to abort.
Sending 5, 100-byte ICMP Echos to 172.22.1.6, timeout is 2 seconds:
.!!!!
```
Success rate is 80 percent (4/5), round-trip min/avg/max = 92/92/92 ms

The crypto engine connection table of Brussels-R1 in Code Listing 6.48 reflects the results of the ping test conducted in Code Listing 6.47. The four outbound ICMP echo packets originating from Brussels-R3 were encrypted, and the returning four inbound ICMP echo-reply packets from London-R6 were decrypted.

Code Listing 6.48:   Brussels-R1 crypto engine connection table

---

```
Brussels-R1#show crypto engine connection active
```

ID	Interface	IP-Address	State	Algorithm	Encrypt	Decrypt
7	no idb	no address	set	DES_56_CBC	0	0
8	Serial0/0	199.1.1.5	set	HMAC_SHA+DES_56_CB	0	**4**
9	Serial0/0	199.1.1.5	set	HMAC_SHA+DES_56_CB	**4**	0

Code Listing 6.49 displays the crypto engine connection table of London-R5. From the perspective of London-R5, the four inbound ICMP echo packets from Brussels-R3 were decrypted, and four outbound ICMP echo-reply packets from London-R6 (responding to Brussels-R3's ICMP echo packets) were encrypted.

Code Listing 6.49:   London-R5 crypto engine connection table

---

```
London-R5#sh crypto engine connection active
```

ID	Interface	IP-Address	State	Algorithm	Encrypt	Decrypt
1	<none>	<none>	set	HMAC_SHA+DES_56_CB	0	0
2000	Serial0/1	200.1.1.5	set	HMAC_SHA+DES_56_CB	0	**4**
2001	Serial0/1	200.1.1.5	set	HMAC_SHA+DES_56_CB	**4**	0

Code Listing 6.50 displays the empty crypto engine connection table of Brussels-R2. Note that the ICMP traffic is not traversing Brussels-R2 at this time.

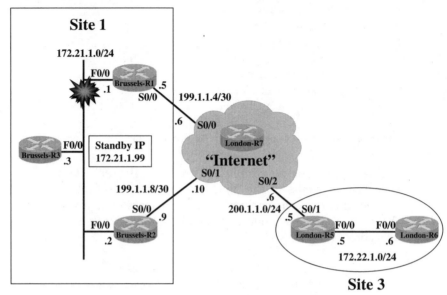

Figure 6.5: Shutting down Brussels-R1 interface FastEthernet0/0.

Code Listing 6.50:   Brussels-R2 crypto engine connection table

---

```
Brussels-R2#show crypto engine connection active

 ID Interface      IP-Address      State  Algorithm       Encrypt  Decrypt
```

Here are the observations before HSRP failover:

- Brussels-R1 initiates IPSec connection with London-R5.
- IPSec connection between Brussels-R1 and London-R5 is validated.
- Traffic between subnets 172.21.1.0/24 and 172.22.1.0/24 are encrypted (and decrypted) properly.

6.5.2.3 Verification after HSRP failover.   As illustrated in Figure 6.5, the HSRP failover is achieved by shutting down interface FastEthernet0/0 on Brussels-R1. Code Listings 6.51 to 6.58 exhibit some of the verification test results after the HSRP failover for the configurations implemented in section 6.5.2.1.

The "show standby" command in Code Listing 6.51 illustrates the HSRP status (after failover) of Brussels-R1 as unknown, indicating that Brussels-R1 is no longer the active router after being disconnected from subnet 172.21.1.0/24.

Code Listing 6.51:   HSRP status of Brussels-R1 after failover

```
Brussels-R1#show standby
FastEthernet0/0 - Group 0
  Local state is Init, priority 110, may preempt
  Hellotime 3 holdtime 10
  Hot standby IP address is 172.21.1.99 configured
  Active router is unknown expired
  Standby router is unknown expired
  Standby virtual mac address is 0000.0c07.ac00
```

The "show standby" command in Code Listing 6.52 illustrates the HSRP status (after failover) of Brussels-R2 as the current active router, signifying that our HSRP configuration is working properly. IP traffic from subnet 172.21.1.0/24 should now be directed to Brussels-R2 instead of to Brussels-R1.

Code Listing 6.52:   HSRP status of Brussels-R2 after failover

```
Brussels-R2#show standby
FastEthernet0/0 - Group 0
  Local state is Active, priority 100, may preempt
  Hellotime 3 holdtime 10
  Next hello sent in 00:00:02.424
  Hot standby IP address is 172.21.1.99 configured
  Active router is local
  Standby router is unknown expired
  Standby virtual mac address is 0000.0c07.ac00
```

The ping test conducted in Code Listing 6.53, however, from Brussels-R3 (172.21.1.3) to London-R6 (172.22.1.6) yields a negative result.

Code Listing 6.53:   Ping test from Brussels-R3 (172.21.1.3) to London-R6 (172.22.1.6)

```
Brussels-R3#ping
Protocol [ip]:
Target IP address: 172.22.1.6
Repeat count [5]:
Datagram size [100]:
Timeout in seconds [2]:
Extended commands [n]: y
Source address or interface: 172.21.1.3
```

```
Type of service [0]:
Set DF bit in IP header? [no]:
Validate reply data? [no]:
Data pattern [0xABCD]:
Loose, Strict, Record, Timestamp, Verbose[none]:
Sweep range of sizes [n]:
Type escape sequence to abort.
Sending 5, 100-byte ICMP Echos to 172.22.1.6, timeout is 2 seconds:
.....
```
Success rate is 0 percent (0/5)

The crypto engine connection table of Brussels-R2 remains empty (Code Listing 6.54), indicating that IPSec has not been initiated and the IPSec tunnels between Brussels-R2 and London-R5 have not been built.

Code Listing 6.54:   Brussels-R2 crypto engine connection table

---

```
Brussels-R2#show crypto engine connection active

  ID Interface      IP-Address      State  Algorithm        Encrypt  Decrypt
```

The "show crypto ipsec sa" command in Code Listing 6.55 sheds more light on the findings in Code Listing 6.54. Note that no IPSec security associations (SAs) between Brussels-R2 and London-R5 have been established. Since the respective SAs are not up, the five ICMP echo packets from Brussels-R3 cannot be sent out appropriately and are reflected as five send errors.

Code Listing 6.55:   Brussels-R2 IPSec security association

```
Brussels-R2#show crypto ipsec sa

interface: Serial0/0
   Crypto map tag: chapter6, local addr. 199.1.1.9

   local  ident (addr/mask/prot/port): (172.21.1.0/255.255.255.0/0/0)
   remote ident (addr/mask/prot/port): (172.22.1.0/255.255.255.0/0/0)
   current_peer: 200.1.1.5
     PERMIT, flags={origin_is_acl,}
    #pkts encaps: 0, #pkts encrypt: 0, #pkts digest 0
    #pkts decaps: 0, #pkts decrypt: 0, #pkts verify 0
    #send errors 5, #recv errors 0
```

```
local crypto endpt.: 199.1.1.9, remote crypto endpt.: 200.1.1.5
path mtu 1500, media mtu 1500
current outbound spi: 0

inbound esp sas:

inbound ah sas:

outbound esp sas:

outbound ah sas:
```

As illustrated in Code Listing 6.56, the ping test conducted from London-R6 (172.22.1.6) to Brussels-R3 (172.21.1.3) also registers a negative result.

Code Listing 6.56:   Ping test from London-R6 (172.22.1.6) to Brussels-R3 (172.21.1.3)

```
London-R6#ping
Protocol [ip]:
Target IP address: 172.21.1.3
Repeat count [5]:
Datagram size [100]:
Timeout in seconds [2]:
Extended commands [n]: y
Source address or interface: 172.22.1.6
Type of service [0]:
Set DF bit in IP header? [no]:
Validate reply data? [no]:
Data pattern [0xABCD]:
Loose, Strict, Record, Timestamp, Verbose[none]:
Sweep range of sizes [n]:
Type escape sequence to abort.
Sending 5, 100-byte ICMP Echos to 172.21.1.3, timeout is 2 seconds:
.....
Success rate is 0 percent (0/5)
```

The crypto engine connection table of London-R5 in Code Listing 6.57 shows that the five outbound ICMP echo packets from London-R6 to Brussels-R3 are encrypted and are still sent out of the IPSec tunnel to Brussels-R1, which is not the favored outcome. Note that the five encrypted outbound packets are aggregated with the four listed in Code Listing 6.49 to reflect a cumulative total of nine in Code Listing 6.57.

Code Listing 6.57:   London-R5 crypto engine connection table

---

```
London-R5#show crypto engine connection active

  ID Interface   IP-Address   State   Algorithm              Encrypt   Decrypt
   1 <none>      <none>       set     HMAC_SHA+DES_56_CB           0         0
2000 Serial0/1   200.1.1.5    set     HMAC_SHA+DES_56_CB           0         4
2001 Serial0/1   200.1.1.5    set     HMAC_SHA+DES_56_CB           9         0
```

The crypto engine connection table of Brussels-R1 in Code Listing 6.58 further consolidates the findings in Code Listing 6.57. The table shows that the five inbound ICMP echo packets from London-R6 to Brussels-R3 are decrypted, but there are no returning ICMP echo-reply packets from Brussels-R3 to encrypt because Brussels-R1 has already been disconnected from subnet 172.21.1.0/24. Note that the five decrypted inbound packets are aggregated with the four listed in Code Listing 6.48 to reflect a cumulative total of nine in Code Listing 6.58.

Code Listing 6.58:   Brussels-R1 crypto engine connection table

---

```
Brussels-R1#show crypto engine connection active

  ID Interface   IP-Address   State   Algorithm              Encrypt   Decrypt
   7 no idb      no address   set     DES_56_CBC                   0         0
   8 Serial0/0   198.1.1.5    set     HMAC_SHA+DES_56_CB           0         9
   9 Serial0/0   198.1.1.5    set     HMAC_SHA+DES_56_CB           4         0
```

Here are the observations after HSRP failover:

- Brussels-R2 attempts an IPSec connection with London-R5 but fails.
- Brussels-R2 IPSec connection request is invalidated at London-R5. This is because the same crypto access list is used in both the crypto map entries on London-R5. The IPSec process at London-R5 will always match the first crypto map entry found; in this case, it will always match Brussels-R1 since Brussels-R2 is not in the first entry.
- Therefore, London-R5 still sends encrypted packets to Brussels-R1 since this IPSec peer is still reachable to London-R5 after the failover.
- IP connectivity (ping test) fails in both directions.

6.5.3   Scenario 2—IPSec backup peers with HSRP

Recall in scenario 1 after the HSRP failover, when Brussels-R2 attempts an IPSec connection with London-R5, the IPSec connection request is invalidated at London-R5. This is because the same crypto access list is used in both the crypto map entries on London-R5. As such, the IPSec process at London-R5 will always match the first crypto map entry found. Therefore, it will always match Brussels-R1 since Brussels-R2 is not in the first entry.

In this scenario, we shall rectify the previous IPSec dual peers configuration on London-R5 by using the IPSec backup peers configuration. Instead of using two crypto map entries, one for each HSRP router, we now use two peer statements in the same crypto map entry. Note that in this setup, Site 3 will be utilizing only a single path to Site 1 at any one time even though Site 1 is reachable by two different paths.

6.5.3.1 Configurations. Code Listing 6.59 illustrates the IPSec backup peers configuration for London-R5. Comparing with Code Listing 6.42, our configuration for London-R5 now has only one crypto map entry with two peers, instead of the previous two crypto map entries with one peer per entry.

Code Listing 6.59:   London-R5 IPSec backup peers configuration

---

```
hostname London-R5
!
<Output Omitted>
!
! --- IPSec backup peers configuration for London-R5 using
! --- one crypto map entry with two peers.
crypto map chapter6 10 ipsec-isakmp
 set peer 199.1.1.5 ! --- Remote IPSec primary peer Brussels-R1.
 set peer 199.1.1.9 ! --- Remote IPSec backup peer Brussels-R2.
 set transform-set chapter6
 match address 110
!
<Output Omitted>
```

6.5.3.2 Verification before HSRP failover. Scenario 2 shares the same verification test results as scenario 1 (see section 6.5.2.2) before the HSRP failover.

6.5.3.3 Verification after HSRP failover. As illustrated in Figure 6.5, the HSRP failover is achieved by shutting down interface FastEthernet0/0 on Brussels-R1. Code Listing 6.60 to 6.69 exhibits some of the verification test results after the HSRP failover for the modified London-R5 IPSec backup peers configuration listed in Code Listing 6.59.

As illustrated in Code Listing 6.60, the ping test conducted from London-R6 (172.22.1.6) to Brussels-R3 (172.21.1.3) registers a negative result.

Code Listing 6.60:   Ping test from London-R6 (172.22.1.6) to Brussels-R3 (172.21.1.3)

```
London-R6#ping
Protocol [ip]:
Target IP address: 172.21.1.3
Repeat count [5]:
Datagram size [100]:
Timeout in seconds [2]:
Extended commands [n]: y
Source address or interface: 172.22.1.6
Type of service [0]:
Set DF bit in IP header? [no]:
Validate reply data? [no]:
Data pattern [0xABCD]:
Loose, Strict, Record, Timestamp, Verbose[none]:
Sweep range of sizes [n]:
Type escape sequence to abort.
Sending 5, 100-byte ICMP Echos to 172.21.1.3, timeout is 2 seconds:
.....
Success rate is 0 percent (0/5)
```

The crypto engine connection table of London-R5 in Code Listing 6.61 shows that the five outbound ICMP echo packets from London-R6 to Brussels-R3 are encrypted and are still sent out of the IPSec tunnel to Brussels-R1, which is not the desired outcome. Note that the five encrypted outbound packets are aggregated with the four listed in Code Listing 6.49 to reflect a cumulative total of nine in Code Listing 6.61.

Code Listing 6.61:   London-R5 crypto engine connection table

---

```
London-R5#show crypto engine connection active
```

ID	Interface	IP-Address	State	Algorithm	Encrypt	Decrypt
1	<none>	<none>	set	HMAC_SHA+DES_56_CB	0	0
2000	Serial0/1	200.1.1.5	set	HMAC_SHA+DES_56_CB	0	4
2001	Serial0/1	200.1.1.5	set	HMAC_SHA+DES_56_CB	9	0

The crypto engine connection table of Brussels-R1 in Code Listing 6.62 further consolidates the findings in Code Listing 6.61. The table shows that the five inbound ICMP echo packets from London-R6 to Brussels-R3 are decrypted, but there are no returning ICMP echo-reply packets from Brussels-R3 to encrypt because Brussels-R1 has already been disconnected from subnet

172.21.1.0/24. Note that the five decrypted inbound packets are aggregated with the four listed in Code Listing 6.48 to reflect a cumulative total of nine in Code Listing 6.62.

Code Listing 6.62:   Brussels-R1 crypto engine connection table

---

```
Brussels-R1#show crypto engine connection active

  ID Interface    IP-Address    State  Algorithm              Encrypt   Decrypt
   7 no idb       no address    set    DES_56_CBC                   0         0
   8 Serial0/0    199.1.1.5     set    HMAC_SHA+DES_56_CB           0         9
   9 Serial0/0    199.1.1.5     set    HMAC_SHA+DES_56_CB           4         0
```

The crypto engine connection table of Brussels-R2 still remains empty (Code Listing 6.63), indicating that IPSec has not been initiated, and the IPSec tunnels between Brussels-R2 and London-R5 have not been built. Up to this point, the results and observations for this scenario are identical to that of scenario 1.

Code Listing 6.63:   Brussels-R2 crypto engine connection table

---

```
Brussels-R2#show crypto engine connection active

  ID Interface    IP-Address     State  Algorithm              Encrypt  Decrypt
```

How about performing the ping test from Brussels-R3 instead? Interestingly, this time the ping test conducted in Code Listing 6.64 from Brussels-R3 (172.21.1.3) to London-R6 (172.22.1.6) exhibits a positive result (80 percent successful).

Code Listing 6.64:   Ping test from Brussels-R3 (172.21.1.3) to London-R6 (172.22.1.6)

```
Brussels-R3#ping
Protocol [ip]:
Target IP address: 172.22.1.6
Repeat count [5]:
Datagram size [100]:
Timeout in seconds [2]:
Extended commands [n]: y
Source address or interface: 172.21.1.3
```

```
Type of service [0]:
Set DF bit in IP header? [no]:
Validate reply data? [no]:
Data pattern [0xABCD]:
Loose, Strict, Record, Timestamp, Verbose[none]:
Sweep range of sizes [n]:
Type escape sequence to abort.
Sending 5, 100-byte ICMP Echos to 172.22.1.6, timeout is 2 seconds:
.!!!!
```
Success rate is 80 percent (4/5), round-trip min/avg/max = 92/92/92 ms

The crypto engine connection table of Brussels-R2 in Code Listing 6.65 reflects the results of the ping test conducted in Code Listing 6.64. The IPSec tunnels to London-R5 have been built. The four outbound ICMP echo packets originating from Brussels-R3 were encrypted, and the returning four inbound ICMP echo-reply packets from London-R6 were decrypted.

Code Listing 6.65:   Brussels-R2 crypto engine connection table

---

```
Brussels-R2#show crypto engine connection active
```

ID	Interface	IP-Address	State	Algorithm	Encrypt	Decrypt
9	no idb	no address	set	DES_56_CBC	0	0
10	Serial0/0	199.1.1.9	set	HMAC_SHA+DES_56_CB	0	**4**
11	Serial0/0	199.1.1.9	set	HMAC_SHA+DES_56_CB	**4**	0

From the crypto engine connection table of London-R5 in Code Listing 6.66, we can see that the IPSec tunnels to Brussels-R2 have been built. The four inbound ICMP echo packets from Brussels-R3 were decrypted, and the four outbound ICMP echo-reply packets from London-R6 (responding to Brussels-R3's ICMP echo packets) were encrypted, all via the newly created IPSec tunnels to Brussels-R2.

Code Listing 6.66: London-R5 crypto engine connection table

```
London-R5#show crypto engine connection active
```

ID	Interface	IP-Address	State	Algorithm	Encrypt	Decrypt
1	<none>	<none>	set	HMAC_SHA+DES_56_CB	0	0
2	<none>	<none>	set	HMAC_SHA+DES_56_CB	0	0
2000	Serial0/1	200.1.1.5	set	HMAC_SHA+DES_56_CB	0	4
2001	Serial0/1	200.1.1.5	set	HMAC_SHA+DES_56_CB	9	0
2002	Serial0/1	200.1.1.5	set	HMAC_SHA+DES_56_CB	0	**4**
2003	Serial0/1	200.1.1.5	set	HMAC_SHA+DES_56_CB	**4**	0

Let's do the ping test again at London-R6. As illustrated in Code Listing 6.67, the ping test conducted from London-R6 (172.22.1.6) to Brussels-R3 (172.21.1.3) now displays a positive result (100 percent successful).

Code Listing 6.67:   Ping test from London-R6 (172.22.1.6) to Brussels-R3 (172.21.1.3)

```
London-R6#ping
Protocol [ip]:
Target IP address: 172.21.1.3
Repeat count [5]:
Datagram size [100]:
Timeout in seconds [2]:
Extended commands [n]: y
Source address or interface: 172.22.1.6
Type of service [0]:
Set DF bit in IP header? [no]:
Validate reply data? [no]:
Data pattern [0xABCD]:
Loose, Strict, Record, Timestamp, Verbose[none]:
Sweep range of sizes [n]:
Type escape sequence to abort.
Sending 5, 100-byte ICMP Echos to 172.21.1.3, timeout is 2 seconds:
!!!!!
Success rate is 100 percent (5/5), round-trip min/avg/max = 92/92/92 ms
```

The crypto engine connection table of London-R5 in Code Listing 6.68 shows a favorable outcome this time. The five outbound ICMP echo packets from London-R6 to Brussels-R3 are encrypted and sent out of the IPSec tunnel to Brussels-R2; the five returning inbound ICMP echo-reply packets from Brussels-R3, received via the IPSec tunnel of Brussels-R2, are decrypted accordingly. Note that the five encrypted outbound and five decrypted inbound packets are aggregated with the four listed in Code Listing 6.66 to reflect a cumulative total of nine in Code Listing 6.68.

Code Listing 6.68: London-R5 crypto engine connection table

```
London-R5#show crypto engine connection active
```

ID	Interface	IP-Address	State	Algorithm	Encrypt	Decrypt
1	<none>	<none>	set	HMAC_SHA+DES_56_CB	0	0
2	<none>	<none>	set	HMAC_SHA+DES_56_CB	0	0
2000	Serial0/1	200.1.1.5	set	HMAC_SHA+DES_56_CB	0	4
2001	Serial0/1	200.1.1.5	set	HMAC_SHA+DES_56_CB	9	0
2002	Serial0/1	200.1.1.5	set	HMAC_SHA+DES_56_CB	0	9
2003	Serial0/1	200.1.1.5	set	HMAC_SHA+DES_56_CB	9	0

The crypto engine connection table of Brussels-R2 in Code Listing 6.69 further consolidates the findings in Code Listing 6.68. The table shows that the five inbound ICMP echo packets from London-R6 to Brussels-R3 are decrypted, and the five outbound ICMP echo-reply packets from Brussels-R3 (responding to London-R6's ICMP echo packets) were encrypted. Note that the five encrypted outbound and five decrypted inbound packets are aggregated with the four listed in Code Listing 6.65 to reflect a cumulative total of nine in Code Listing 6.69.

Code Listing 6.69:   Brussels-R2 crypto engine connection table

---

```
Brussels-R2#show crypto engine connection active

  ID Interface   IP-Address   State  Algorithm           Encrypt   Decrypt
   9 no idb      no address   set    DES_56_CBC                0         0
  10 Serial0/0   199.1.1.9    set    HMAC_SHA+DES_56_CB        0         9
  11 Serial0/0   199.1.1.9    set    HMAC_SHA+DES_56_CB        9         0
```

Here are the observations after HSRP failover:

- Initial ping test from London-R6 to Brussels-R3 still fails because the IPSec process on Brussels-R2 has not been initiated yet and the IPSec tunnels to London-R5 have not been built. Therefore, London-R5 still sends encrypted packets to Brussels-R1 since this IPSec peer is still reachable to London-R5 after the failover.

- However, the IP connectivity is restored explicitly via the ICMP echo packets from Brussels-R3 as this helps to fire up the IPSec tunnels on Brussels-R2 to London-R5.

- In other words, connectivity is lost when packets are sent to the IPSec peer (Brussels-R1) that cannot complete the final delivery, and connectivity is restored only when a packet is received from the IPSec peer (Brussels-R2) that can complete the final delivery.

6.5.4   GRE tunnel overview

6.5.4.1   Tunneling preliminaries.   Tunneling provides an alternative to encapsulate packets within a transport protocol and is implemented as a point-to-point virtual interface. The tunnel interface is not tied to specific payload or delivery protocols but rather is a mechanism designed to provide the services necessary to implement any standard point-to-point encapsulation scheme. Because tunnels are point-to-point links, a separate tunnel (virtual interface) is required for each link.

In short, tunneling has the following three main components:

1. *Payload protocol*: The protocol to be encapsulated (for example, connection-less network service [CLNS], IP, or Internetwork packet exchange [IPX].

2. *Carrier protocol*: Tunneling protocols such as the generic routing encapsulation (GRE) or IPSec (tunnel mode).

3. *Delivery protocol*: Protocols such as IP, which is used to carry the encapsulated protocol.

In the following sections, we discuss how GRE can be used in conjunction with IPSec to provide VPN resiliency.

6.5.4.2 IPSec VPN resiliency with GRE tunnels.   GRE (RFC 1701) is a mechanism for encapsulating arbitrary packets within an arbitrary transport protocol. Note that when IP is used as the delivery protocol (RFC 1702), GRE packets that are encapsulated within IP will use IP protocol type 47. In addition, when IP is used as the payload protocol (RFC 1702), the IP packets will be encapsulated with a protocol type field of 0x800.

GRE is capable of handling the delivery of multiprotocol and IP multicast traffic between two sites, which only have IP unicast connectivity. GRE tunnels also enable the use of private network addressing (RFC 1918) across a service provider's backbone without the need for running the NAT feature.

The importance of using GRE tunnels in a VPN environment is based on the fact that IPSec encryption works only on IP unicast frames. GRE can be used in conjunction with IPSec to pass routing updates between sites on an IPSec VPN. GRE encapsulates the clear text packet, and IPSec (in transport or tunnel mode) encrypts the packet. This packet flow of IPSec over GRE enables routing updates, which are generally multicast, to be passed over an encrypted link. IPSec alone cannot achieve this, because it does not support multicast.

Network resiliency in a VPN environment is an important factor to consider in the decision to use GRE tunnels, IPSec tunnels, or tunnels that utilize IPSec over GRE. For VPN resilience, the remote site should be configured with two GRE tunnels, one to the primary router and the other to the backup router. If the GRE tunnels are secured with IPSec, each tunnel has its own IKE SA and a pair of IPSec SAs. Because GRE can carry multicast and broadcast traffic, it is now possible to configure a routing protocol for these tunnels. With the routing protocol in place, the failover mechanism works automatically.

On loss of connectivity to the primary router, routing protocol will discover the failure and route to the backup router, thereby providing network resiliency. Since the backup GRE tunnel is already up and secured, the failover time is determined by the convergence time of the routing protocol. Besides providing a failover mechanism, GRE tunnels provide the ability to encrypt multicast and broadcast packets and non-IP protocols with IPSec. They also provide enhanced performance and scalability for site-to-site VPN services. In the next section, we look at how GRE can be incorporated with IPSec to provide a reliable HSRP failover.

6.5.5   Scenario 3—IPSec and GRE tunnels with HSRP

In scenario 2, we managed to achieve full IP connectivity between Brussels-R3 and London-R5 during HSRP failover by configuring IPSec backup peers on London-R5. However, for this method to work effectively, a packet must first be received by London-R5 from the IPSec peer (Brussels-R2) that can complete the final delivery to fire up the appropriate IPSec tunnels. Therefore, the IPSec backup peer technique is not a foolproof solution and is not totally resilient.

In this section, we shall incorporate GRE tunnels discussed in section 6.5.4 into the IPSec implementation, making our network more robust and resilient to HSRP failover. The whole idea is to run a routing protocol (static routing or dynamic routing) over the GRE tunnels. The routing updates then control which GRE tunnels to use before and after HSRP failover. Note that in this setup, Site 3 will be utilizing the two paths to Site 1.

6.5.5.1   Configurations. Code Listings 6.70 to 6.72 illustrate the IPSec and GRE tunnels for Brussels-R1, Brussels-R2, and London-R5 depicted in Figure 6.6. In the configurations, comments (in *italics*) precede certain configuration lines to explain them.

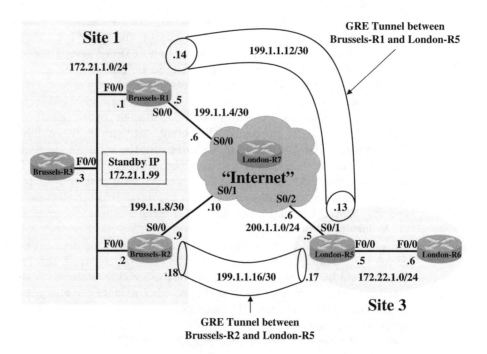

Figure 6.6:   GRE tunnel layout for scenario 3.

Code Listing 6.70: Brussels-R1 IPSec and GRE tunnel configuration

---

```
hostname Brussels-R1
!
<Output Omitted>
!
crypto ipsec transform-set chapter6 esp-des esp-sha-hmac
! --- Even though the default IPSec tunnel mode
! --- is implemented here, the reader can choose
! --- between tunnel or transport mode, taking into
! --- consideration that the IPSec transport mode reduces
! --- the overall packet size as compared with the IPSec
! --- tunnel mode, thus improving the effective throughput.
!
! --- Specify the crypto map chapter6 where we define
! --- our remote peer London-R5 (200.1.1.5),
! --- transform set chapter6, and our crypto access list 110.
 crypto map chapter6 10 ipsec-isakmp
 set peer 200.1.1.5
 set transform-set chapter6
 match address 110
!
! --- Create GRE tunnel Tunnel0 from Brussels-R1 to London-R5.
interface Tunnel0 ! --- Tunnel0 is a virtual interface.
 ip address 199.1.1.14 255.255.255.252
 tunnel source Serial0/0 ! --- Tunnel starts from local interface Serial0/0.
 tunnel destination 200.1.1.5 ! --- Tunnel ends at London-R5 (200.1.1.5).
! --- Since the GRE tunnel is a unique interface,
! --- it is assigned its own crypto map.
 crypto map chapter6 ! --- Apply crypto map chapter6 to Tunnel0.
!
<Output Omitted>
!
interface Serial0/0
 ip address 199.1.1.5 255.255.255.252
! --- Apply crypto map chapter6 to Serial0/0
! --- to activate crypto engine.
 crypto map chapter6
!
! --- Routing from Brussels-R1 to subnet 172.22.1.0/24
! --- is via a static route pointing out through Tunnel0.
ip route 172.22.1.0 255.255.255.0 Tunnel0

! --- Routing from Brussels-R1 to other Internet destinations
! --- is via a default route to Internet gateway London-R7
```

```
! --- with a next-hop address of 199.1.1.6.
ip route 0.0.0.0 0.0.0.0 199.1.1.6
!
! --- The crypto access list for crypto map chapter6
! --- is encrypting GRE traffic between
! --- the two tunnel endpoints 199.1.1.5 (Brussels-R1)
! --- and 200.1.1.5 (London-R5) as the IP traffic is now
! --- encapsulated within the GRE packets.
access-list 110 permit gre host 199.1.1.5 host 200.1.1.5
```

Code Listing 6.71:   Brussels-R2 IPSec and GRE tunnel configuration

---

```
hostname Brussels-R2
!
<Output Omitted>
!
crypto ipsec transform-set chapter6 esp-des esp-sha-hmac
! --- Even though the default IPSec tunnel mode
! --- is implemented here, the reader can choose
! --- between tunnel or transport mode, taking into
! --- consideration that the IPSec transport mode reduces
! --- the overall packet size as compared with the IPSec
! --- tunnel mode, thus improving the effective throughput.
!
! --- Specify the crypto map chapter6 where we define
! --- our remote peer London-R5 (200.1.1.5),
! --- transform set chapter6, and our crypto access list 110.
 crypto map chapter6 10 ipsec-isakmp
 set peer 200.1.1.5
 set transform-set chapter6
 match address 110
!
! --- Create GRE tunnel Tunnel0 from Brussels-R2 to London-R5.
interface Tunnel0 ! --- Tunnel0 is a virtual interface.
 ip address 199.1.1.18 255.255.255.252
 tunnel source Serial0/0 ! --- Tunnel starts from local interface Serial0/0.
 tunnel destination 200.1.1.5 ! --- Tunnel ends at London-R5 (200.1.1.5).
! --- Since the GRE tunnel is a unique interface,
! --- it is assigned its own crypto map.
 crypto map chapter6 ! --- Apply crypto map chapter6 to Tunnel0.
!
<Output Omitted>
!
interface Serial0/0
```

```
 ip address 199.1.1.9 255.255.255.252
! --- Apply crypto map chapter6 to Serial0/0
! --- to activate crypto engine.
 crypto map chapter6
!
! --- Routing from Brussels-R2 to subnet 172.22.1.0/24
! --- is via a static route pointing out through Tunnel0.
ip route 172.22.1.0 255.255.255.0 Tunnel0

! --- Routing from Brussels-R2 to other Internet destinations
! --- is via a default route to Internet gateway London-R7
! --- with a next-hop address of 199.1.1.10.
ip route 0.0.0.0 0.0.0.0 199.1.1.10
!
! --- The crypto access list for crypto map chapter6
! --- is encrypting GRE traffic between the two tunnel endpoints
! --- 199.1.1.9 (Brussels-R2) and 200.1.1.5 (London-R5) as the
! --- IP traffic is now encapsulated within the GRE packets.
access-list 110 permit gre host 199.1.1.9 host 200.1.1.5
```

Code Listing 6.72:   London-R5 IPSec and GRE tunnel configuration

---

```
hostname London-R5
!
<Output Omitted>
!
crypto ipsec transform-set chapter6 esp-des esp-sha-hmac
! --- Even though the default IPSec tunnel mode
! --- is implemented here, the reader can choose between
! --- tunnel or transport mode, taking into consideration that
! --- the IPSec transport mode reduces the overall packet size
! --- as compared with the IPSec tunnel mode,
! --- thus improving the effective throughput.
!
! --- Specify the crypto map chapter6 sequence number 10,
! --- where we define our first remote peer
! --- Brussels-R1 (199.1.1.5), transform set chapter6,
! --- and a different crypto access list 110.
crypto map chapter6 10 ipsec-isakmp
 set peer 199.1.1.5
 set transform-set chapter6
 match address 110

! --- Specify the crypto map chapter6 sequence number 20,
! --- where we define our second remote peer
```

```
! --- Brussels-R2 (199.1.1.9), transform set chapter6,
! --- and a different crypto access list 120.
crypto map chapter6 20 ipsec-isakmp
 set peer 199.1.1.9
 set transform-set chapter6
 match address 120
!
! --- Create GRE tunnel Tunnel0 from London-R5 to Brussels-R1.
interface Tunnel0 ! --- Tunnel0 is a virtual interface.
 ip address 199.1.1.13 255.255.255.252
 tunnel source Serial0/1 ! --- Tunnel starts from local interface Serial0/1.
 tunnel destination 199.1.1.5 ! --- Tunnel ends at Brussels-R1 (199.1.1.5).
! --- Since the GRE tunnel is a unique interface,
! --- it is assigned its own crypto map.
 crypto map chapter6 ! --- Apply crypto map chapter6 to Tunnel0.
!
! --- Create GRE tunnel Tunnel1 from London-R5 to Brussels-R2.
interface Tunnel1 ! --- Tunnel1 is a virtual interface.
 ip address 199.1.1.17 255.255.255.252
 tunnel source Serial0/1 ! --- Tunnel starts from local interface Serial0/1.
 tunnel destination 199.1.1.9 ! --- Tunnel ends at Brussels-R2 (199.1.1.9).
! --- Since the GRE tunnel is a unique interface,
! --- it is assigned its own crypto map.
 crypto map chapter6 ! --- Apply crypto map chapter6 to Tunnel1.
!
interface FastEthernet0/0
 ip address 172.22.1.5 255.255.255.0
!
interface Serial0/1
 bandwidth 64
 ip address 200.1.1.5 255.255.255.252
! --- Apply crypto map chapter6 to Serial0/1
! --- to activate crypto engine.
 crypto map chapter6
!
! --- Routing from London-R5 to subnet 172.21.1.0/24
! --- is via a static route pointing out through Tunnel0
! --- and a static route pointing out through Tunnel1.
! --- Thus, we can load-balance packets over both tunnels.
ip route 172.21.1.0 255.255.255.0 Tunnel0
ip route 172.21.1.0 255.255.255.0 Tunnel1

! --- Routing from London-R5 to other Internet destinations
! --- is via a default route to Internet gateway London-R7
! --- with a next-hop address of 200.1.1.6.
```

```
ip route 0.0.0.0 0.0.0.0 200.1.1.6
!
! --- Crypto access list 110 is encrypting GRE traffic
! --- between the two tunnel endpoints
! --- 200.1.1.5 (London-R5) and 199.1.1.5 (Brussels-R1) since
! --- the IP traffic is now encapsulated within the GRE packets.
access-list 110 permit gre host 200.1.1.5 host 199.1.1.5


! --- Crypto access list 120 is encrypting GRE traffic
! --- between the two tunnel endpoints 200.1.1.5 (London-R5)
! --- and 199.1.1.9 (Brussels-R2) since the IP traffic is now
! --- encapsulated within the GRE packets.
access-list 120 permit gre host 200.1.1.5 host 199.1.1.9
```

6.5.5.2 Verification before HSRP failover. Code Listings 6.73 to 6.83 illustrate some of the verification test results before the HSRP failover for the IPSec and GRE tunnel configurations implemented in the previous section.

Code Listing 6.73 displays the IP routing table of Brussels-R1. From the table, we can see that subnet 172.22.1.0/24 is reachable via a static route pointing out through Tunnel0 (GRE tunnel).

Code Listing 6.73: Brussels-R1 IP routing table

```
Brussels-R1#show ip route
<Output Omitted>

Gateway of last resort is 199.1.1.6 to network 0.0.0.0

     172.21.0.0/24 is subnetted, 1 subnets
C       172.21.1.0 is directly connected, FastEthernet0/0
     172.22.0.0/24 is subnetted, 1 subnets
S       172.22.1.0 is directly connected, Tunnel0
     199.1.1.0/30 is subnetted, 2 subnets
C       199.1.1.4 is directly connected, Serial0/0
C       199.1.1.12 is directly connected, Tunnel0
S*   0.0.0.0/0 [1/0] via 199.1.1.6
```

Similarly, Code Listing 6.74 displays the IP routing table of Brussels-R2. From the table, we can see that subnet 172.22.1.0/24 is reachable via a static route pointing out through Tunnel0 (GRE tunnel).

Code Listing 6.74: Brussels-R2 IP routing table

```
Brussels-R2#show ip route
<Output Omitted>


Gateway of last resort is 199.1.1.10 to network 0.0.0.0

     172.21.0.0/24 is subnetted, 1 subnets
C        172.21.1.0 is directly connected, FastEthernet0/0
     172.22.0.0/24 is subnetted, 1 subnets
S        172.22.1.0 is directly connected, Tunnel0
     199.1.1.0/30 is subnetted, 2 subnets
C        199.1.1.8 is directly connected, Serial0/0
C        199.1.1.16 is directly connected, Tunnel0
S*   0.0.0.0/0 [1/0] via 199.1.1.10
```

Code Listing 6.75 displays the IP routing table of London-R5. From the table, we can see that subnet 172.21.1.0/24 is reachable via a static route pointing out through Tunnel0 and a static route pointing out through Tunnel1, thus indicating that packets are load balanced over both GRE tunnels. Incorporating load balancing is worthwhile as it reduces the overall routing convergence time significantly. Note that the scenario uses static routing for simplicity reasons and optimum convergence time. The reader might want to use dynamic routing protocols (e.g., RIPv2 or OSPF) for better manageability and scalability. The reader should also keep in mind that path selection converges as fast as the chosen routing protocol.

Code Listing 6.75: London-R5 IP routing table

```
London-R5#show ip route
<Output Omitted>


Gateway of last resort is 200.1.1.6 to network 0.0.0.0

     200.1.1.0/30 is subnetted, 1 subnets
C        200.1.1.4 is directly connected, Serial0/1
     172.21.0.0/24 is subnetted, 1 subnets
S        172.21.1.0 is directly connected, Tunnel0
                    is directly connected, Tunnel1
     172.22.0.0/24 is subnetted, 1 subnets
C        172.22.1.0 is directly connected, FastEthernet0/0
     199.1.1.0/30 is subnetted, 2 subnets
C        199.1.1.12 is directly connected, Tunnel0
C        199.1.1.16 is directly connected, Tunnel1
S*   0.0.0.0/0 [1/0] via 200.1.1.6
```

As illustrated in Code Listing 6.76, the ping test is conducted successfully (60 percent) from Brussels-R3 (172.21.1.3) to London-R6 (172.22.1.6). Brussels-R3 sends out five ICMP echo packets and receives three ICMP echo-reply packets from London-R6.

Code Listing 6.76:   Ping test from Brussels-R3 (172.21.1.3) to London-R6 (172.22.1.6)

```
Brussels-R3#ping
Protocol [ip]:
Target IP address: 172.22.1.6
Repeat count [5]:
Datagram size [100]:
Timeout in seconds [2]:
Extended commands [n]: y
Source address or interface: 172.21.1.3
Type of service [0]:
Set DF bit in IP header? [no]:
Validate reply data? [no]:
Data pattern [0xABCD]:
Loose, Strict, Record, Timestamp, Verbose[none]:
Sweep range of sizes [n]:
Type escape sequence to abort.
Sending 5, 100-byte ICMP Echos to 172.22.1.6, timeout is 2 seconds:
..!!!
Success rate is 60 percent (3/5), round-trip min/avg/max = 104/105/108 ms
```

From the crypto engine connection table of Brussels-R1 in Code Listing 6.77, four outbound ICMP echo packets originating from Brussels-R3 are encrypted, and there are no returning inbound ICMP echo-reply packets from London-R6.

Code Listing 6.77:   Brussels-R1 crypto engine connection table

```
Brussels-R1#show crypto engine connection active

  ID Interface   IP-Address    State  Algorithm           Encrypt   Decrypt
  24 no idb      no address    set    DES_56_CBC               0         0
  25 Serial0/0   199.1.1.5     set    HMAC_SHA+DES_56_CB       0         0
  26 Serial0/0   199.1.1.5     set    HMAC_SHA+DES_56_CB       4         0
```

The "show crypto ipsec sa" command in Code Listing 6.78 reflects four encrypted packets, concurring with the findings in Code Listing 6.77. There is one send error, indicating that one of the original five ICMP echo packets is bad.

Code Listing 6.78:   Brussels-R1 IPSec security association

```
Brussels-R1#show crypto ipsec sa

<Output Omitted>

   local   ident (addr/mask/prot/port): (199.1.1.5/255.255.255.255/47/0)
   remote ident (addr/mask/prot/port): (200.1.1.5/255.255.255.255/47/0)
   current_peer: 200.1.1.5
     PERMIT, flags={origin_is_acl,parent_is_transport,}
    #pkts encaps: 4, #pkts encrypt: 4, #pkts digest 4
    #pkts decaps: 0, #pkts decrypt: 0, #pkts verify 0
    #send errors 1, #recv errors 0

<Output Omitted>
```

The crypto engine connection table of London-R5 in Code Listing 6.79 reflects another set of peculiar results: four inbound ICMP echo packets originating from Brussels-R3 and received from the IPSec/GRE tunnel between Brussels-R1 and London-R5 are decrypted; three returning outbound ICMP echo-reply packets originating from London-R6 are encrypted and sent out of the IPSec/GRE tunnel between Brussels-R2 and London-R5.

Code Listing 6.79:   London-R5 crypto engine connection table

---

```
London-R5#show crypto engine connection active
```

ID	Interface	IP-Address	State	Algorithm	Encrypt	Decrypt
1	\<none\>	\<none\>	set	HMAC_SHA+DES_56_CB	0	0
2	\<none\>	\<none\>	set	HMAC_SHA+DES_56_CB	0	0
2000	Serial0/1	200.1.1.5	set	HMAC_SHA+DES_56_CB	0	4
2001	Serial0/1	200.1.1.5	set	HMAC_SHA+DES_56_CB	0	0
2002	Serial0/1	200.1.1.5	set	HMAC_SHA+DES_56_CB	0	0
2003	Serial0/1	200.1.1.5	set	HMAC_SHA+DES_56_CB	3	0

The "show crypto ipsec sa" command in Code Listing 6.80 helps shed more light on the findings in Code Listing 6.79. Four encrypted inbound packets received from IPSec peer Brussels-R1 (199.1.1.5) are decrypted; three outbound packets are encrypted and sent to IPSec peer Brussels-R2 (199.1.19) with one send error.

The four decrypted packets concur with the four encrypted ICMP echo packet that originated from Brussels-R3 and were sent out of the IPSec/GRE tunnel between Brussels-R1 and London-R5 (see Code Listing 6.77). Once the four decrypted ICMP echo packets reach London-R6, four returning ICMP echo-reply packets are generated. Of the four return packets, three are encrypted and sent out of the IPSec/GRE tunnel between London-R5 and Brussels-R2 (see Code Listing 6.79), and one bad packet is dropped.

Code Listing 6.80:   London-R5 IPSec security association

```
London-R5#show crypto ipsec sa

<Output Omitted>

   local   ident (addr/mask/prot/port): (200.1.1.5/255.255.255.255/47/0)
   remote ident (addr/mask/prot/port): (199.1.1.5/255.255.255.255/47/0)
   current_peer: 199.1.1.5
     PERMIT, flags={origin_is_acl,}
   #pkts encaps: 0, #pkts encrypt: 0, #pkts digest 0
   #pkts decaps: 4, #pkts decrypt: 4, #pkts verify 4
   #pkts compressed: 0, #pkts decompressed: 0
   #pkts not compressed: 0, #pkts compr. failed: 0, #pkts decompress
    failed: 0
   #send errors 0, #recv errors 0

<Output Omitted>

   local   ident (addr/mask/prot/port): (200.1.1.5/255.255.255.255/47/0)
   remote ident (addr/mask/prot/port): (199.1.1.9/255.255.255.255/47/0)
   current_peer: 199.1.1.9
     PERMIT, flags={origin_is_acl,}
   #pkts encaps: 3, #pkts encrypt: 3, #pkts digest 3
   #pkts decaps: 0, #pkts decrypt: 0, #pkts verify 0
   #pkts compressed: 0, #pkts decompressed: 0
   #pkts not compressed: 0, #pkts compr. failed: 0, #pkts decompress
    failed: 0
   #send errors 1, #recv errors 0

<Output Omitted>
```

From the crypto engine connection table of Brussels-R2 in Code Listing 6.81, three inbound packets are decrypted. These three decrypted packets correspond with the three encrypted ICMP echo-reply packets that originated from London-R6 and were sent out of the IPSec/GRE tunnel between London-R5 and Brussels-R2 (see Code Listing 6.79).

Code Listing 6.81:   Brussels-R2 crypto engine connection table

---

```
Brussels-R2#show crypto engine connection active

ID Interface   IP-Address   State  Algorithm          Encrypt   Decrypt
18 no idb       no address   set    DES_56_CBC              0         0
19 Serial0/0    199.1.1.9    set    HMAC_SHA+DES_56_CB      0         3
20 Serial0/0    199.1.1.9    set    HMAC_SHA+DES_56_CB      0         0
```

To further substantiate and consolidate all the above findings, let's perform a traceroute test at both Brussels-R3 and London-R6.

In Code Listing 6.82, a traceroute is performed from Brussels-R3 (172.21.1.3) to London-R6 (172.22.1.6). From the command output, we can see that path taken to reach London-R6 (172.22.1.6) is via Brussels-R1 (172.21.1.1) and the GRE tunnel between Brussels-R1 and London-R5 (199.1.1.13).

Code Listing 6.82:   Traceroute test from Brussels-R3 (172.21.1.3) to London-R6 (172.22.1.6)

```
Brussels-R3#traceroute
Protocol [ip]:
Target IP address: 172.22.1.6
Source address: 172.21.1.3
Numeric display [n]:
Timeout in seconds [3]:
Probe count [3]:
Minimum Time to Live [1]:
Maximum Time to Live [30]:
Port Number [33434]:
Loose, Strict, Record, Timestamp, Verbose[none]:
Type escape sequence to abort.
Tracing the route to 172.22.1.6

  1 172.21.1.1 0 msec 4 msec 0 msec
  2 199.1.1.13 80 msec 80 msec 80 msec
  3 172.22.1.6 84 msec *   80 msec
```

In Code Listing 6.83, a traceroute is performed from London-R6 (172.22.1.6) to Brussels-R3 (172.21.1.3). From the command output, we can see that the path taken to reach Brussels-R3 (172.21.1.3) is via London-R5 (172.22.1.5) and the GRE tunnel between London-R5 and Brussels-R2 (199.1.1.18).

Code Listing 6.83:   Traceroute test from London-R6 (172.22.1.6) to Brussels-R3 (172.21.1.3)

```
London-R6#traceroute
Protocol [ip]:
Target IP address: 172.21.1.3
Source address: 172.22.1.6
Numeric display [n]:
Timeout in seconds [3]:
Probe count [3]:
Minimum Time to Live [1]:
Maximum Time to Live [30]:
Port Number [33434]:
Loose, Strict, Record, Timestamp, Verbose[none]:
Type escape sequence to abort.
Tracing the route to 172.21.1.3


 1 172.22.1.5 0 msec 4 msec 0 msec
 2 199.1.1.18 72 msec 76 msec 76 msec
 3 172.21.1.3 76 msec *  72 msec
```

Code Listings 6.82 and 6.83 together demonstrate that per-destination load balancing (default for Cisco routers) is in place. In fact, this is an example of asymmetric routing. Therefore, traffic from source subnet 172.21.1.0/24 to destination subnet 172.22.1.0/24 will traverse the GRE tunnel between Brussels-R1 and London-R5; traffic from source subnet 172.22.1.0/24 to destination subnet 172.21.1.0/24 will traverse the GRE tunnel between London-R5 and Brussels-R2. This also explains why the ICMP echo packets from Brussels-R3 to London-R6 traverse the GRE tunnel between Brussels-R1 and London-R5 and the ICMP echo-reply packets from London-R6 to Brussels-R3 traverse the GRE tunnel between London-R5 and Brussels-R2.

6.5.5.3 Verification after HSRP failover.   As illustrated in Figure 6.5, the HSRP failover is achieved by shutting down interface FastEthernet0/0 on Brussels-R1. Code Listing 6.84 to 6.88 exhibits some of the verification test results after the HSRP failover for the IPSec and GRE tunnel configurations listed in section 6.5.5.1.

As illustrated in Code Listing 6.84, the ping test conducted from London-R6 (172.22.1.6) to Brussels-R3 (172.21.1.3) registers a positive result (100 percent successful). London-R6 sends out five ICMP echo packets and receives five ICMP echo-reply packets from Brussels-R3.

Code Listing 6.84:   Ping test from London-R6 (172.22.1.6) to Brussels-R3 (172.21.1.3)

```
London-R6#
London-R6#ping
Protocol [ip]:
Target IP address: 172.21.1.3
Repeat count [5]:
Datagram size [100]:
Timeout in seconds [2]:
Extended commands [n]: y
Source address or interface: 172.22.1.6
Type of service [0]:
Set DF bit in IP header? [no]:
Validate reply data? [no]:
Data pattern [0xABCD]:
Loose, Strict, Record, Timestamp, Verbose[none]:
Sweep range of sizes [n]:
Type escape sequence to abort.
Sending 5, 100-byte ICMP Echos to 172.21.1.3, timeout is 2 seconds:
!!!!!
Success rate is 100 percent (5/5), round-trip min/avg/max = 104/104/108 ms
```

The crypto engine connection table of London-R5 in Code Listing 6.85 depicts the desirable outcome. The five outbound ICMP echo packets from London-R6 to Brussels-R3 are encrypted and sent out of the IPSec/GRE tunnel between London-R5 and Brussels-R2. Likewise, the five encrypted returning inbound ICMP echo-reply packets from Brussels-R3 to London-R6 received through the IPSec/GRE tunnel between Brussels-R2 and London-R5 are decrypted. Note that the five encrypted outbound packets are aggregated with the three listed in Code Listing 6.79 to reflect a cumulative total of eight in Code Listing 6.85.

Code Listing 6.85:   London-R5 crypto engine connection table

---

```
London-R5#sh crypto engine connection active
```

ID	Interface	IP-Address	State	Algorithm	Encrypt	Decrypt
1	<none>	<none>	set	HMAC_SHA+DES_56_CB	0	0
2	<none>	<none>	set	HMAC_SHA+DES_56_CB	0	0
2000	Serial0/1	200.1.1.5	set	HMAC_SHA+DES_56_CB	0	4
2001	Serial0/1	200.1.1.5	set	HMAC_SHA+DES_56_CB	0	0
2002	Serial0/1	200.1.1.5	set	HMAC_SHA+DES_56_CB	0	**5**
2003	Serial0/1	200.1.1.5	set	HMAC_SHA+DES_56_CB	**8**	0

The crypto engine connection table of Brussels-R2 in Code Listing 6.86 further consolidates the findings in Code Listing 6.85. The table shows that the five encrypted inbound ICMP echo packets from London-R6 to Brussels-R3 are decrypted, and the five returning outbound ICMP echo-reply packets from Brussels-R3 to London-R6 are encrypted. Note that the five decrypted inbound packets are aggregated with the three listed in Code Listing 6.81 to reflect a cumulative total of eight in Code Listing 6.86.

Code Listing 6.86:   Brussels-R2 crypto engine connection table

---

```
Brussels-R2#show crypto engine connection active

ID Interface    IP-Address    State  Algorithm              Encrypt   Decrypt
18 no idb       no address    set    DES_56_CBC                0         0
19 Serial0/0    199.1.1.9     set    HMAC_SHA+DES_56_CB        0         8
20 Serial0/0    199.1.1.9     set    HMAC_SHA+DES_56_CB        5         0
```

In Code Listing 6.87, a traceroute is performed from London-R6 (172.22.1.6) to Brussels-R3 (172.21.1.3). From the command output, we can see that path taken to reach Brussels-R3 (172.21.1.3) is via London-R5 (172.22.1.5) and the GRE tunnel between London-R5 and Brussels-R2 (199.1.1.18).

Code Listing 6.87:   Traceroute test from London-R6 (172.22.1.6) to Brussels-R3 (172.21.1.3)

```
London-R6#trace
Protocol [ip]:
Target IP address: 172.21.1.3
Source address: 172.22.1.6
Numeric display [n]:
Timeout in seconds [3]:
Probe count [3]:
Minimum Time to Live [1]:
Maximum Time to Live [30]:
Port Number [33434]:
Loose, Strict, Record, Timestamp, Verbose[none]:
Type escape sequence to abort.
Tracing the route to 172.21.1.3

  1 172.22.1.5 4 msec 0 msec 0 msec
  2 199.1.1.18 76 msec 76 msec 72 msec
  3 172.21.1.3 76 msec *   72 msec
```

In Code Listing 6.88, a traceroute is performed from Brussels-R3 (172.21.1.3) to London-R6 (172.22.1.6). From the command output, we can see that path taken to reach London-R6 (172.22.1.6) is via Brussels-R2 (172.21.1.2) and the GRE tunnel between Brussels-R2 and London-R5 (199.1.1.17).

Code Listing 6.88:   Traceroute test from Brussels-R3 (172.21.1.3) to London-R6 (172.22.1.6)

```
Brussels-R3#traceroute
Protocol [ip]:
Target IP address: 172.22.1.6
Source address: 172.21.1.3
Numeric display [n]:
Timeout in seconds [3]:
Probe count [3]:
Minimum Time to Live [1]:
Maximum Time to Live [30]:
Port Number [33434]:
Loose, Strict, Record, Timestamp, Verbose[none]:
Type escape sequence to abort.
Tracing the route to 172.22.1.6

  1 172.21.1.2 0 msec 0 msec 4 msec
  2 199.1.1.17 80 msec *   80 msec
  3 172.22.1.6 80 msec *   80 msec
```

The traceroute results in Code Listings 6.87 and 6.88 indicate that the GRE tunnel between London-R5 and Brussels-R1 is no longer a valid path to reach subnets 172.21.1.0/24 from London-R6, and neither is it a valid path for Brussels-R3 to reach subnet 172.22.1.0/24. This is because Brussels-R1 is now disconnected from subnet 172.21.1.0/24 in the HSRP failover.

To conclude Case Study 6.2, incorporating GRE tunnels with IPSec provides the best solution to support a fail-safe HSRP switchover, since it gives us full control of the usage of IPSec/GRE tunnels and HSRP routers.

6.6. IPSec VPN Performance Optimization

In this section, we discuss some of the key factors that can affect the performance of an IPSec VPN:

- Fragmentation
- IKE SA lifetimes
- IKE keepalives

We will touch on these factors in greater detail in the subsequent sections.

6.6.1   Fragmentation

Packet fragmentation will eventually lead to packet reassembly, which is a resource (CPU and memory allocation) intensive process. As a result, the overall network performance decreases. Typically, fragmentation occurs when a packet is sent over a tunnel and the encapsulated packet becomes too large to fit on the smallest link on the tunnel path. One way to overcome this problem is to use the path maximum transmission unit discovery (PMTUD). PMTUD will determine the maximum transfer unit or MTU that a host can use to send a packet through the tunnel without triggering fragmentation. To allow PMTUD in the network, ICMP message Type 3 (Destination Unreachable), Code 4 (Fragment Required but DF [Don't Fragment] bit is set) should not be filtered or blocked. If this is not possible, then we will have to resort either to manually setting the MTU to a lower value on the VPN termination device and allow PMTUD locally or to clear the DF bit explicitly to force fragmentation.

6.6.2   IKE SA lifetimes

The IKE SA lifetime (see Chapter 2) determines how long the IKE SA will be retained by each peer until it expires. Before an SA expires, it can be reused by subsequent IKE negotiations, thereby reducing the time required to setup new IPSec SAs. Therefore, to optimize the IPSec VPN performance, we can configure a longer IKE SA lifetime. However, the reader should also take into consideration that the longer the lifetime, the less secure the IKE negotiation is likely to be.

6.6.3   IKE keepalives

IKE keepalives are used for detecting loss of connectivity between two IPSec peers and can be incorporated into an IPSec VPN to provide network resiliency. The keepalive packets are sent every 10 seconds by default. An IPSec termination point will conclude that it has lost connectivity with its peers once three keepalive packets are missed consecutively. To restore connectivity, a new tunnel is established with a predefined backup or secondary peer. From the configuration perspective, the IPSec termination point must have at least two IPSec peer addresses in its crypto map statement (see Code Listing 6.59). During the connectivity reestablishment, the IPSec termination point will go into main mode (MM) and quick mode (QM) negotiations with the second peer in its list. IKE keepalives should be used with devices that do not support GRE.

Recall that in the scenario depicted in section 6.5.3, we did not really use IKE keepalives to establish a new tunnel with the IPSec backup peer. Instead we explicitly bring up the IPSec backup peer with an ICMP echo packet. To establish the new tunnel with the IPSec backup peer, the IPSec primary peer must no longer be reachable; only then can we detect the lost of connectivity through the IKE keepalives. However, in the scenario the IPSec primary peer is still reachable after the failover and therefore using IKE keepalives does not help in this case.

In terms of VPN performance, a longer interval between keepalives will help to reduce excessive CPU usage, thus increasing overall network performance. However, similar to IKE SA lifetimes, the longer the interval, the longer it will take to detect a loss of connectivity.

6.7 Summary

Chapter 6 embarks upon the more advanced topics that affect interoperability, reliability, and performance of the network when deploying IPSec VPN in large enterprises. Specifically, the chapter focuses on incorporating IPSec with NAT, HSRP, GRE, and performance optimization parameters such as fragmentation, IKE SA lifetimes, and IKE keepalives.

IPSec with NAT is presented with two scenarios in Case Study 6.1, and IPSec with HSRP is portrayed with three scenarios in Case Study 6.2. The chapter concludes by discussing some of the performance parameters that can be used to fine-tune and optimize the performance of an IPSec VPN.

Deployment of MPLS VPN for Service Providers

Chapter

7

MPLS Concepts

7.1 Introduction

Multiprotocol label switching (MPLS) is the most elementary building block that is mandatory when implementing MPLS VPN. It is therefore crucial to understand the MPLS inner workings before covering MPLS VPN itself. This chapter gives an overview of all the MPLS main concepts and core technologies, ensuring that the reader has the necessary MPLS knowledge when deploying MPLS VPN. The MPLS unicast IP routing model is used throughout the chapter to illustrate these concepts and technologies.

MPLS has two modes of operation: frame mode and cell mode. The focus of this chapter is on the MPLS operation that applies to routers using frame-mode MPLS. Cell-mode (ATM-based) MPLS is beyond the scope of this book. Nevertheless, concepts that are related to cell mode will be mentioned briefly.

7.2 Terminologies

This section gives a general conceptual overview of the terms that appear in this chapter. Some of them are more specifically defined in later sections of the chapter.

- *Forwarding equivalence class (FEC)*: a group of IP packets that are forwarded in the same manner over the same path with the same forwarding treatment.

- *Label*: a short fixed-length physically contiguous identifier that is used to identify an FEC, usually of local significance.

- *Label swap*: the basic forwarding operation, consisting of looking up an incoming label to determine the outgoing label, encapsulation, port, and other data-handling information.

- *Label switched path (LSP)*: the path through one or more label switch routers (LSRs) at one level of the hierarchy, which is followed by a packet in a particular FEC.

- *Label switch router (LSR)*: an MPLS node that is capable of forwarding native layer 3 packets.

- *Label stack*: an ordered set of labels.

- *MPLS domain*: a contiguous set of nodes in a single routing or administrative domain that operate MPLS routing as well as forwarding.

- *MPLS label*: a label that is carried in a packet header and that represents the packet's FEC.

- *MPLS node*: a node that is running MPLS. An MPLS node will be aware of MPLS control protocols, will operate one or more layer 3 routing protocols, and will be capable of forwarding packets based on labels. An MPLS node may optionally be also capable of forwarding native layer 3 packets.

- *Virtual circuit*: a circuit used by a connection-oriented layer 2 technology, such as ATM or frame relay, requiring the maintenance of state information in layer 2 switches.

7.3   Conventional IP Forwarding Versus MPLS Forwarding

As an IP packet traverses from one router to the next, each router, as illustrated in Figure 7.1, makes an independent forwarding decision for that packet. Conventionally, each router analyzes the packet's header and runs a network layer routing algorithm. Each router separately selects a next-hop for the packet, based on its analysis of the packet's header and the outcome of running its routing algorithm. In other words, destination-based routing lookup is required on each hop. In some cases, the routers may require full routing information and therefore increases the size of the forwarding table as well as the

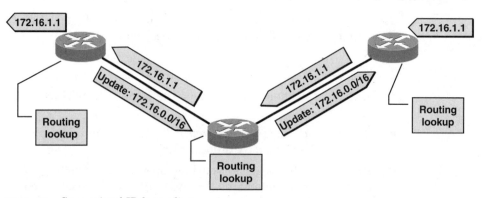

Figure 7.1: Conventional IP forwarding.

forwarding complexity. This can become a significant scalability issue when the number of routes grows.

Packet headers contain considerably more information than is needed to choose the next-hop. Basically, selecting the next-hop involves two functions. The first function partitions the entire set of possible packets into a set of forwarding equivalence classes (FECs). The second maps each FEC to a next-hop. An FEC is a set of packets that are forwarded over the same path through a network even if their final destinations are different.

In conventional IP forwarding, a particular router will typically consider two packets to be in the same FEC if there is some address prefix PREFIX_A (for example, 172.16.0.0/16) in that router's routing tables such that PREFIX_A is the longest match for each packet's destination address. As the packet traverses the network, each hop in turn reexamines the packet and assigns it to an FEC.

Multiprotocol label switching (MPLS) emerged from the IETF's effort to standardize a number of proprietary multilayer switching solutions that were initially proposed in the mid-1990s. MPLS technologies are applicable to any network layer protocol. However, in this chapter, the focus is on the use of IP as the network layer protocol.

In MPLS, the assignment of a particular packet to a particular FEC is done just once (by the ingress router) when the packet enters the network (MPLS domain). The FEC to which the packet is assigned is encoded as a short fixed-length value known as a label. The packets are labeled before they are forwarded. When a packet is forwarded to its next-hop, the label is propagated along with it.

At subsequent hops (core routers), there is no further analysis of the packet's network layer header. Instead the label is used as an index into a label-forwarding table that specifies the next-hop and a corresponding new label. The old label is swapped (or replaced) with the new label, and the packet is forwarded to its next-hop. When the labeled packet arrives at the egress router, the egress router discards the label and forwards the packet using the conventional longest-match IP forwarding.

In the MPLS forwarding paradigm, once a packet is assigned to an FEC, subsequent packet forwarding is driven by label inspection, and the subsequent routers (core routers) do not need to perform additional header analysis. In other words, only the ingress or egress router performs a routing lookup and label assignment/removal, while the core routers swap labels (label swapping) and forward packets based on simple label lookups. This simplifies the forwarding process and reduces the forwarding overhead on the core routers. Figure 7.2 shows an example of the MPLS forwarding operation.

MPLS forwarding has several advantages over conventional IP forwarding:

- MPLS forwarding can be done by switches that are capable of doing label lookup and replacement but are not capable of analyzing the network layer headers.

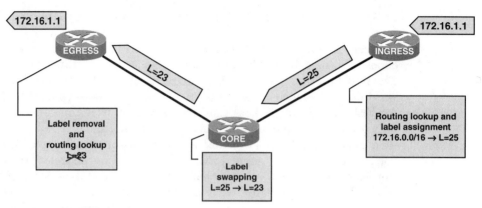

Figure 7.2:   MPLS forwarding example.

- Since a packet is assigned to an FEC when it enters the network, in determining the packet classification assignment the ingress router may use any information it has about the packet. For example, packets arriving on different ports may be assigned to different FECs. Conventional IP forwarding, on the other hand, can consider only information that travels with the packet in the packet header.

- The same packet entering the network at different ingress routers can be labeled differently; consequently, forwarding decisions that rely on the ingress router can be made easily. This cannot be achieved with conventional IP forwarding because the identity of a packet's ingress router does not travel with the packet.

- The complexity of assigning a packet to an FEC is confined to the ingress/egress routers only. This has no impact on the core routers that merely forward labeled packets.

- Sometimes it might be desirable to force a packet to choose an explicit route (possibly due to link underutilization) before entering the network. In conventional forwarding, this requires that the packet carry an encoding of the explicit route (source routing). In MPLS, a label can be used to represent the route, so that the identity of the explicit route need not be carried with the packet.

In the next section, we examine MPLS in two different dimensions or planes to ensure a better understanding of the MPLS inner workings or architecture.

7.4   MPLS Architecture

MPLS has two main building blocks:

1. The control plane is in charge of the routing information exchange and label exchange between adjacent devices. The control plane uses standard routing

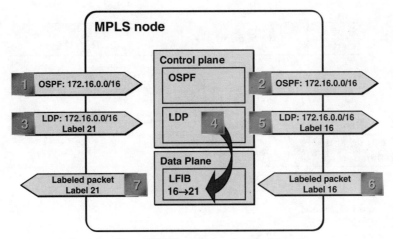

Figure 7.3: Functionality of MPLS control and data planes.

protocols, such as open shortest path first (OSPF), intermediate system-to-intermediate system (IS-IS), and border gateway protocol (BGP) to exchange information with other routers to build an IP forwarding table or forwarding information base (FIB; see section 7.6.2). The control plane also requires label distribution (label swapping) protocols such as label distribution protocol (LDP), BGP, and resource reservation protocol (RSVP) to exchange labels and populate the label-forwarding information base (LFIB; see section 7.6.2).

2. The data plane is in charge of forwarding of packets according to the destination address or label. The data plane is a simple label-based forwarding engine that is independent of the type of routing protocol or label distribution protocol. A label-forwarding information base (LFIB) is used to forward packets based on labels. The control plane populates and maintains the contents of the LFIB table.

Figure 7.3 illustrates the functionality of the MPLS control and data planes in which OSPF is the routing protocol and LDP (RFC 3036) is the label exchange protocol. Both protocols reside in the control plane. In the example, the MPLS node running OSPF receives the routing update about network 172.16.0.0/16, enters this prefix into the routing table, and forwards the routing information to its upstream neighbors. At the same time, LDP receives label 21 to be used for packets associated with a destination address 172.16.X.X. A local label of value 16 is generated and sent to upstream neighbors so that these neighbors can label packets with the appropriate next-hop label. LDP also inserts an entry into the LFIB table located in the data plane where local label 16 is mapped to label 21. The data plane then forwards all received packets with label 16 through the appropriate interfaces and replaces the label with label 21.

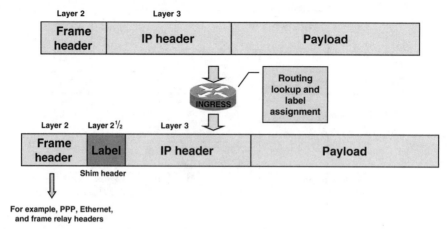

For example, PPP, Ethernet,
and frame relay headers

Figure 7.4: Frame-mode MPLS.

7.5 Frame-Mode MPLS

MPLS is designed for use on most layer 1 media and layer 2 encapsulation. For frame-based encapsulations such as PPP (typically implemented by Packet over Sonet/SDH [PoS]), frame relay, and Ethernet, MPLS inserts a 4-byte label (shim header) between the layer 2 and layer 3 headers as illustrated in Figure 7.4. This is known as frame-mode (or frame-based) MPLS. In frame-mode MPLS, when the ingress router receives a regular IP packet, it performs the following tasks:

- Do a routing table lookup to determine the outgoing interface.
- Assign and impose a label (shim header) between the layer 2 frame header and the layer 3 packet header if the outgoing interface is enabled for MPLS and a next-hop label for the destination exists.
- Forward out the labeled packet.

Note that the rest of the MPLS-enabled (core) routers in the network core simply forward the packet based on the label. In the following section, we discuss the physical components, or the set of nodes commonly referred as label switch routers (LSRs), which form an MPLS domain.

7.6 Label Switch Routers

A router that supports MPLS is known as a label switch router (LSR). There are two types of LSRs: core and edge. Core LSRs and edge LSRs are typically devices (routers) that are capable of doing both label switching and IP routing. Devices that have all their interfaces enabled for MPLS are called core LSRs because they are located at the core of an MPLS domain. Devices that have only

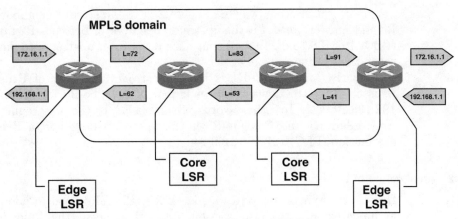

Figure 7.5: Placement of core LSR and edge LSR in an MPLS domain.

Figure 7.6: Upstream and downstream LSRs.

some of their interfaces enabled for MPLS are called edge LSRs because they are typically located at the edge of an MPLS domain.

Core LSRs primarily forward packets based on labels (label swapping). They swap a label on the top of a label stack with a next-hop label or a stack of labels in the core. Edge LSRs not only do label imposition (push operation) and disposition (pop operation), they also forward packets based on destination IP addresses and label them if the outgoing interface is enabled for MPLS. Edge LSRs are further categorized into ingress and egress LSRs. The ingress LSR receives IP packets, performs packet classifications (grouping packets into FECs), assigns labels, and forwards the labeled IP packets into the MPLS domain. The egress LSR removes the labels from the IP packets before forwarding them out of the MPLS domain. Figure 7.5 illustrates the placement of the core LSR and the edge LSR in an MPLS domain.

7.6.1 Upstream and downstream LSRs

In the MPLS world, upstream and downstream are relative. They always refer to a prefix, or more appropriately an FEC. Figure 7.6 further illustrates this.

In Figure 7.6, Cologne-R3 is the downstream neighbor of Brussels-R2 and Brussels-R2 is the downstream neighbor of Amsterdam-R1 for destination (FEC) 192.168.23.0/24. The LSRs know their downstream neighbors through the IP routing protocol, and the next-hop (NH) address is the downstream neighbor. Put another way, Brussels-R2 is the upstream neighbor of Cologne-R3 and Amsterdam-R1 is the upstream neighbor of Brussels-R2 for destination (FEC) 192.168.23.0/24. Note that data flows from upstream to downstream to reach the destination network.

Similarly, Amsterdam-R1 is the downstream neighbor of Brussels-R2 and Brussels-R2 is the downstream neighbor of Cologne-R3 for destination (FEC) 192.168.12.0/24. In other words, Brussels-R2 is the upstream neighbor of Amsterdam-R1 and Cologne-R3 is the upstream neighbor of Brussels-R2 for destination (FEC) 192.168.12.0/24.

7.6.2 Architecture of LSRs

Despite the type (core LSR or edge LSR), all LSRs need to adopt the MPLS architecture discussed in section 7.4 and perform the following tasks as illustrated in Figure 7.7:

- Exchange routing information
- Exchange labels
- Forward packets

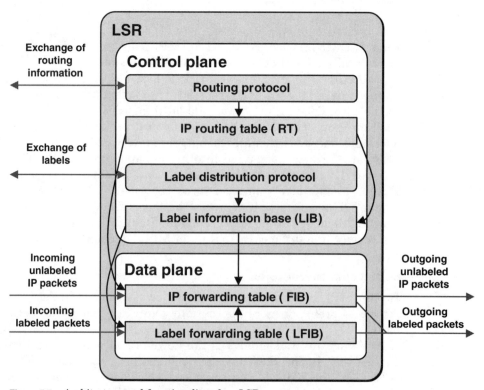

Figure 7.7: Architecture and functionality of an LSR.

The first two tasks belong to the control plane, and the last task is part of the data plane. The control plane consists of a routing protocol that exchanges routing information and maintains the contents of the main IP routing table (RT). The IP forwarding table (FIB) in the data plane is based on the IP routing table and is used to forward IP packets with no labels. A packet is labeled if a next-hop label is available for a particular destination IP network; otherwise, it is not labeled.

The control plane also includes a label distribution protocol that exchanges labels and stores them in a label information base (LIB). The LIB in the control plane is used by the label distribution protocol, where an IP prefix or network is allocated a locally generated (also locally significant) label that is mapped to a next-hop label learned (via the label distribution protocol) from the down-stream neighbor. The locally generated label is then propagated to the upstream neighbors, where it might be used as the next-hop label by these neighbors. The LIB is also used in the data plane to provide the following functionalities:

- A label is added to the FIB to map an IP prefix to a next-hop label.

- The previous locally generated label is added to the label-forwarding infor-mation base (LFIB) and also mapped to a next-hop label. The LFIB in the data plane is used to forward IP packets with labels. Local labels that were previously announced to upstream neighbors are mapped to next-hop labels that were previously received from the downstream neighbors.

Note that every LSR builds its LIB, LFIB, and FIB based on received labels. Figure 7.8 illustrates an example of MPLS Unicast IP forwarding. The figure shows an LSR where OSPF is used to exchange IP routing information and LDP (RFC 3036) is used to exchange labels.

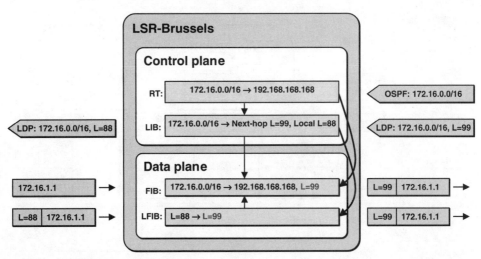

Figure 7.8: MPLS unicast IP forwarding example.

In this example, LSR-Brussels learns about the destination IP network 172.16.0.0/16 as well as the next-hop address (192.168.168.168) to this network from its downstream neighbor via OSPF. This routing information is stored in the IP routing table (RT).

Meanwhile, LSR-Brussels also receives the next-hop label 99 from its downstream neighbor via LDP. Besides generating a local label 88 for the network learned via OSPF, the LSR also propagates this local label to the upstream neighbors. Together with the information gathered in the RT, the local label 88 is mapped to the next-hop label 99 received from the downstream neighbor, and this label binding information is entered into the LIB. From the label-binding information found in the LIB, the next-hop label 99 is added to the FIB (derived from the RT) as an IP network to a next-hop label map entry. The local label 88 is also bound to the next-hop label 99 in the LFIB.

Now that the FIB and LFIB are in place, an incoming unlabeled IP packet with a destination IP address of 172.16.1.1 is forwarded using the FIB table, where a next-hop label indicates that the outgoing packet should be labeled with label 99. Likewise, an incoming labeled (label 88) packet with the same destination IP address is forwarded using the LFIB table, where the incoming label 88 is swapped with the next-hop label 99.

7.6.3 MPLS Frame-mode forwarding operation

In the previous section, we examined the MPLS frame-based forwarding only from the perspective of a single LSR. In this section, we look at the MPLS frame-mode forwarding operation again, but this time from a domain-wide perspective.

The example in Figure 7.9 shows a domain-wide MPLS frame-mode forwarding. An IP routing protocol, for example, OSPF or IS-IS, is used within

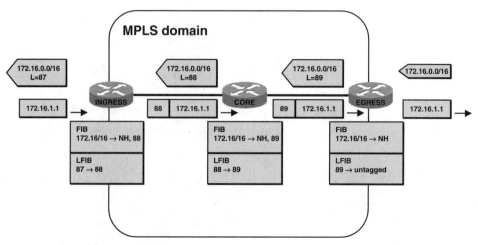

Figure 7.9: MPLS frame-mode forwarding example.

the MPLS domain to exchange routing information, such as determining the best path to network 172.16.0.0/16 (the FEC in this example) between the LSRs and a label distribution protocol, for instance, LDP is used to exchange label/FEC binding-information between adjacent neighbors.

In this case, the egress LSR generates local label 89 for FEC 172.16.0.0/16 and propagates this label/FEC binding information to its upstream neighbor. The core LSR (egress LSR's upstream neighbor) then uses label 89 as the next-hop label for FEC 172.16.0.0/16. The core LSR in turn generates local label 88 for the FEC and propagates this label/FEC binding information to its upstream neighbor. The next-hop label 89 is added to core LSR's FIB as an IP network to a next-hop label map entry, and the local label 88 is bound to the next-hop label 89 in its LFIB.

The ingress LSR (core LSR's upstream neighbor) then uses label 88 as the next-hop label for the FEC. Likewise, the ingress LSR generates local label 87 for the FEC, and propagates this label/FEC binding information to its upstream neighbor. The next-hop label 88 is added to ingress LSR's FIB as an IP network to a next-hop label map entry, and the local label 87 is bound to the next-hop label 88 in its LFIB.

On the network ingress, with reference to the FIB, the ingress LSR assigns label 88 to the incoming unlabeled IP packet with a destination address of 172.16.1.1 and forwards it to its downstream neighbor. When this labeled IP packet arrives at the core LSR (ingress LSR's downstream neighbor), based on the contents of its LFIB the core LSR swaps label 88 with label 89 and forwards it to its downstream neighbor. When this label-swapped IP packet arrives at the egress LSR (core LSR's downstream neighbor), based on the contents of its LFIB the egress LSR removes (pops) label 89 (penultimate hop popping is discussed in section 7.12), and a regular routing lookup is performed via the FIB to forward the IP packet.

7.7 MPLS Labels

Labels are stacked on top of the layer 3 packets. A label is a 4-octet (32-bit) fixed-length and locally significant identifier that is used to identify an FEC. The label that is imposed on a particular packet represents the FEC to which that packet is assigned.

A label (see Figure 7.10) comprises the following fields:

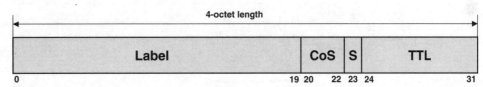

Figure 7.10: MPLS label format.

- *20-bit label*: Theoretically, the range of label values is from 0 through [2^{20} − 1]. Because the use of globally unique labels would impose a management burden as well as limit the number of usable labels, labels have local significance only and therefore change on every hop.

- *3-bit CoS (Class of service)*: This field is used to define a class of service by reflecting the IP precedence value (also 3-bit in length) of the encapsulated IP packet and can affect the queuing and discard algorithms applied to the packet as it is transmitted through the network.

- *1-bit bottom-of-stack indicator*: This bit supports a hierarchical label stack and indicates whether this is the last label in the label stack before the IP header. It is set to 1 for the last entry in the label stack (bottom of the stack) and 0 for all other label stack entries.

- *8-bit TTL (time-to-live)*: This field provides conventional IP TTL functionality. The TTL field is used to prevent forwarding loops and is decremented by a value of 1 on every hop. (See section 7.9 for more details on MPLS TTL processing.)

7.8   MPLS Label Stack

The MPLS label is inserted between the data link layer (layer 2) header and the network layer (layer 3) header. There can be more than one label forming a label stack (Figure 7.11). The top of the label stack appears first in the packet, and the bottom appears last. The network layer immediately follows the last label in the label stack.

MPLS devices (nodes) forward packets based on the label value and not the IP information. When an MPLS node receives a labeled packet, only the label value at the top of the stack is looked up. Consequently, if the lookup is successful, the MPLS node will be able to determine:

- The next-hop to which the packet is to be forwarded.

- The operation to be performed on the label stack before forwarding. This operation may be to replace the top label stack entry with another, or to pop an entry off the label stack, or to replace the top label stack entry, and then to push one or more additional entries on the label stack.

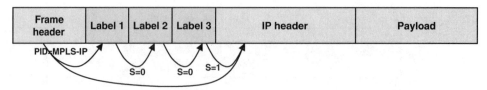

Figure 7.11:   MPLS label stack.

Besides learning the next-hop and the label stack operation, the MPLS node may also learn the outgoing data link encapsulation, for instance, a LAN encapsulation frame. When transporting labeled packets over LAN media, exactly one labeled packet is carried in each LAN frame. The label stack entries immediately precede the network layer header and follow any data link layer headers.

An MPLS label does not contain any information about the layer 3 protocol being carried in a packet. To indicate that the payload starts with a label or labels and is followed by an IP header (see Figure 7.11), a protocol identifier (PID [for example, an EtherType value]) for MPLS-capable layer 3 protocols need to be specified in the layer 2 header. The EtherType value 0x8847 indicates that a frame is carrying an MPLS unicast IP packet, and the EtherType value 0x8848 indicates that a frame is carrying an MPLS multicast packet. These EtherType values can be used to carry labeled packets with either the Ethernet encapsulation or the IEEE 802.3 LLC/SNAP encapsulation.

Typically, only one label is assigned per packet. However, additional labels may be added to packets by applications such as MPLS VPNs (see section 7.13). An MPLS VPN uses two labels: the top label points to the egress LSRs and the second label identifies the VPN.

Note that label values 0–15 are reserved and are defined as follows:

- A value of 0 represents the "IPv4 Explicit NULL Label." This label value is legitimate only at the bottom of the label stack. It indicates that the label stack must be popped and the forwarding of the packet must then be based on the IPv4 header.

- A value of 1 represents the "router alert label." This label value is legitimate anywhere in the label stack except at the bottom. When a received packet contains this label value at the top of the label stack, it is delivered to a local software module for processing. The label beneath it in the stack determines the actual forwarding of the packet. However, if the packet is to be forwarded further, the router alert label should be pushed back onto the label stack before forwarding. The use of this label is analogous to the use of the "router alert option" in IP packets (for example, ping with record route option).

- A value of 2 represents the "IPv6 Explicit NULL Label." This label value is legitimate only at the bottom of the label stack. It indicates that the label stack must be popped and the forwarding of the packet must then be based on the IPv6 header.

- A value of 3 represents the "Implicit NULL Label" (see penultimate popping in section 7.12). This is a label that an LSR may assign and distribute but that never actually appears in the encapsulation. When an LSR replaces the label at the top of the stack with a new label and the new label is "Implicit NULL," the LSR will pop the stack instead of doing the replacement.

- Values 4–15 are reserved for future use.

When the last label is popped from a packet's label stack (resulting in the stack being emptied), further processing of the packet is based on the packet's network layer header. Since the label stack does not contain any field that explicitly identifies the network layer protocol, the LSR that pops the last label off the stack must therefore be able to identify the packet's network layer protocol.

7.9   MPLS TTL Processing

The TTL field in IPv4 header or the hop limit field in IPv6 header is used for loop control. It can be used to prevent forwarding (routing) loops due to faulty configurations. Because an LSR does not examine the IP header, the TTL field is included in the label so that the TTL function can still be supported. The following rules are applied when processing the TTL field in the MPLS label:

- When an IP packet arrives at an ingress LSR of an MPLS domain, a single label stack entry is added to the packet. The TTL value of this label stack entry is copied from the IP TTL value. When this packet next encounters a core LSR in the MPLS domain, the TTL value in the top label stack entry is decremented by a value of 1:
  —If this TTL value is zero, the packet is not forwarded any further and is silently discarded.
  —If this TTL value is positive, it is placed in the TTL field of the top label stack entry for the outgoing packet, and the packet is forwarded.
- When the packet arrives at the egress LSR of the MPLS domain, the TTL value in the single stack entry is decremented by a value of 1 and the label is popped (resulting in an empty label stack):
  —If this TTL value is zero, the packet is not forwarded any further and is silently discarded.
  —If this TTL value is positive, it is copied into the TTL field of the IP header and the IP packet is forwarded using regular IP routing. Note that the IP header checksum must be modified to reflect the new TTL value before forwarding.

7.10   MPLS Label Assignment, Distribution, and Retention

The MPLS architecture specifies several modes/parameters for label assignment, distribution, and retention, such as:

- Per-platform or per-interface label space
- Unsolicited downstream or on-demand downstream label distribution
- Independent or ordered label distribution control
- Liberal or conservative label retention

In the following sections, we describe these modes in more detail.

7.10.1 Label space

There are two possible alternatives when assigning labels to networks:

1. *Per-platform label space*: per-platform label space is used with frame-mode MPLS, where one single label is assigned to a destination network and announced to all neighbors. The label must be locally unique and valid on all incoming interfaces. This label can be used on any incoming interface. In other words, the LFIB on an LSR does not contain an incoming interface, which means the label is not bound to the interface where the packet was received. Therefore, per-platform label space is less secure because it is susceptible to label spoofing, whereby an adjacent router can send a labeled packet with a label that has not been previously allocated to this router.

2. *Per-interface label space*: local labels are assigned to IP destination prefixes on a per-interface basis. The label assigned to an input interface can be reused on another interface with a different IP destination. These labels must be unique for a specific input interface. In this case, the LFIB on an LSR contains an incoming interface. That is, the labels are bound to the interface where the packet was received. Therefore, per-interface label space is more secure since LSRs cannot send packets with labels that were not assigned to them. In other words, per-interface label space will foil any label spoofing attempt.

Note that per-platform label space is used by routers (edge and core LSRs) implementing frame-mode MPLS, and per-interface label space is used by ATM switches (ATM core LSRs) as well as routers with ATM interfaces (ATM edge LSRs) when implementing cell-mode MPLS.

7.10.2 Label distribution

There are two available label distribution schemes:

- *Unsolicited downstream distribution*: the label for a network prefix is asynchronously allocated and advertised to all neighbors, regardless of whether the neighbors are upstream or downstream LSRs for the destination. There is no control mechanism to govern the distribution of labels in an ordered fashion. Typically, frame-mode MPLS uses unsolicited downstream distribution.

- *On-demand downstream distribution*: requires each LSR to specifically request a label from its downstream neighbor, and the downstream neighbor will distribute the label on request.

Note that unsolicited downstream distribution is used by routers (edge and core LSRs) implementing frame-mode MPLS, and on-demand downstream distribution is used by ATM switches (ATM core LSRs) as well as routers with ATM interfaces (ATM edge LSRs) when implementing cell-mode MPLS.

7.10.3 Label distribution control

There are two types of label distribution control:

- *Independent control mode*: is typically combined with unsolicited downstream label distribution (see section 7.10.2), whereby labels can be created and distributed independent of any other LSR. When independent control mode is applied, an LSP might encounter an incoming labeled packet where there is no corresponding outgoing label in the LFIB table. Therefore, an LSR using independent control mode must have layer 3 capabilities, that is, it must be able to perform full layer 3 lookups. In other words, independent control mode can be applied only on edge LSRs. Typically, independent control mode is applied to edge LSRs using frame-mode MPLS.

- *Ordered control mode*: is typically combined with on-demand downstream label distribution (see section 7.10.2), whereby labels can be created and distributed by an LSR for a particular FEC only if it is the egress LSR for that FEC or it has already received the next-hop label from its next-hop LSR; otherwise, it must request a label from its next-hop LSR. This requirement results in an ordered sequence of downstream requests until an LSR is found that already has a next-hop label or an edge LSR is reached that uses independent control mode, for example, an egress LSR.

Note that independent control mode is used by routers (edge and core LSRs) implementing frame-mode MPLS. Routers (ATM edge LSRs) implementing cell-mode MPLS also use the independent control mode since they are the endpoints of the ATM virtual circuits. Ordered control mode is used by ATM switches (ATM core LSRs) when implementing cell-mode MPLS.

7.10.4 Label retention

There are two available label retention modes:

- *Liberal retention mode*: LSR retains labels from all neighbors even if these neighbors are not the downstream peers for a specific destination network. Liberal retention mode improves convergence time, when next-hop is again available after IP convergence. However, it requires more memory and label space. Typically, liberal retention mode is applied to LSRs using frame-mode MPLS.

- *Conservative retention mode*: LSR retains labels only from next-hops neighbors; all other labels are ignored, that is, it discards all labels for FECs without next-hop. Conservative retention mode is implicitly achieved through on-demand downstream label distribution (see section 7.10.2), whereby no label is received unless it is requested. It also frees up memory and label space.

Note that liberal retention mode is used by routers (edge and core LSRs) implementing frame-mode MPLS, and conservative retention mode is used by

ATM switches (ATM core LSRs) as well as routers with ATM interfaces (ATM edge LSRs) when implementing cell-mode MPLS.

7.10.5 Label distribution protocol

A label distribution protocol is required to perform the operations that were described in sections 7.10.1 to 7.10.4. Specifically, a label distribution protocol is a set of procedures by which one LSR informs another of the label/FEC bindings it has made. Two LSRs that use a label distribution protocol to exchange label/FEC binding information are known as label distribution peers with respect to the binding information they exchange. If two LSRs are label distribution peers, there is a label distribution adjacency between them. Note that two LSRs may be label distribution peers with respect to some set of bindings but not with respect to some other set of bindings. In addition, the label distribution protocol also encompasses any negotiations in which two label distribution peers need to engage to learn of each other's MPLS capabilities.

The MPLS architecture does not assume a single label distribution protocol but provides for multiple such protocols. Existing protocols such as BGP (RFC 3107—Carrying Label Information in BGP-4) and RSVP (RFC 3209—RSVP-TE: Extensions to RSVP for LSP tunnels) have been extended so that label distribution can be piggybacked on them. New protocols such as LDP (Label Distribution Protocol [RFC 3036]) and CR-LDP (Constraint-based LDP [RFC 3212]) have also been defined for the explicit purpose of distributing labels.

LDP is used for mapping unicast IP destinations into labels; RSVP and CR-LDP are used in traffic engineering; and BGP is used for distributing external labels in MPLS VPNs. Note that in this chapter, we try to use the acronym LDP to refer specifically to the protocol defined in RFC 3036; when discussing label distribution protocols in general, we try to avoid the acronym.

7.11 Label Switched Paths

Before the routing and delivery of packets in a given FEC, a path through the network, known as a label switched path (LSP), must be established. An LSP is a sequence of LSRs that forward packets for a specific FEC. Each LSR independently selects the next-hop for each FEC, and swaps the top label in a packet that is traversing the LSP. Put another way, an LSP is functionally equivalent to a virtual circuit because it defines an ingress-to-egress path through a network that is followed by all packets assigned to a particular FEC. Note that the terms LSP and LSP tunnel are interchangeable.

So far, the examples that we have used in this chapter are all based on MPLS unicast IP forwarding. In MPLS unicast IP forwarding, the FECs are determined by the destination IP networks found in the main routing table, and an LSP is created for each entry found in the routing table. Each LSP is created over the best path selected by the IGP (interior gateway protocol), toward the

destination network. In this case, an IGP (for example, OSPF or IS-IS) is used
to propagate routing information to all routers in an MPLS domain to determine the best path to specific destination networks, and a label distribution
protocol (such as LDP) is used to propagate labels for these networks as well
as to build the LSPs. LSPs are unidirectional. This means the return traffic will
use a different LSP, which is usually the reverse path since most IGPs provide
symmetrical routing.

LSP may diverge from IGP best path through explicit routing with MPLS
traffic engineering (TE), where a single LSR, usually the ingress or egress,
specifies some (loose explicit routing) or all (strict explicit routing) of the LSRs
in the LSP for a given FEC. MPLS traffic engineering is beyond the scope of
this book.

7.12 Penultimate Hop Popping

Figure 7.12 illustrates how labels are forwarded and used in a typical frame-
mode MPLS domain. In this example, the egress LSR must do a dual lookup:
one lookup in the LFIB table to determine whether the label must be removed
and another lookup in the IP FIB table to determine the longest-match route.
Dual lookup is not an optimal way of forwarding labeled packets. Penultimate
hop popping eases the requirement for a dual lookup on the egress LSR.

In penultimate hop popping (PHP) as illustrated in Figure 7.13, the label at
the top of the stack is removed (popped) by the immediate upstream neighbor
of the egress LSR. In other words, a label can be popped one hop earlier on the
router before the last hop (egress LSR) within an MPLS domain. The egress
LSR requests the pop operation through the label distribution protocol by
advertising a pop or implicit null label (encoded using a label value of 3 [see

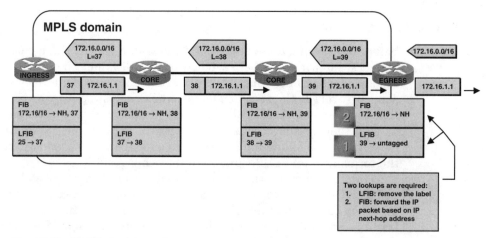

Figure 7.12: Dual-lookup scenario.

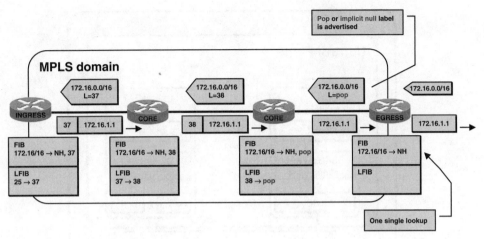

Figure 7.13: Penultimate hop popping.

section 7.8]). This label instructs upstream routers to remove the label instead of swapping it with the next-hop label. In this case, the egress LSR will not receive a labeled packet. Hence, it need not do a LFIB lookup and remove the label itself. Therefore, penultimate hop popping optimizes MPLS performance to a certain extent because one less LFIB lookup is required on the egress LSR. However, the egress LSR will still need to do an IP FIB lookup for determining the more specific (longest-match) route.

7.13 MPLS Applications

Besides reducing the amount of per-packet processing required at each router in an IP network, MPLS also provides new capabilities in applications such as traffic engineering (TE), class of service (CoS), and virtual private networks (VPNs). We will discuss only MPLS VPNs in this section. Traffic engineering and class of service are beyond the scope of this book.

Regardless of the available MPLS applications, the functionality is always separated into the control plane and the data plane. The applications differ only in the control plane in which appropriate mechanisms are used to exchange routing information and labels. Each application may use a different routing protocol to exchange routing information and a different (or additional) label distribution protocol to exchange label/FEC binding information. While in the data plane, all the applications use a common label-forwarding engine. In other words, MPLS provides the flexibility to evolve control functionality without changing the forwarding mechanism.

Recall that MPLS unicast IP routing is the MPLS application that we have been using throughout the chapter to illustrate all the MPLS concepts and technologies. In MPLS unicast IP routing, two mechanisms are required on the control plane: IP routing protocol (for example, OSPF, IS-IS) and label distri-

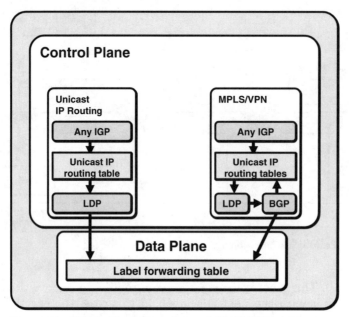

Figure 7.14: MPLS unicast IP routing and MPLS VPN architecture.

bution protocol (for example, LDP). The FEC is a destination network that is stored in the IP routing table.

In MPLS VPNs, networks are learned via an IGP (for example, OSPF, RIPv2) from a customer router or via BGP from other internal routers. In this instance, two labels are used:

1. Top label indicates the BGP next-hop. This label is distributed through LDP.

2. Second-level label indicates the outgoing interface on the egress router. This label is propagated via MP-BGP (multiprotocol BGP [RFC 2858]).

In MPLS VPNs, the FEC is a destination VPN or VPN site descriptor. Figure 7.14 illustrates the simplified architecture for MPLS unicast IP routing and MPLS VPN. Further details of the MPLS VPN architecture are discussed in Chapter 8.

7.14 Summary

One of the most important benefits of MPLS is that it allows service providers to deliver new services, such as MPLS VPNs, that cannot be readily supported by conventional IP routing techniques. This is because MPLS has the flexibility to evolve control functionality without altering the forwarding mechanism. For service providers to offer a robust VPN service to its customers, they have to solve the issues of data privacy and support the use of non-unique, private

IP addresses within a VPN. MPLS allows service providers to conquer these challenges by providing a simple, flexible, and yet powerful tunneling mechanism. We shall explore this further in Chapter 8.

Chapter 7 discussed a number of MPLS concepts and technologies in detail so that in the later MPLS VPN chapters the reader will have the fundamental MPLS concepts to build on and will also be better equipped with the necessary MPLS know-how to support the actual deployment of MPLS VPNs.

MPLS VPN Architecture

8.1 Introduction

The general MPLS concepts were discussed in Chapter 7. Chapter 8 is an extension of that subject, and the focus is now on one of the most important and popular MPLS applications—MPLS VPN. Before the actual deployment of MPLS VPN, it is important to have a thorough understanding of its related components, implementation mechanisms, and operational processes.

Hence, in this chapter we take the reader through the inner workings of the MPLS VPN architecture by examining the MPLS VPN functional as well as architectural building blocks, the MPLS VPN routing model, and the MPLS VPN forwarding mechanisms. Comprehending the functionalities and operations of the MPLS VPN architecture will give the reader the extra leverage required to deploy a fully operational MPLS VPN with ease.

8.2 MPLS VPN Terminologies

The following list is a general conceptual overview of the terms used in this chapter. Some of them are more specifically defined in later sections of the chapter.

- *Border router*: PE router that interfaces to other P networks.

- *Customer edge router (CE router)*: The CE router forms part of the C network and interfaces with a PE router.

- *Customer network (C-Network)*: The network that is under customer control.

- *Extended community*: In MPLS VPN context, BGP extended community attribute is used to identify a route origin, a route target, or a site of origin.

- *Provider edge router (PE router)*: PE router serves as the edge router (or edge-LSR) of the P network and interfaces with CE routers.

- *Provider Network (P-Network)*: The MPLS VPN backbone or core network under the control of a service provider.

- *Provider router (P router)*: The core router (or core-LSR) in the P network that provides transit transport across the provider or MPLS VPN backbone but has no knowledge of VPN.

- *Route distinguisher (RD)*: A 64-bit attribute prepended to each VPN route to uniquely identify prefixes among VPNs. An RD is VRF based and not VPN based.

- *Route target (RT)*: A 64-bit BGP extended community attribute used to identify routers that should receive the route.

- *Site*: Contiguous portion of the C network. A site is connected to the MPLS VPN backbone through one or more PE/CE links.

- *Site of origin (SOO)*: A 64-bit BGP extended community attribute used to identify the customer site that originates the route.

- *VPN-IPv4 address*: An address that includes the 64-bit RD and the 32-bit IPv4 address.

- *VPN routing and forwarding instance (VRF)*: Routing table and FIB table populated by routing protocol contexts.

- *VPN-aware network*: A provider backbone in which MPLS VPN is deployed.

8.3   MPLS VPN Functional Components

MPLS VPN describes a mechanism that allows service providers to use their IP backbone to provide VPN services to their customers. An MPLS-based VPN network is divided into three main functional components:

1. CE routers

2. PE routers

3. P routers

Figure 8.1 illustrates the placement of these fundamental building blocks. The details of these functional components are described in the following sections.

8.3.1   Customer edge routers

A customer edge (CE) router provides customer access to the service provider over a data link to one or more provider edge (PE) routers. The CE router establishes an adjacency with its directly connected PE routers. After the adjacency is established, the CE router advertises the site's local VPN routes to the PE router and learns remote VPN routes from the PE router.

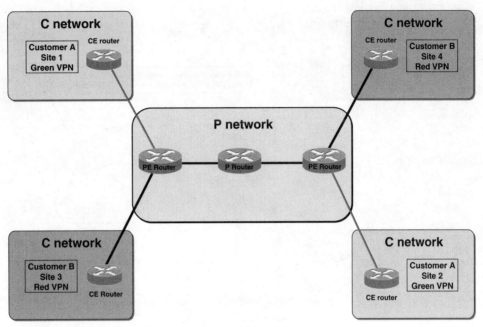

Figure 8.1: Fundamental building blocks of an MPLS VPN.

8.3.2 Provider edge routers

In MPLS VPN, the architecture of a PE router is similar to that in the dedicated PE router peer-to-peer model (see section 1.4.2.2) except that the whole architecture is condensed into one physical router. PE routers exchange routing information with CE routers using static routing, RIPv2, OSPF, or EBGP, and each customer is assigned an independent VPN routing table known as a VPN routing and forwarding (VRF) table (more details on VRF tables in section 8.5.3). In other words, the PE router implements isolation between customers via VRF tables. Each PE router maintains a VRF for each of its directly connected customer sites. Multiple interfaces on a PE router can be associated with a single VRF if these sites belong to the same VPN. In addition, PE routers use MPLS forwarding with the P routers and conventional IP forwarding with the CE routers.

Although a PE router maintains VPN routing information, it is required to maintain VPN routes only for those VPNs to which it is directly connected. This eliminates the need for PE routers to maintain all of the service provider's VPN routes, thus improving overall scalability. After learning local VPN routes advertised from the CE routers, the PE routers exchange VPN routing information with other PE routers using IBGP, and only routes relevant to the PE router's VRFs are populated.

As the number of customer routes is expected to be huge, the only well-known routing protocol with the required scalability is BGP. Therefore, BGP is used

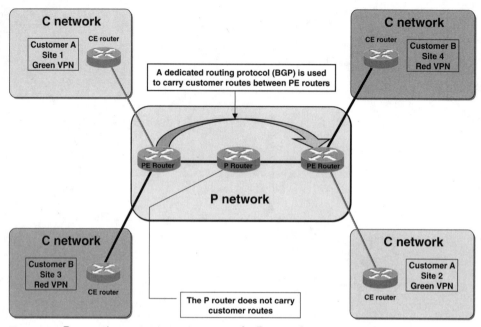

Figure 8.2: Propagating customer routes across the P network.

in MPLS VPN to propagate customer routes directly between PE routers (Figure 8.2). The PE routers must establish any-to-any IBGP sessions with each other to exchange the VPN routing information; however, this IBGP fully meshed requirement can be alleviated with BGP route reflectors (see Chapter 11).

In MPLS VPN, it is important for the reader to differentiate the functions between BGP and MPLS. Succinctly, BGP is used to distribute VPN routing information across the provider's backbone, and MPLS is used to forward VPN traffic from one VPN site to another.

8.3.3   Provider routers

A provider (P) router is any router in the provider's network that does not connect directly to CE routers. In the dedicated PE router peer-to-peer model (see section 1.4.2.2), the P router contains all customer routes and filters routing updates between different PE routers using BGP communities. In MPLS VPN, P routers function as MPLS transit (core) LSRs when forwarding VPN data traffic between PE routers. Since traffic is forwarded across the MPLS backbone using a two-level MPLS label stack (see section 7.13), P routers are required to maintain routes only to the PE routers with a common IGP; they are not required to maintain a specific VPN routing information for each customer site. In others words, the P routers need not carry customer routes (see Figure 8.2).

8.4   MPLS Architectural Components

The MPLS VPN architecture has two groundbreaking components: route distinguisher (RD) and route target (RT). RDs enable overlapping customer address space, and RTs allow the implementation of complex VPN topologies that were hard to implement with other VPN architectures. RDs and RTs are discussed in the following sections.

8.4.1   Route distinguisher

In Section 8.3.2, we mentioned that BGP is the dedicated routing protocol that is used to carry customer routes between the PE routers. One important issue arises: How can BGP propagate several identical prefixes, belonging to different customers, between PE routers?

The MPLS VPN architecture supports overlapping customer address space by prepending the customer's IP prefix with a unique 64-bit prefix called the route distinguisher (RD). The RD is used to convert a non-unique 32-bit customer IPv4 address into a unique 96-bit VPNv4 address (also known as an VPN-IPv4 address). VPNv4 addresses are exchanged only between PE routers; they are transparent to the customer. Therefore, BGP between PE routers must support the exchange of both traditional IPv4 and VPNv4 prefixes. BGP that supports address families besides IPv4 addresses is known as multiprotocol BGP (MP-BGP [RFC 2858]). The content of an MP-BGP update is discussed in greater length in section 8.5.4. Figure 8.3 illustrates the application of the RD in an MPLS VPN.

Note that if two or more VPNs have a common site (overlapping VPNs), the address space must be unique between these VPNs even with a distinctive RD. Overlapping VPNs are discussed in Chapter 10.

8.4.2   Route target

Besides transforming overlapping IPv4 addresses into globally unique VPNv4 addresses, the RD has no special role in the MPLS VPN architecture. RD is VRF-based and not VPN-based. There has to be a unique one-to-one mapping between RDs and VRFs, which makes RD the VRF identifier. However, simple VPN topologies may require only one RD per customer; in such cases, the RD can also be considered a VPN identifier. In other words, RD is configured in the PE router for each VRF, and it may or may not be related to a site or a VPN.

The problem arises in the implementation of more complex VPN topologies (overlapping VPNs) when a customer site can belong to more than one VPN. The RD cannot identify a site that participates in multiple VPNs. Therefore, a different method is required in which a set of VPN identifiers can be attached to a route to indicate its membership in different VPNs.

Route targets (RT) are additional attributes (extended BGP communities) attached to a VPNv4 route (carried in an MP-BGP update) to indicate its VPN

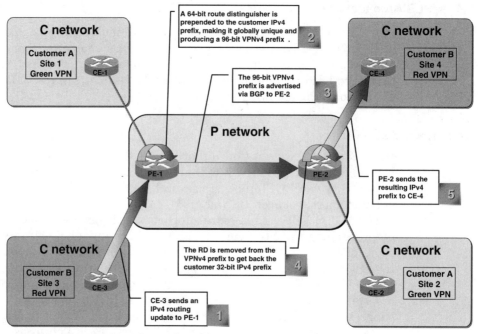

Figure 8.3: Route distinguisher application process.

membership, thus fulfilling the requirements of complex VPN topologies. Note that any number of RTs can be attached to a single route, and extended BGP communities are used to encode these attributes. More details on extended communities with respect to MP-BGP will be found in section 8.5.4.2.

Export RTs are appended to a customer route when it is converted from an IPv4 route to a VPNv4 route. Export RTs are used to identify the set of sites to which a particular route should be exported and are configured separately for each VRF table in a PE router.

When VPNv4 routes are advertised to other PE routers, these routers need to select which routes to import into their VRF tables. This selection is based on import RTs. Each VRF table in a PE router can have various import RTs configured, indicating the set of VPNs from which this particular VRF table is accepting routes. In short, RTs allow the implementation of complex VPN topologies with ease, which can be a difficult task to achieve in traditional peer-to-peer VPN architectures.

8.4.3   MPLS VPN connectivity model

RDs and RTs (and site of origin; to be discuss in Section 8.5.4.2) together help to construct the MPLS VPN connectivity model. The MPLS VPN connectivity model can be thought of as a community of interest or closed user group (CUG). Since MPLS VPN supports complex VPN topologies, the traditional definition of a VPN needs to be redefined. In other words, the MPLS VPN connectivity model is simply a collection of sites sharing common routing information. Through this sharing of routing information, a site may belong to more than one VPN, resulting in differing routing requirements for sites that belong to different sets of VPNs. These routing requirements are supported with multiple VRF tables on the PE routers.

If sites with different requirements are associated with the same VRF table, some of these sites might be able to access destinations that should not be accessible to them. To avoid this security vulnerability, a single VRF table can be used only for sites with similar connectivity requirements in simple VPN topologies, while multiple VRF tables for each VPN are required for dissimilar connectivity requirements in complex VPN topologies.

8.5   MPLS VPN Routing Model

Recall that in Chapter 7 we discussed the MPLS architecture being split into two planes: control and data forwarding. MPLS VPN is still based on the typical MPLS architecture except that it introduces more variety of control mechanisms, such as:

- LDP and IGP for normal MPLS unicast IP forwarding (as illustrated in Chapter 7). LDP and IGP are required to forward VPN packets (based on LDP labels) among the P and PE routers across the P network. More details on MPLS VPN packet forwarding appear in Section 8.6.

- PE-CE routing protocols between the PE and CE routers for individual VRF tables and MP-BGP for propagating VPNv4 updates tagged with a VPN label among PE routers. More details on PE-CE routing and MP-BGP are in subsequent sections. MPLS VPN packet forwarding related to MP-BGP is discussed in Section 8.6.1.

It is worth mentioning that the routing function plays a very crucial role in the entire MPLS VPN technology implementation. The MPLS VPN routing model can be defined as having the following routing requirements:

- CE routers should not be aware of the MPLS VPN, and its routing should be based only on traditional IP routing protocols.

- For scalability reasons, P routers should not participate in MPLS VPN routing and should not carry any VPN routes.

- PE routers should support both MPLS VPN services and traditional Internet services.

8.5.1   CE routing

CE routers run traditional IP routing protocols and exchange routing updates with PE routers, which appear to them as any other routers in the C network.

This exchange is also known as PE-CE routing, and the choice of routing protocols that can be run between the PE and CE routers includes static routes, RIP version 2 (RIPv2), open shortest path first (OSPF), and external BGP (EBGP). The implementation examples based on these PE-CE routing protocols are illustrated in Chapter 9.

From the customer's perspective, the MPLS VPN backbone looks like an ordinary BGP backbone, with the PE routers performing route redistribution between individual sites and the core backbone. In this case, the P routers are concealed from the customer view, which makes the interior topology of the BGP backbone invisible to the customer.

8.5.2   P routing

From the P router perspective, the MPLS VPN backbone appears even simpler, since P routers do not participate in MPLS VPN routing and do not carry VPN routes. Besides running a common backbone IGP with other P routers as well as PE routers, and exchange routing information regarding core subnets (which includes core links and BGP next-hop addresses [usually the loopback addresses of PE routers]), P routers also run a common label distribution protocol (for example, LDP [RFC 3036]) with the other P and PE routers to learn about the MPLS labels for the BGP next-hop addresses (see section 8.6.1).

For scalability reasons, BGP deployment on P routers is not necessary with the MPLS VPN operation. Only the border routers (PE routers) are required to run BGP. This also implies that the P routers are not required to participate in Internet routing to support Internet connectivity. As such, the size of routing table (memory usage) on a P router will be relatively smaller, which further improves scalability.

8.5.3   PE routing

The main bulk of MPLS VPN routing lies in the PE routers because they are the only routers that see all the routing portions of the MPLS VPN (Figure 8.4):

- They exchange VPN routes with the CE routers via various per-VRF routing protocols (mentioned in section 8.5.1).
- They exchange VPNv4 routes via MP-BGP sessions with other PE routers.
- They exchange backbone routes with P and other PE routers via a common backbone IGP.

The PE routers maintain separate routing tables:

- *Global IP routing table*: The global IP routing table is the usual IP routing table found in a router even if it is not running MPLS VPN. In this case, it contains all the non-VPN backbone routes (PE and P routes) populated by the backbone IGP and Internet (IPv4) routes populated from the global IPv4

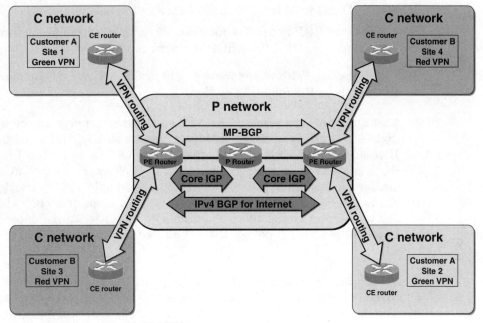

Figure 8.4: All aspects of MPLS VPN routing.

BGP table. In other words, PE routers are also required to support Internet connectivity, and this is achieved through the normal BGP protocol (IPv4 BGP).

- *VRF* (VPN Routing and Forwarding) *table*: A VRF table is associated with one or more directly connected sites (CE routers). It can be correlated with any type of interface, whether it is a logical or physical interface, and these interfaces may share the same VRF table if the connected sites share the same routing requirements. VRF routing tables contain VPN-specific (or CE) routes populated locally by PE-CE routing protocols and VPNv4 routes populated through MP-BGP (VPNv4 BGP) sessions from other PE routers.

Note that MP-BGP sessions are really IBGP sessions, except that MP-BGP is an extension of the BGP protocol that carries routing information about other protocols and address families (in this instance, VPNv4 addresses). MP-BGP sessions can also be considered VPNv4 BGP sessions. In the next section, we examine in detail the content of an MP-BGP update.

8.5.4 MP-BGP update

An MP-BGP update contains the following:

- VPNv4 address
- BGP extended communities, such as route targets and site of origin (optional)

- MPLS label used for VPN packet forwarding (see section 8.6.1)
- Any other BGP attributes for example, AS-path, local preference, multi-exit discriminator (MED), or BGP standard community

We discuss VPNv4 addresses and BGP extended communities more thoroughly in the following sections.

8.5.4.1 VPNv4 addresses. The VPNv4 address propagated in the MP-BGP update comprises a 64-bit RD (see Section 8.4.1) and a 32-bit customer IPv4 address. The RD is configured in the VRF table on the PE router. The RD has a one-to-one relationship with a VPN when all sites in a VPN have similar routing requirements, for example, in simple VPN topologies. However, RD is associated with a specific site instead of a specific VPN when each site has a different connectivity requirement. In other words, depending on the implemented VPN topology, the RD may or may not be related to a VPN or a site.

8.5.4.2 BGP extended communities. Typical BGP standard communities are 32-bit length and are the predecessors of BGP extended communities. The extended community is a transitive optional BGP attribute, with BGP type code 16. BGP extended communities are 64-bit length. This increased range is used to support a wide variety of applications, such as MPLS VPN routing updates, without worrying about overlapping. The inclusion of a type field provides structure for the community space that allows the usage of policy based on the application for which the community value will be used. BGP extended communities (minimum RTs) are always attached to the VPNv4 routes in MP-BGP updates. The BGP extended community attribute has two presentation formats:

- **<16-bit extended community type>: <16-bit Autonomous-System>: <32-bit number>**—In this display format, the high-order 16 bits represent the extended community type to be used in a particular application (we shall cover more on the BGP extended community types that are associated with MPLS VPN later in this section). The next 16 bits represent the public (registered) AS (autonomous system) number of the service provider defining the community, and the lower 32-bit field represents any 32-bit number. Both the 16-bit AS-number and the 32-bit number are configurable. This display format is encoded as 0x00 or 0x40 in the high-order octet of the 16-bit type field to ascertain consistent formatting across all routers taking part in an MPLS VPN.
- **<16-bit extended community type>: <32-bit IPv4-Address>: <16-bit number>**—In this display format, the high-order 16 bits has the same meaning as the previous display format. The next 32 bits represent a public (registered) IPv4 address belonging to the service provider defining the community, and the lower 16-bit field represents any 16-bit number. Both the 32-

bit IPv4 address and the 16-bit number are configurable. This display format is encoded as 0x01 or 0x41 in the high-order octet of the 16-bit type field to ascertain consistent formatting across all routers taking part in an MPLS VPN.

Returning to BGP extended community types, types 0 through 0x7FFF inclusive are assigned by IANA (Internet Assigned Numbers Authority), and types 0x8000 through 0xFFFF inclusive are vendor specific. The extended community types that are associated with MPLS VPN include:

- **Route origin**: Identifies one or more routers that inject a set of routes (that carry this community) into BGP. The type field for the route origin community is 0x0001 or 0x0101. Note that the route origin community is similar to the site-of-origin (SOO) community.

- **Route target**: Identifies one or more routers that may receive a set of routes (that carry this community) advertised by MP-BGP or identifies the set of routers to which the route must be propagated. Put another way, RT is used to indicate the VPN membership of a customer route and also to support the population of customer routes between different VRF tables (see section 8.4.2). The type field for the route target community is 0x0002 or 0x0102.

- **Site of origin (SOO)**: Identifies the customer site that originates the route. The SOO community is used to prevent loops in MPLS VPNs with multi-homed sites where the AS-path attribute cannot be used. The type field for SOO is 0x0003 or 0x0103. SOO will be touched on again in Chapter 9 with an illustrative example.

- **OSPF domain identifier, OSPF route type, and OSPF router ID**: Refer to IETF document draft-rosen-vpns-ospf-bgp-mpls-06.txt.

8.5.5 MPLS VPN routing example

In this section, we will consolidate all the concepts that we have covered earlier in Sections 8.5.1 to 8.5.4 into an illustrative example as shown in Figure 8.5. Typically, the entire MPLS VPN routing operation is accomplished in three main stages:

Stage 1: PE routers receive IPv4 routing updates from CE routers and populate these routes into the appropriate VRF table.

Stage 2: PE routers export (redistribute) VPN routes from VRF tables into MP-BGP and propagate them as VPNv4 routes via MP-IBGP to other PE routers. A full-mesh of MP-IBGP sessions is required between the PE routers, or BGP route reflectors could be used to mitigate the full mesh IBGP requirement.

Stage 3: The PE routers receiving MP-BGP updates will import the incoming VPNv4 routes into their VRF tables when the RTs attached to these

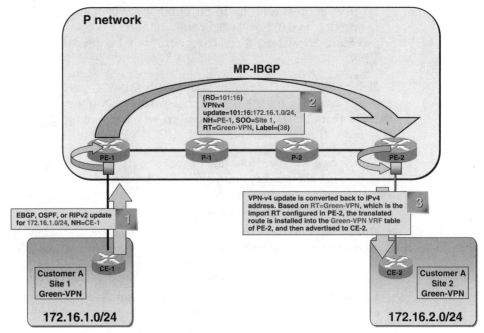

Figure 8.5: MPLS VPN routing operation.

routes match the import RTs configured in the VRF tables. The VPNv4 routes installed in VRF tables are then converted back to IPv4 routes and propagated to the CE routers. The SOO attribute (optional) attached to the VPNv4 route controls the IPv4 route advertisement to the CE routers. A route installed into a VRF is not advertised to a CE router if the SOO attached to the route is identical to the SOO attribute associated with the CE router.

In Figure 8.5, PE-1 receives IPv4 customer route 172.16.1.0/24 (with an IPv4 source address 172.16.1.1) from CE-1 and populates them into the Green-VPN VRF table. PE-1 then does the following tasks:

- Prepends the 32-bit IPv4 customer route with an RD attribute value of 101:16 to obtain the VPNv4 route 101:16:172.16.1.0/24.

- Assigns RT = Green-VPN and SOO = Site 1.

- Rewrites next-hop attribute from NH = CE-1 to NH = PE-1 (PE-1's loopback).

- Assigns a VPN label for the VPNv4 route based on VRF table and/or outgoing interface (see section 8.6.1).

- Compiles all the above-mentioned data into an MP-IBGP update and sends this update to PE-2.

After PE-2 has received the MP-IBGP update from PE-1, it translates the 96-bit VPNv4 route back to the original 32-bit IPv4 customer route and installs the route into the VRF table identified by the RT attribute (based on PE-2 configuration). Consequently, the label of value 38 associated with the VPNv4 route 101:16:172.16.1.0/24 will be set on packets forwarded toward destination subnet 172.16.1.0/24 (more about this in section 8.6.4).

8.6   MPLS VPN Packet Forwarding

In the previous section, we covered the MPLS VPN routing model, looking at how MP-BGP is used to distribute VPN routing information across the provider's network. This section is a continuation of that section, and we shall describe how MPLS is used with MP-BGP to forward the actual VPN traffic from one VPN site to another.

8.6.1   MPLS VPN label stack and label propagation

In MPLS VPN packet forwarding, VPN packets are labeled with a label stack (see section 7.13) by the ingress PE router. The ingress PE router has two label-stack entries associated with a remote VPN route. The top label is the LDP label that is bound to the BGP next-hop assigned by the next-hop P router. This label is distributed via LDP (see sections 7.6.2 and 7.6.3) and is taken from the local label information base (LIB). Note that the BGP next-hop must be reachable via the common backbone IGP running between the P and PE routers.

The bottom label is the VPN label. The egress PE routers allocate a VPN label to every VPN route received from attached CE routers and to every summary route summarized within the PE router. These locally assigned VPN labels are then propagated via MP-BGP to all other PE routers together with the VPNv4 prefixes. Both of these label stack entries are stored in the VRF table of the ingress PE router.

The VPN label is bound to an outgoing interface whenever the CE router is the next-hop of the VPN route; in this case, the egress PE router need only perform label lookup on the VPN packet. However, the VPN label (aggregate label) is bound to a VRF table for aggregate VPN routes, VPN routes pointing to a null interface, and routes for directly connected VPN interfaces. In this case, the egress PE router would first perform a label lookup to find the target VRF table and then perform an IP lookup (FIB lookup) within the VRF table to determine the correct layer 2 encapsulation.

Figure 8.6 illustrates the VPN label propagation between PE routers. In this example, a VPN label of value 38 is assigned to the route 172.16.1.0/24 received from CE-1 by the egress PE router, PE-1. The VPN label is propagated to PE-2 in an MP-IBGP update, and a label stack is built in PE-2's VRF table. The PE and P routers learned the reachable BGP next-hop (PE-1 with loopback address 192.168.1.1) through the backbone IGP, and LDP labels are

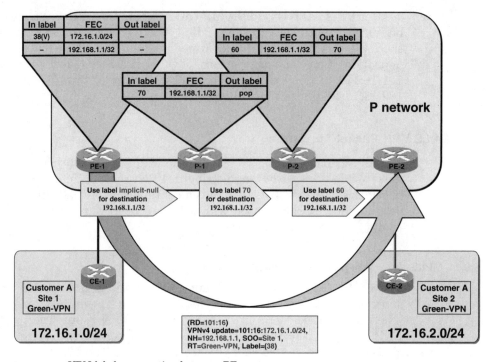

Figure 8.6: VPN label propagation between PE routers.

distributed through LDP corresponding to the BGP next-hop address 192.168.1.1.

8.6.2 MPLS VPN label forwarding

The two-level MPLS label stack discussed in the previous section satisfies all MPLS VPN forwarding requirements:

- The ingress PE router receives a normal IP packet from the CE router, does an "IP longest match" in the VRF table, finds the BGP next-hop, and imposes the MPLS label stack (top LDP label + bottom VPN label) on this packet.

- The P routers perform label switching based on the top LDP label and forward the packet toward the egress PE router.

- The egress PE router's upstream LDP neighbor (penultimate hop) removes the top LDP label (MPLS VPN penultimate hop popping [see section 8.6.3]) on the packet before forwarding it to the egress PE router.

- When the egress PE router receives the packet, it does a label lookup on the remaining bottom VPN label to determine an outgoing interface or VRF table, removes the label, and either forwards the packet toward the CE router through the outgoing interface or sends the packet out to the destination VPN site after performing another IP lookup in the VRF table to determine the layer 2 encapsulation.

8.6.3 MPLS VPN penultimate hop popping

As discussed in section 7.12, penultimate hop popping (PHP) can be performed in frame-based MPLS networks to optimize MPLS performance to a certain extent, because one less lookup is required on the egress PE router. In MPLS VPN PHP, the last P router (P-1) in the LSP tunnel "pops" the top LDP label, which was previously requested via the egress PE router (PE-1) through LDP by advertising an implicit-null label as illustrated in Figure 8.6.

The egress PE router (PE-1) then receives a labeled packet that contains only the bottom VPN label, uses the VPN label to select which VPN/CE to forward the packet to, removes the VPN label, and routes the packet toward the VPN site.

8.6.4 MPLS VPN packet forwarding operation

In Sections 8.6.1 to 8.6.3, we discussed the functional procedures associated with MPLS VPN label propagation and label forwarding. In this section, we examine an example of the operation of the MPLS VPN packet forwarding as illustrated in Figure 8.7. Note that the example shown in Figure 8.7 is a continuation of the example portrayed in Figure 8.6.

In the example, the ingress PE router, PE-2 receives a normal IP packet with a destination address of 172.16.1.1. PE-2 does an "IP longest match" lookup from its Green-VPN VRF table, finds the BGP next-hop PE-1 (with loopback address 192.168.1.1), and imposes the MPLS label stack <top LDP label, bottom VPN label> = <60,38> derived previously in Figure 8.6. In this case, the BGP next-hop address 192.168.1.1 is reachable through the common backbone IGP running among the P and PE routers with an associated next-hop LDP label 60.

Next, the packet is forwarded to P router P-2, which proceeds to switch the packet based solely on the top LDP label. In this instance, LDP label 60 is swapped with another next-hop LDP label 70 and is then forwarded to P router P-1. P-1 is the penultimate hop, which does a PHP to remove the top LDP label 70.

The egress PE router PE-1 then receives a labeled packet that contains only the bottom VPN label 38 (which was previously assigned by itself to route 172.16.1.0/24 received from CE-1 illustrated in Figure 8.6). Based on this VPN label, PE-1 determines the outgoing interface to forward the packet to CE router CE-1 (which is the next-hop of VPN route 172.16.1.0/24). Finally, the VPN label is removed and the packet is routed toward CE-1.

8.6.5 Caveats for MPLS VPN packet forwarding

For MPLS VPN packet forwarding to work properly, it is important to remember the following criteria:

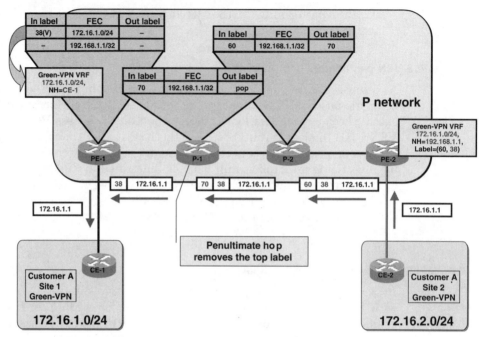

Figure 8.7: MPLS VPN packet-forwarding example.

- The VPN label must be allocated by the BGP next-hop, and this BGP next-hop must be the PE router. This implies that the BGP next-hop should not be modified during the process of propagating the MP-BGP update.
- There must be an unbroken LSP between the ingress and egress PE routers because the VPN label can be identified only by the egress PE router that originates it:
  — Since LDP is bound to the IGP table rather than the BGP table, LDP labels are not assigned to BGP routes. If a BGP next-hop is propagated as a BGP route, it will not be allocated any LDP label by LDP, and the MPLS VPN label stack will not be built properly. As a result, the LSP will be broken. Therefore, BGP next-hops should not be advertised as BGP routes; instead they should be announced by the common backbone IGP between the P and PE routers. In other words, the BGP next-hop address (PE address) must be uniquely known in the backbone IGP.
  — BGP next-hops advertised by IGP should not be summarized in the P network by the P routers because this will break and disrupt the LSP at the P router that does the summarization. In other words, instead of terminating at the egress PE router, the LSP will now end at the summarizing P router. Recall that VPN labels are assigned by the egress PE router and not by the P routers. Therefore, when the summarizing P

router receives the VPN label, it will not know how to interpret it and the VPN packet will be discarded.

8.7 Summary

Chapter 8 takes the reader through the various functional and architectural building blocks of the MPLS VPN architecture, such as route distinguisher (RD), route target (RD), and site of origin (SOO). RD is used to prevent address overlap when more than one VPN site have identical IPv4 address spaces, while RT and SOO are used to implement the MPLS VPN closed user group (CUG) concept. CUG is really a VPN community.

Next, in the MPLS VPN routing model, we looked at how multiprotocol BGP (MP-BGP) can be used to advertise reachable VRF information to all members of a VPN community. MP-BGP peering (fully meshed MP-IBGP sessions) needs to be configured in all PE routers within a VPN community; otherwise, BGP route reflectors could be used to mitigate the full mesh IBGP requirement. Last but not least, in MPLS VPN packet forwarding, we discussed how MPLS can be used to transport all traffic between all VPN community members across a MPLS VPN service provider network.

Chapters 9 to 11 cover deploying MPLS VPN in a service provider environment, including

- Using different PE-CE routing protocols, such as static routing, RIPv2, and EBGP;

- Implementing some of the more complex MPLS VPN topologies, such as overlapping VPNs, central service VPNs, hybrid VPNs, and hub-and-spoke VPNs; and

- Scaling MPLS VPN.

Note that the device configurations found in the case studies from Chapters 9 to 11 have been developed and tested using Cisco IOS software release 12.2(11T) and Cisco 2621 (CE) as well as Cisco 3640 (PE) routers.

Building MPLS VPN

9.1 Introduction

This chapter describes how to build an MPLS VPN by first going through the design considerations, followed by the implementation preliminaries, and finally the actual implementation. The chapter uses three different case studies to illustrate the actual deployment of MPLS VPN. Case Study 9.1 gives the reader good hands-on coverage of how to design and implement MPLS VPN practically from scratch. The PE-CE routing protocol adopted by this first case study is EBGP. Case Study 9.2 is an extension of Case Study 9.1 and deploys static routes and RIPv2 as the alternative PE-CE routing protocols besides EBGP. Finally, Case Study 9.3 shows the reader how to prevent potential BGP loops by using the site-of-origin (SOO) extended BGP community attribute when the standard AS-path–based BGP loop prevention has been compromised.

9.2 Design and Implementation Preliminaries

Before embarking on the design and implementation details, let us do a quick recap of VPN-aware PE-CE routing protocols and VRF tables.

9.2.1 VPN-aware PE-CE routing protocols

Recall that in Chapter 8, we mentioned that each VPN, or more specifically each VRF table, needs to be maintained by a separate and isolated VPN-aware PE-CE routing process. The rationale behind this is to prevent unwanted route leakage between VPNs. In addition, overlapping IP address spaces would create chaotic routing conditions if all VRF tables share identical copies of the PE-CE routing protocol.

Therefore, the concept of routing contexts was introduced to support the need for separate isolated copies of PE-CE routing protocols. Routing contexts are simply separate routing processes (in the case of OSPF) or separate isolated instances of the same routing protocols (for example, EBGP, RIPv2, and static routes). In this case, each routing process or instance will have its own autonomous routing protocol parameters and will maintain its own routing table. With routing contexts, it is now possible to have one routing process per VRF table.

Note that the current Cisco IOS software implementation limits the overall number of routing protocols in a router to 32. Two routing methods are pre-defined (static and connected), and two routing protocols are required for proper MPLS VPN backbone operation (BGP and backbone IGP). Therefore, the number of PE-CE routing processes is restricted to 28.

9.2.2   VRF tables

A VRF is a routing and forwarding instance that can be used for a single VPN site or for multiple VPN sites connected to the same PE router provided these sites share identical connectivity requirements. Multiple VRFs are required for VPN sites that do not share the same connectivity requirements (see Chapter 10).

The routes received from routing contexts are installed into the IP routing table contained within the VRF table. The per-VRF forwarding table (FIB) is built from the per-VRF routing table and is used to forward all packets received through the interfaces associated with the VRF. Any number of interfaces can be associated with a VRF, regardless of whether it is a physical or logical inter-face. However, since the router has to uniquely identify which FIB is to be used for packets received over an interface, each interface can be associated with only one VRF.

In Cisco implementations, these per-VRF interfaces have to support Cisco Express Forwarding (CEF). Succinctly, CEF uses a complete IP switching table, or rather an FIB table, which holds identical information to the IP routing table. The CEF FIB is essentially a replacement for the standard IP routing table. The generation of route entries in the FIB is change triggered. In other words, when there is a change in the standard routing table, this change is also reflected in the FIB. Furthermore, the FIB does not keep information about the outgoing interface and the corresponding layer 2 header. This information is stored in an adjacency table, which is identical to the ARP (address resolution protocol) cache except that it holds the layer 2 header instead of destination MAC (media access control) addresses.

Other MPLS VPN attributes that are associated with a VRF include:

- The route distinguisher (RD), which is prepended to all routes exported from the VRF into the global BGP VPNv4 table.

- A set of export route targets (RTs), which are attached to any route exported from the VRF.

- A set of import route targets (RTs). VPNv4 updates are imported into the VRF when the RTs attached to these VPNv4 updates match with the import RTs configured for the VRF.

9.3   Design Considerations

When designing MPLS VPN, we need to consider the following:

- VPN topology
- Selecting the backbone IGP
- Autonomous system numbering
- Selecting an RD
- Selecting an RT
- Selecting an SOO (optional)
- PE-CE addressing scheme
- Selecting the PE-CE routing protocol

9.3.1   VPN topology

In MPLS VPN, the VPN topology is driven by the connectivity requirement and site membership. The VPN connectivity requirement and VPN site membership are customer oriented. In other words, the customer's business environment determines the ultimate VPN topology.

In this chapter, we examine simple MPLS VPN topologies that support any-to-any connectivity between customer sites associated with identical VPN membership (which is similar to the fully meshed overlay VPN networks). In this case, each CE router will advertise its own address space, and each site will have full routing knowledge of all other sites with the same VPN membership. MP-BGP is used to propagate VPNv4 updates between the PE routers, and optimal routing is achieved in the backbone because each route will have the BGP next-hop that is nearest to the destination VPN site. In simple VPN topology, no site is used as the central point for connectivity. More complex VPN topologies such as overlapping VPN topology, central services VPN topology, hybrid VPN topology, and hub-and-spoke VPN topology are discussed in Chapter 10.

9.3.2   Selecting the backbone interior gateway protocol

In this chapter, we discuss only integrated IS-IS as the backbone IGP since it is running between the three PE routers—Brussels-PE, London-PE, and Paris-PE.

9.3.2.1   Integrated IS-IS overview.   Integrated IS-IS (RFC 1195) is an extended version of IS-IS (intermediate system to intermediate system) for mixed OSI (with the ISO Connectionless Network Service [CLNS]) and IP environment. It allows for three types of routing domains (OSI, IP, dual), and represents a more scalable option than OSPF in the IP world. We briefly explain some of the important features and functionalities of this routing protocol in the subsequent sections.

9.3.2.2 IS-IS addressing. As IS-IS originated from the open system interconnect (OSI) protocol stack, it also adopts the OSI network-layer addressing scheme. OSI network layer addressing is implemented with network service access points (NSAPs).

An IS-IS NSAP is divided into three parts: 1 to 13 octets for area address (which includes the authority and format identifier [AFI]), 6 octets for system ID, and 1 octet for NSEL (N-Selector or service identifier). The total length of NSAP is from 8 (minimum) to 20 (maximum) octets. Network entity title (NET) is NSAP with a service identifier (NSEL) of 0x00 and is used in routers. Official NSAP prefixes are required for connecting to a public OSI network. For private address space, an AFI of value 49 can be used. A router requires the NSAP to determine its area address (for level 2 routing) and system ID (for Level 1 routing).

The area address uniquely identifies the routing area and the system ID identifies each routing node (also referred to as an intermediate system [IS]). All routers within an area must use the same area address. A nonrouting network node (also referred to as an end system [ES]) may be adjacent to a level 1 router only if they both have a common area address. The system ID can be a MAC address for CLNS implementation or a router loopback IP address for IP implementation. The system ID has to be unique both within an area and within level 2 routers forming the routing domain. The typical recommendation is to use a unique domain-wide system ID for all the routers (or ISs).

9.3.2.3 IS-IS levels. IS-IS is a hierarchical routing protocol that uses a two-level area structure to denote its routing domain. The two-level hierarchy comprises level 2 (backbone) and level 1 (nonbackbone) areas.

Level 1 routing occurs between routers within the same area and is implemented with level 1 (L1) routers (also referred to as station routers). L1 routers constitute an area and keep one copy of the link-state database (LSDB) with intra-area information only. They also enable any ESs to communicate with the world via the most optimal exit from the network segment that they reside on.

Level 2 routing happens between different areas within the same routing domain and is implemented with level 2 (L2) routers (also referred to as area routers). L2 routers interconnect areas and keep one copy of the LSDB that stores interarea information only.

In IS-IS, the area borders lie on the links. In other words, each IS-IS router belongs to exactly one area. Therefore, only one link-state packet is required per IS-IS router per area (including redistributed prefixes) when advertising a link-state change. This also allows a more flexible approach to extending the backbone with the help of level 1–2 (L1/L2) routers acting as area border routers. L1/L2 routers keep two separate copies of the LSDBs, one for level 1 intra-area information and the other for level 2 inter-area information. L1/L2 routers perform both L1 and L2 routing. They also inform L1 routers that it is a potential exit point for the area (see section 9.3.2.6). Put another way, in an IS-IS routing domain a level 1 area is a collection of L1 as well as L1/L2 routers, and the level 2 (backbone) area is a contiguous collection of L1/L2 as well as L2 routers from various areas.

9.3.2.4   IS-IS routing metrics.   In IS-IS, router-specific or link-state information such as routing metrics are encoded in TLV (type/length/value) variables (maximum size per variable is 255 bytes). IS-IS routing metric is associated with an outgoing interface. There are four possible metrics: delay, default, expense, and error. The delay, expense, and error metrics are optional, and they are meant for type-of-service (ToS) routing. Cisco routers use only the default metric in IS-IS routing implementation.

The default metric (arbitrary cost, not based on bandwidth) is a 6-bit integer ranging between 0 and 63 (default 10) with the total path metric limited to 1023. However, with the introduction of new extended IS-IS TLV variables (defined in the IETF document "draft-ietf-isis-traffic-04.txt"), it is now possible to carry a 24-bit-wide metric with the total path metric increased to 32 bits.

9.3.2.5   IS-IS routing algorithms.   The shortest path first (SPF) algorithm is used to determine the SPF topology (based on NETs) to destination networks. L1 or L2 routers require only one SPF calculation. However, separate route calculations are performed for L1 and L2 areas in L1/L2 routers. Whichever way, a full SPF recalculation is still invoked when the SPF topology changes.

In IS-IS, IP subnets and ESs are represented as leaf objects, which do not participate in SPF. Therefore, a partial route calculation (PRC) is run whenever a router receives a link-state packet (LSP) where only IP or ES information has changed. In other words, changes of IP (or ESs) information do not lead to full SPF recalculation, and as a result, the CPU utilization in the router is optimized. PRC is also invoked after completing SPF to insert IP prefix information in the LSDB. The IP forwarding table is thus built with PRC.

9.3.2.6   IS-IS IP routing.   There are two types of IS-IS routing: Level 1 intra-area routing and level 2 interarea routing. With an L1 router, IP routing is based on IP reachability using the same path as OSI routing (system ID). In the event the destination address is outside the area, the L1 router will send a packet to a nearest active L1/L2 router based on the attached (ATT) bit. The ATT bit is set in the L1 LSP by a L1/L2 router indicating to the L1 routers that it is a potential exit point of the area. This results in default routing in L1 routers in which they will point a default route to the closest L1/L2 router. The packet then travels via L2 routing toward the destination area where the best L1 path is used to reach the destination network. Level 2 routing is more straightforward. IP routing with a L2 router is based on L2 IP prefixes that consider the total path cost.

To place a route into the IP forwarding table, the router requires information about the IP next-hop address, but L1 or L2 link-state packet (LSP) advertises only the IP prefix, subnet, and cost. Therefore, the IP next-hop address for L1 or L2 routes must be derived from somewhere else, and in this case, it is learned from the CLNS adjacency table. To determine the IP next-hop address, routers would need to establish CLNS adjacencies and use CLNS packets. This explains why CLNS configurations (such as NET) are still required even when integrated IS-IS is used only for IP routing

9.3.2.7   Simple IS-IS design.   The main objective of using IS-IS is to provide a scalable internal IP routing design. This section gives the reader some simple design tips on how to incorporate IS-IS as the backbone IGP for an MPLS VPN implementation:

- Use one IS-IS area for the core.
- Use wide metrics (see section 9.3.2.4).
- Use level 2 routing wherever possible.
- Implementing a level 2 area first is advantageous because it will be easier to incorporate a level 1 area later to scale the network as it grows.
- Running L1/L2 routing is strongly discouraged because L1/L2 routers are required to have two LSDBs (see section 9.3.2.3), which in turn run two SPF calculations (see section 9.3.2.5). This can exhaust valuable memory and CPU cycles fairly quickly.
- Keep in mind that the maximum number of routers in one area is not absolute. It depends on the IS-IS topology database size, neighbors, and the number of changes in the network.

9.3.3   Autonomous system numbering

The Internet Assigned Numbers Authority (IANA) is responsible for the assignment of a public autonomous system number (ASN), which is 16 bits in length. In MPLS VPNs, an ASN combined with an arbitrary number creates a route distinguisher (RD), which uniquely identifies a VPN routing and forwarding (VRF) instance in the PE router configuration. An RD is used with an IPv4 prefix to create a globally unique prefix for a VRF instance.

Typically, a service provider (SP) assigns public ASNs to customers that need to connect to the PE routers using EBGP. Alternatively, the SP can allocate private ASNs, which range from 64,512 to 65,535. Using private ASNs for VPN sites allows easier configuration and saves public ASNs.

There might be cases when the same ASN is allocated to several VPNs. The PE router receiving an MP-BGP update will verify whether any of the ASNs in the AS-path attribute is equivalent to the ASN of the corresponding VPN location. If the numbers are identical, the ASNs in the AS-path attribute are replaced by the SP's ASN using the BGP AS-override feature. We shall illustrate more on this specific BGP feature in section 9.7.2.

9.3.4 Selecting a route distinguisher

A route distinguisher (see section 8.4.1 for more details) is an 8-byte (64-bit) value added to the beginning of the customer's IPv4 prefixes to change them into globally unique VPNv4 prefixes. The format of a route distinguisher (RD) can be either of the following:

- IP-Address:nn—This format comprises an IP address and an arbitrary number, which can be used only when the MPLS VPN uses a private ASN and the VPNv4 addresses are propagated beyond the private AS (for example, when VPN routes are exchanged between different SPs).

- ASN:nn—This format comprises an ASN and an arbitrary number. Use ASN.nn with an AS number that is assigned by the IANA to ensure that a route distinguisher is unique among SPs.

9.3.5 Selecting a route target

A route target (see section 8.4.2 for more details) is a 64-bit BGP extended community attribute attached to a VPNv4 route to indicate its VPN membership. There are two types of route targets (RTs):

1. Import RT —When VPNv4 routes are advertised to other PE routers, these routers need to select which routes to import into their VRF tables. This selection is based on import RTs. Each VRF table in a PE router can have various import RTs configured, indicating the set of VPN sites that this particular VRF table is accepting routes from.

2. Export RT —Export RTs are appended to a customer route when it is converted from an IPv4 route to a VPNv4 route. Export RTs are used to identify the set of sites to which a particular route should be exported and are configured separately for each VRF table in a PE router.

Note that any number of RTs can be attached to a single VPNv4 route. Complex VPN topologies have always been difficult to achieve with traditional peer-to-peer VPN architectures. However, in MPLS VPN, with the help of RTs, complex VPN topologies can now be implemented easily (see Chapter 10 for further illustrations). Similar to RDs, the RTs can be specified in one of two formats: (1) 32-bit IP address followed by a 16-bit decimal number (IP-Address:nn) or (2) 16-bit ASN followed by a 32-bit decimal number (AS:nn).

9.3.6 Selecting a site of origin

The AS-path–based BGP loop prevention is bypassed when either the AS-override or the Allowas-in BGP feature is used (AS-override is covered in section 9.7.2 and Allowas-in is discussed in chapter 10). The site of origin (SOO) is an optional extended BGP community attribute that can be used to prevent loops in these scenarios. SOO is required only for multihomed customer sites, and it must be unique for each site. SOO is typically denoted as: *<Customer Private-AS-Number>:<Customer Site-Number>*.

SOO is used to identify network prefixes that originated from a site to prevent the advertisement of these prefixes back to the same site again. When a network prefix is advertised to a CE router for a particular site, the attached SOO is verified against the SOO associated with the site to which the CE router belongs. Identical SOO values indicate that a loop has occurred, and the network prefix will not be advertised further.

9.3.7 PE-CE addressing scheme

Various PE-CE addressing schemes can be used:

- Customer addresses —This approach is recommended if the SP is not managing the CE or an unmanaged CE service exists.

- Public addresses —This approach uses a comparatively large number of public addresses for PE-CE numbering, but it is also the most straightforward addressing scheme.

- Private addresses (RFC 1918)—This is a relatively efficient addressing scheme. The only precautionary step is for the SP to ensure that its customer is not using the same private address block as that allocated for PE-CE numbering.

9.3.8 Selecting the PE-CE routing protocol

Both static and dynamic routing protocols can be configured to distribute IP prefixes between a customer edge (CE) router at a VPN site and the associated provider edge (PE) router. The routes statically defined or dynamically learned over a specific VRF interface are installed into the associated VRF tables within the PE router, and not into the global routing table or other unassociated VRF tables (unless the routes are to be imported using the associated route target).

For ease of administration, it would be appropriate to standardize on a PE-CE routing protocol. The PE-CE routing context can be implemented with the following routing protocols:

- Static routing
- RIPv2
- EBGP
- OSPF

We shall discuss static routing, RIPv2, and EBGP in the following sections. OSPF is not within the scope of this book.

9.3.8.1 Static routing. Static routes can be used in environments where a customer site (stub site) has a single connection to the P network and uses a single

IP prefix. Static routes are recommended in sites where the SP is required to enforce control over customer routing for some specific central services. However, if new or changing routes occur frequently within the site, static routing is not recommended.

With static routing, a default route is configured on the CE routers and specific static routes are configured on the PE routers. Static routes must be redistributed into MP-BGP to notify other PE routers of the remote networks originating from the customer's VPN site.

9.3.8.2 RIP version 2 routing. RIPv2 allows for dynamic routing between the PE and CE routers. This option is typically used in stub sites where the SP does not manage the CE routers, where every CE router needs to know all routes from other CE routers with the same VPN membership, or where there is more than one IP prefix per customer site.

The RIP metric can be preserved by copying it into the BGP multi-exit discriminator (MED) attribute via the ingress PE router during redistribution from RIP into MP-BGP. The value of the MED attribute (with the original RIP hop count) is copied into the RIP metric when the MP-BGP route is redistributed back into RIP. This allows transparency of the RIP metric through the MPLS VPN backbone, and the entire MPLS VPN backbone appears as a single hop to the CE routers.

9.3.8.3 EBGP routing. EBGP is typically used for multihomed sites where optimal routing is mandatory. Deploying EBGP as the PE-CE routing protocol allows the continuity of BGP policies between customer sites since the BGP attributes set by one customer site are transparently advertised to other customer sites. A standard BGP session is established between the PE and CE routers. Updates received from neighboring CE routers end up in the appropriate address family (routing context) of the BGP table. Since the same routing protocol is used throughout the network, redistribution is unnecessary. However, exporting from the VRF table is still required to prepend an RD to the IPv4 prefix and to affix the RT(s) to the resultant VPNv4 route.

9.4 MPLS VPN Configuration Checklist

Before going into the case studies, it is appropriate to briefly familiarize the reader with the MPLS VPN configuration checklist:

- Identify the PE routers.
- Configure VRF instance for each VPN.
- Determine the RD and RT assignments.
- Configure MP-BGP between the PE routers (fully meshed or using route reflector).
- Identify interfaces connecting to the VPN sites.

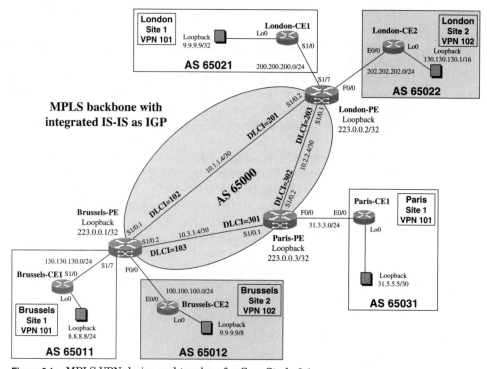

Figure 9.1:   MPLS VPN design and topology for Case Study 9.1.

- Allocate interfaces to the VRF instance.
- Select the PE-CE routing protocol.
- Configure the necessary redistribution between MP-BGP and PE-CE routing protocol.

9.5   Case Study 9.1: EBGP as PE-CE Routing Protocol

9.5.1   Case overview

This case study provides a sample configuration of an MPLS VPN over frame relay when EBGP is present on the customer side as the PE-CE routing protocol. The MPLS VPN deployment uses Cisco routers and spans five different customer sites: Brussels Site 1, Brussels Site 2, London Site 1, London Site 2, and Paris Site 1. The MPLS VPN design and topology for Case Study 9.1 are shown in Figure 9.1. Note that, for simplicity, P routers are omitted in all our case studies.

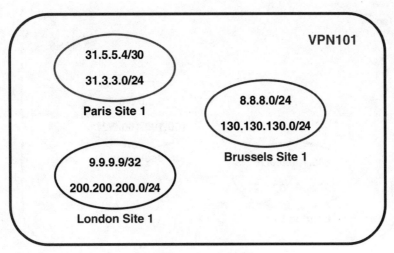

Figure 9.2: Sites and subnets associated with VPN101.

9.5.2 Network design

9.5.2.1 VPN topology. Two VPNs are defined in this case study and they are VPN101 and VPN102. VPN101 encompasses the following site members and their associated subnets (see Figure 9.2):

- Brussels Site 1: With subnets 130.130.130.0/24 and 8.8.8.0/24
- London Site 1: With network 200.200.200.0/24 and host route 9.9.9.9/32
- Paris Site 1: With subnets 31.3.3.0/24 and 31.5.5.4/30

The above three sites are solely the site members of VPN101; they do not belong to any other VPNs. As such, the connectivity requirement is for all the subnets between all three sites to be reachable within VPN101 only.

VPN102 encompasses the following site members and their associated subnets (see Figure 9.3):

- Brussels Site 2: With subnet 100.100.100.0/24 and network 9.0.0.0/8.
- London Site 2: With networks 202.202.202.0/24 and 130.130.0.0/16.

The above two sites are solely the site members of VPN102; they do not belong to any other VPNs. As such, the connectivity requirement is for all the subnets between the two sites to be reachable within VPN102 only.

Since the sites that belong to VPN101 are mutually exclusive from those that belong to VPN102, it is possible to have overlapping address spaces between the two VPNs. Nevertheless, site members belonging to the same VPN would still be required to use nonduplicating address ranges.

Notice that there are two occurrences of overlapping address space, one at Brussels Site 1 (in VPN101) and London Site 2 (in VPN102) for subnet

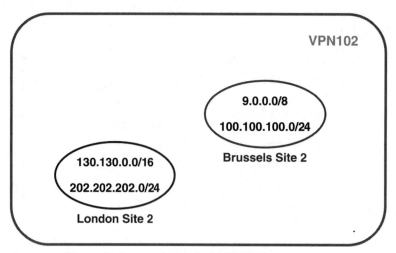

Figure 9.3:   Sites and subnets associated with VPN102.

130.130.130.0/24 and network 130.130.0.0/16, respectively. The other is at London Site 1 (in VPN101) and Brussels Site 2 (in VPN102) for host route 9.9.9.9/32 and network 9.0.0.0/8, respectively. (This is an illustration that over-lapping address spaces are indeed supported in MPLS VPNs.)

9.5.2.2   Backbone IGP.   Integrated IS-IS is selected as the backbone IGP for the MPLS VPN. It is implemented in Brussels-PE, London-PE, and Paris-PE to ensure full IP connectivity between these PE routers. The CLNS NETs defined for the three PE routers are:

- Brussels-PE: 49.0001.0000.0000.0001.00
- London-PE: 49.0001.0000.0000.0002.00
- Paris-PE: 49.0001.0000.0000.0003.00

We also employ wide metrics and a level 1 only area in the case study. The level 1 area is used for illustrative purposes. The reader might want to consider using a level 2 area instead (see section 9.3.2.7).

9.5.2.3   Autonomous system numbering.   Private ASNs are used for the five different sites:

- Brussels Site 1 uses ASN 65011.
- Brussels Site 2 uses ASN 65012.
- London Site 1 uses ASN 65021.
- London Site 2 uses ASN 65022.
- Paris Site 1 uses ASN 65031.

9.5.2.4 Selecting the RD and RT. In this case study, a unique RD is used for each VPN (rather than each VRF). The VRF of each PE router connecting the sites belonging to VPN101 is configured with RD 65000:101, and the VRF of each PE router connecting sites belonging to VPN102 is configured with RD 65000:102. Implementing the same RD for VRFs used for the same VPN helps to conserve memory.

Based on our VPN topology illustrated in section 9.5.2.1, only one RT is required. Therefore, we let the import and export RTs adopt the same value as the RD. Note that each RT will consume at least 64 bits per routing update. Using the same RD and RT helps to ease management, monitoring, and troubleshooting of the MPLS VPN.

9.5.2.5 PE-CE addressing scheme. Customer address spaces are applied for the PE-CE numbering for the five sites.

9.5.2.6 PE-CE routing protocol. EBGP is selected as the PE-CE routing protocol for all five sites.

9.5.3 Configuring the CE routers

Code Listings 9.1 to 9.5 illustrate the EBGP (the PE-CE routing protocol) configurations for the five CE Routers—Brussels-CE1, Brussels-CE2, London-CE1, London-CE2, and Paris-CE1—portrayed in Figure 9.1.

Code Listing 9.1 illustrates the EBGP configuration for Brussels-CE1, which is in AS 65011. Brussels-CE1 peers to its EBGP neighbor Brussels-PE in AS 65000 and advertises subnet 8.8.8.0/24 to Brussels-PE.

Code Listing 9.1:   EBGP configuration for Brussels-CE1

```
hostname Brussels-CE1
!
interface Loopback0
 ip address 8.8.8.8 255.255.255.0
!
interface Serial1/0
 bandwidth 64
 ip address 130.130.130.2 255.255.255.0
!
! Regular BGP Configuration
!
router bgp 65011
 no synchronization
 network 8.8.8.0 mask 255.255.255.0 ! Advertise subnet 8.8.8.0/24
! Peer to EBGP neighbor Brussels-PE in AS 65000
 neighbor 130.130.130.1 remote-as 65000
 no auto-summary ! Disable auto-summarization to propagate subnets unchanged
```

Code Listing 9.2 illustrates the EBGP configuration for Brussels-CE2, which is in AS 65012. Brussels-CE2 peers to its EBGP neighbor Brussels-PE in AS 65000 and advertises network 9.0.0.0/8 to Brussels-PE.

Code Listing 9.2:   EBGP configuration for Brussels-CE2

```
hostname Brussels-CE2
!
interface Loopback0
 ip address 9.9.9.9 255.0.0.0
!
interface Ethernet0/0
 ip address 100.100.100.2 255.255.255.0
!
! Regular BGP Configuration
!
router bgp 65012
 no synchronization
 network 9.0.0.0 ! Advertise network 9.0.0.0/8
! Peer to EBGP neighbor Brussels-PE in AS 65000
 neighbor 100.100.100.1 remote-as 65000
 no auto-summary ! Disable auto-summarization to propagate subnets unchanged
```

Code Listing 9.3 illustrates the EBGP configuration for London-CE1, which is in AS 65021. London-CE1 peers to its EBGP neighbor London-PE in AS 65000 and advertises host route 9.9.9.9/32 to London-PE.

Code Listing 9.3:   EBGP configuration for London-CE1

```
hostname London-CE1
!
interface Loopback0
 ip address 9.9.9.9 255.255.255.255
!
interface Serial1/0
 bandwidth 64
 ip address 200.200.200.2 255.255.255.0
!
! Regular BGP Configuration
!
router bgp 65021
 no synchronization
```

```
network 9.9.9.9 mask 255.255.255.255 ! Advertise host route 9.9.9.9/32
! Peer to EBGP neighbor London-PE in AS 65000
 neighbor 200.200.200.1 remote-as 65000
 no auto-summary ! Disable auto-summarization to propagate subnets unchanged
```

Code Listing 9.4 illustrates the EBGP configuration for London-CE2, which is in AS 65022. London-CE2 peers to its EBGP neighbor London-PE in AS 65000 and advertises network 130.130.0.0/16 to London-PE.

Code Listing 9.4:   EBGP configuration for London-CE2

```
hostname London-CE2
!
interface Loopback0
 ip address 130.130.130.1 255.255.0.0
!
interface Ethernet0/0
 ip address 202.202.202.2 255.255.255.0
!
! Regular BGP Configuration
!
router bgp 65022
 no synchronization
 network 130.130.0.0 ! Advertise network 130.130.0.0/16
! Peer to EBGP neighbor London-PE in AS 65000
 neighbor 202.202.202.1 remote-as 65000
 no auto-summary ! Disable auto-summarization to propagate subnets unchanged
```

Code Listing 9.5 illustrates the EBGP configuration for Paris-CE1, which is in AS 65031. Paris-CE1 peers to its EBGP neighbor Paris-PE in AS 65000 and advertises subnet 31.5.5.4/30 to Paris-PE.

Code Listing 9.5:   EBGP configuration for Paris-CE1

```
hostname Paris-CE1
!
interface Loopback0
 ip address 31.5.5.5 255.255.255.252
!
interface Ethernet0/0
 ip address 31.3.3.2 255.255.255.0
!
```

```
! Regular BGP Configuration
!
router bgp 65031
 no synchronization
 network 31.5.5.4 mask 255.255.255.252 ! Advertise subnet 31.5.5.4/30
 ! Peer to EBGP neighbor Paris-PE in AS 65000
 neighbor 31.3.3.1 remote-as 65000
 no auto-summary ! Disable auto-summarization to propagate subnets unchanged
```

9.5.4 Configuring the PE Routers

Code Listings 9.6 to 9.8 illustrate the MPLS VPN configurations for the three PE Routers—Brussels-PE, London-PE, and Paris-PE—portrayed in Figure 9.1. In the configurations, self-explanatory comments (in *italics*) are embedded between the configuration lines to provide clarification to the configuration commands and procedures.

In short, the MPLS VPN configuration for the PE router can be divided into five parts:

Part 1: Enable MPLS

Part 2: Configure backbone IGP

Part 3: Configure VRF tables and assign PE-CE interfaces to specified VRFs

Part 4: Configure MP-BGP

Part 5: Configure per-VRF PE-CE routing protocol

Code Listing 9.6:   MPLS VPN configuration for Brussels-PE

```
hostname Brussels-PE
!
! --- Step 1.1: Create VRF vpn101 for VPN101.
! --- VRF names have only local significance and are case-sensitive.
! --- VRF remains nonfunctional until the RD is configured.
ip vrf vpn101
 rd 65000:101 ! --- Step 1.2: Assign route distinguisher 65000:101 to VRF
vpn101.

! --- Step 1.3: Specify RT 65000:101 to be attached
! --- to every route from this VRF to MP-BGP.
 route-target export 65000:101

! --- Step 1.4: Specify RT 65000:101 to be used as an import filter.
! --- Only routes matching this RT are imported into VRF vpn101.
 route-target import 65000:101
```

```
!
!
ip vrf vpn102 ! --- Step 2.1: Create VRF vpn102 for VPN102.
 rd 65000:102 ! --- Step 2.2: Assign route distinguisher 65000:102 to VRF
vpn102.

! --- Step 2.3: Specify RT 65000:102 to be attached
! --- to every route from this VRF to MP-BGP.
 route-target export 65000:102

! --- Step 2.4: Specify RT 65000:102 to be used as an import filter.
! --- Only routes matching this RT are imported into VRF vpn102.
 route-target import 65000:102
!
ip cef ! --- Step 3.1: Enable Cisco Express Forwarding.
!
interface Loopback0
 ip address 223.0.0.1 255.255.255.255
 ip router isis  ! --- Step 4.5: Enable IS-IS for IP routing on Loopback0.
!
interface FastEthernet0/0
 ip vrf forwarding vpn102 ! --- Step 2.5: Associate FastEthernet0/0 with
VRF vpn102.
 ip address 100.100.100.1 255.255.255.0
!
interface Serial1/0   ·
 encapsulation frame-relay
 frame-relay lmi-type ansi
!
interface Serial1/0.1 point-to-point
 bandwidth 64
 ip address 10.1.1.5 255.255.255.252
 ip router isis ! --- Step 4.6: Enable IS-IS for IP routing on Serial1/0.1.
 tag-switching ip ! --- Step 3.2: Enable MPLS on subinterface Serial1/0.1.
 frame-relay interface-dlci 102
!
interface Serial1/0.2 point-to-point
 bandwidth 64
 ip address 10.3.3.5 255.255.255.252
 ip router isis ! --- Step 4.7: Enable IS-IS for IP routing on Serial1/0.2.
 tag-switching ip ! --- Step 3.3: Enable MPLS on subinterface Serial1/0.2.
 frame-relay interface-dlci 103
!
interface Serial1/7
 bandwidth 64
```

```
ip vrf forwarding vpn101 ! --- Step 1.5: Associate Serial1/7 with VRF vpn101.
 ip address 130.130.130.1 255.255.255.0
 clockrate 64000
!
! --- Step 4.1: Configure Integrated IS-IS as the Backbone IGP.
router isis

! --- Step 4.2: Configure Brussels-PE with CLNS parameter NET
! --- of value 0x49.0001.0000.0000.0001.00.
 net 49.0001.0000.0000.0001.00

! Step 4.3: --- Configure IS-IS level 1 routing globally on Brussels-PE.
 is-type level-1 1

! --- Step 4.4: Use new style of TLVs to carry wider metric.
! --- 24 bits for interface metric and 32 bits for total path metric.
 metric-style wide
!
! --- Step 5.1: Configure MP-BGP for Brussels-PE in AS 65000.
! --- All MP-BGP neighbors are configured under the global BGP configura-
tion.
! --- MP-IBGP sessions are established between loopback interfaces.
router bgp 65000
 no synchronization

! --- Step 5.2: Declare London-PE as MP-IBGP neighbor.
 neighbor 223.0.0.2 remote-as 65000

! --- Step 5.3: Inform London-PE to establish MP-IBGP session to Loopback0.
 neighbor 223.0.0.2 update-source Loopback0

! --- Step 5.4: Declare Paris-PE as MP-IBGP neighbor.
 neighbor 223.0.0.3 remote-as 65000

! --- Step 5.5: Inform Paris-PE to establish MP-IBGP session to Loopback0.
 neighbor 223.0.0.3 update-source Loopback0
 no auto-summary
 !
! --- Step 7.1: Enter the per-VRF BGP routing context for VRF vpn102.
 address-family ipv4 vrf vpn102

! --- Step 7.2: Redistribute directly connected interface FastEthernet0/0
! --- into the per-VRF BGP routing context for VRF vpn102.
 redistribute connected

! --- Step 7.3: Declare Brussels-CE2 in AS 65012 as CE EBGP neighbor.
 neighbor 100.100.100.2 remote-as 65012
```

```
neighbor 100.100.100.2 activate ! --- Step 7.4: Activate Brussels-CE2.
no auto-summary
no synchronization
exit-address-family
 !
! --- Step 6.1: Enter the per-VRF BGP routing context for VRF vpn101.
! --- CE EBGP neighbors have to be specified within the
! --- per-VRF BGP routing context, and not in the global BGP configuration.
! --- All non-BGP per-VRF routes have to be redistributed into the
! --- per-VRF BGP routing context to be propagated by MP-BGP to other
! --- PE routers. The per-VRF BGP routing context also has
! --- auto-summarization and synchronization disabled by default.
address-family ipv4 vrf vpn101

! --- Step 6.2: Redistribute directly connected interface Serial1/7
! --- into the per-VRF BGP routing context for VRF vpn101.
 redistribute connected

! --- Step 6.3: Declare Brussels-CE1 in AS 65011 as CE EBGP neighbor.
neighbor 130.130.130.2 remote-as 65011
neighbor 130.130.130.2 activate ! --- Step 6.4: Activate Brussels-CE1.
no auto-summary
no synchronization
exit-address-family
 !
! --- Step 5.6: Enter the VPNv4 configuration mode to configure
! --- VPNv4 specific parameters on the respective MP-IBGP neighbors.
address-family vpnv4
! --- Step 5.7: Activate London-PE for VPNv4 route exchange.
neighbor 223.0.0.2 activate
! --- Step 5.8: Activate Paris-PE for VPNv4 route exchange.
neighbor 223.0.0.3 activate

! --- Step 5.9: Disable the default IBGP next-hop processing
! --- for VPNv4 route exchange.
! --- If the default next-hop processing is not disabled,
! --- the IPv4 address of a BGP-speaking
! --- CE router might become the VPNv4 BGP next-hop,
! --- causing connectivity across the
! --- MPLS VPN backbone to be broken.
neighbor 223.0.0.2 next-hop-self
neighbor 223.0.0.3 next-hop-self

! --- Step 5.10: Propagate extended BGP communities
! --- to the respective MP-IBGP neighbors.
! --- Extended BGP communities attached to VPNv4 prefixes
```

```
! --- have to be exchanged between MP-IBGP neighbors
! --- for proper MPLS VPN operation.
 neighbor 223.0.0.2 send-community extended
 neighbor 223.0.0.3 send-community extended
 no auto-summary
 exit-address-family
```

Code Listing 9.6 illustrates the MPLS VPN configuration for Brussels-PE. In the configuration:

- Steps 1.1 to 1.5 configure VRF table vpn101 and assign the specified VRF to the corresponding PE-CE interfaces (Serial1/7) that interconnect sites (Brussels Site 1) belonging to vpn101.

- Steps 2.1 to 2.5 configure VRF table vpn102 and assigns the specified VRF to the corresponding PE-CE interfaces (FastEthernet0/0) that interconnects sites (Brussels Site 2) belonging to vpn102.

- Steps 3.1 to 3.3 enable MPLS.

- Steps 4.1 to 4.7 configure the backbone IGP (Integrated IS-IS).

- Steps 5.1 to 5.10 configure MP-BGP.

- Steps 6.1 to 6.4 configure the per-VRF PE-CE routing protocol (EBGP) for vpn101.

- Steps 7.1 to 7.4 configure the per-VRF PE-CE routing protocol (EBGP) for vpn102.

Code Listing 9.7:   MPLS VPN configuration for London-PE

```
hostname London-PE
!
ip vrf vpn101 ! --- Step 1.1: Create VRF vpn101 for VPN101.
 rd 65000:101 ! --- Step 1.2: Assign route distinguisher 65000:101 to VRF vpn101.

! --- Step 1.3: Specify RT 65000:101 to be attached
! --- to every route from this VRF to MP-BGP.
 route-target export 65000:101

! --- Step 1.4: Specify RT 65000:101 to be used as an import filter.
! --- Only routes matching this RT are imported into VRF vpn101.
 route-target import 65000:101
 !
 !
```

```
ip vrf vpn102 ! --- Step 2.1: Create VRF vpn102 for VPN102.
 rd 65000:102 ! --- Step 2.2: Assign route distinguisher 65000:102 to VRF
vpn102.

! --- Step 2.3: Specify RT 65000:102 to be attached
! --- to every route from this VRF to MP-BGP.
 route-target export 65000:102

! --- Step 2.4: Specify RT 65000:102 to be used as an import filter.
! --- Only routes matching this RT are imported into VRF vpn102.
 route-target import 65000:102
!
ip cef ! --- Step 3.1: Enable Cisco Express Forwarding.
!
interface Loopback0
 ip address 223.0.0.2 255.255.255.255
 ip router isis ! --- Step 4.5: Enable IS-IS on Loopback0.
!
interface FastEthernet0/0
 ip vrf forwarding vpn102 ! --- Step 2.5: Associate FastEthernet0/0 with
VRF vpn102.
 ip address 202.202.202.1 255.255.255.0
!
interface Serial1/0
 encapsulation frame-relay
 frame-relay lmi-type ansi
!
interface Serial1/0.1 point-to-point
 ip address 10.2.2.5 255.255.255.252
 ip router isis ! --- Step 4.6: Enable IS-IS on Serial1/0.1.
 tag-switching ip ! --- Step 3.2: Enable MPLS on subinterface Serial1/0.1.
 frame-relay interface-dlci 203
!
interface Serial1/0.2 point-to-point
 ip address 10.1.1.6 255.255.255.252
 ip router isis ! --- Step 4.7: Enable IS-IS for IP routing on Serial1/0.2.
 tag-switching ip ! --- Step 3.3: Enable MPLS on subinterface Serial1/0.2.
 frame-relay interface-dlci 201
!
interface Serial1/7
 bandwidth 64
 ip vrf forwarding vpn101 ! --- Step 1.5: Associate Serial1/7 with VRF vpn101.
 ip address 200.200.200.1 255.255.255.0
 clockrate 64000
!
```

```
! --- Step 4.1: Configure Integrated IS-IS as the Backbone IGP.
router isis


! --- Step 4.2: Configure London-PE with CLNS parameter NET
! --- of value 0x49.0001.0000.0000.0002.00.
 net 49.0001.0000.0000.0002.00


! --- Step 4.3: Configure IS-IS Level-1 routing globally on London-PE.
 is-type level-1


! --- Step 4.4: Use new style of TLVs to carry wider metric.
! --- 24 bits for interface metric and 32 bits for total path metric.
 metric-style wide
!
! --- Step 5.1: Configure MP-BGP for London-PE in AS 65000.
router bgp 65000
 no synchronization


! --- Step 5.2: Declare Brussels-PE as MP-IBGP neighbor.
 neighbor 223.0.0.1 remote-as 65000


! --- Step 5.3: Inform Brussels-PE to establish MP-IBGP session to Loop-
back0.
 neighbor 223.0.0.1 update-source Loopback0


! --- Step 5.4: Declare Paris-PE as MP-IBGP neighbor.
 neighbor 223.0.0.3 remote-as 65000


! --- Step 5.5: Inform Paris-PE to establish MP-IBGP session to Loopback0.
 neighbor 223.0.0.3 update-source Loopback0
 no auto-summary
 !
! --- Step 7.1: Enter the per-VRF BGP routing context for VRF vpn102.
 address-family ipv4 vrf vpn102


! --- Step 7.2: Redistribute directly connected interface FastEthernet0/0
! --- into the per-VRF BGP routing context for VRF vpn102.
 redistribute connected


! --- Step 7.3: Declare London-CE2 in AS 65022 as CE EBGP neighbor.
 neighbor 202.202.202.2 remote-as 65022
 neighbor 202.202.202.2 activate ! --- Step 7.4: Activate London-CE2.
 no auto-summary
 no synchronization
 exit-address-family
```

```
!
! --- Step 6.1: Enter the per-VRF BGP routing context for VRF vpn101.
 address-family ipv4 vrf vpn101

! --- Step 6.2: Redistribute directly connected interface Serial1/7
! --- into the per-VRF BGP routing context for VRF vpn101.
 redistribute connected

! --- Step 6.3: Declare London-CE1 in AS 65021 as CE EBGP neighbor.
 neighbor 200.200.200.2 remote-as 65021
 neighbor 200.200.200.2 activate ! --- Step 6.4: Activate London-CE1.
 no auto-summary
 no synchronization
 exit-address-family
 !
! --- Step 5.6: Enter the VPNv4 configuration mode to configure
! --- VPNv4 specific parameters on the respective MP-IBGP neighbors.
 address-family vpnv4
! --- Step 5.7: Activate Brussels-PE for VPNv4 route exchange.
 neighbor 223.0.0.1 activate
! --- Step 5.8: Activate Paris-PE for VPNv4 route exchange.
 neighbor 223.0.0.3 activate

! --- Step 5.9: Disable the default IBGP next-hop processing
! --- for VPNv4 route exchange.
 neighbor 223.0.0.1 next-hop-self
 neighbor 223.0.0.3 next-hop-self

! --- Step 5.10: Propagate extended BGP communities
! --- to the respective MP-IBGP neighbors.
 neighbor 223.0.0.1 send-community extended
 neighbor 223.0.0.3 send-community extended
 no auto-summary
 exit-address-family
```

Code Listing 9.7 illustrates the MPLS VPN configuration for London-PE. In the configuration:

- Steps 1.1 to 1.5 configure VRF table vpn101 and assign the specified VRF to the corresponding PE-CE interfaces (Serial1/7) that interconnects sites (London Site 1) belonging to vpn101.

- Steps 2.1 to 2.5 configure VRF table vpn102 and assign the specified VRF to the corresponding PE-CE interfaces (FastEthernet0/0) that interconnects sites (London Site 2) belonging to vpn102.

- Steps 3.1 to 3.3 enable MPLS.

- Steps 4.1 to 4.7 configure the backbone IGP (Integrated IS-IS).
- Steps 5.1 to 5.10 configure MP-BGP.
- Steps 6.1 to 6.4 configure the per-VRF PE-CE routing protocol (EBGP) for vpn101.
- Steps 7.1 to 7.4 configure the per-VRF PE-CE routing protocol (EBGP) for vpn102.

Code Listing 9.8:   MPLS VPN configuration for Paris-PE

```
hostname Paris-PE
!
ip vrf vpn101 ! --- Step 1.1: Create VRF vpn101 for VPN101.
 rd 65000:101 ! --- Step 1.2: Assign route distinguisher 65000:101 to VRF
vpn101.

! --- Step 1.3: Specify RT 65000:101 to be attached
! --- to every route from this VRF to MP-BGP.
 route-target export 65000:101

! --- Step 1.4: Specify RT 65000:101 to be used as an import filter.
! --- Only routes matching this RT are imported into VRF vpn101.
 route-target import 65000:101
!
ip cef ! --- Step 3.1: Enable Cisco Express Forwarding.
!
interface Loopback0
 ip address 223.0.0.3 255.255.255.255
 ip router isis ! --- Step 4.5: Enable IS-IS for IP routing on Loopback0.
!
interface FastEthernet0/0
 ip vrf forwarding vpn101 ! --- Step 1.5: Associate FastEthernet0/0 with
VRF vpn101.
 ip address 31.3.3.1 255.255.255.0
!
interface Serial1/0
 encapsulation frame-relay
 frame-relay lmi-type ansi
!
interface Serial1/0.1 point-to-point
 ip address 10.3.3.6 255.255.255.252
 ip router isis ! --- Step 4.6: Enable IS-IS for IP routing on Serial1/0.1.
 tag-switching ip ! --- Step 3.2: Enable MPLS on subinterface Serial1/0.1.
 frame-relay interface-dlci 301
```

```
!
interface Serial1/0.2 point-to-point
 ip address 10.2.2.6 255.255.255.252
 ip router isis ! --- Step 4.7: Enable IS-IS for IP routing on Serial1/0.2.
 tag-switching ip ! --- Step 3.3: Enable MPLS on subinterface Serial1/0.2.
 frame-relay interface-dlci 302
!
! --- Step 4.1: Configure Integrated IS-IS as the backbone IGP.
router isis

! --- Step 4.2: Configure Paris-PE with CLNS parameter NET
! --- of value 0x49.0001.0000.0000.0003.00.
 net 49.0001.0000.0000.0003.00

! --- Step 4.3: Configure IS-IS level 1 routing globally on Paris-PE.
 is-type level-1

! --- Step 4.4: Use new style of TLVs to carry wider metric.
! --- 24 bits for interface metric and 32 bits for total path metric.
 metric-style wide
!
! --- Step 5.1: Configure MP-BGP for Paris-PE in AS 65000.
router bgp 65000
 no synchronization

! --- Step 5.2: Declare Brussels-PE as MP-IBGP neighbor.
 neighbor 223.0.0.1 remote-as 65000

! --- Step 5.3: Inform Brussels-PE to establish MP-IBGP session to Loop-
back0.
 neighbor 223.0.0.1 update-source Loopback0

! --- Step 5.4: Declare London-PE as MP-IBGP neighbor.
 neighbor 223.0.0.2 remote-as 65000

! --- Step 5.5: Inform London-PE to establish MP-IBGP session to Loopback0.
 neighbor 223.0.0.2 update-source Loopback0
 no auto-summary
 !
! --- Step 6.1: Enter the per-VRF BGP routing context for VRF vpn101.
 address-family ipv4 vrf vpn101

! --- Step 6.2: Redistribute directly connected interface FastEthernet0/0
! --- into the per-VRF BGP routing context for VRF vpn101.
 redistribute connected
```

```
! --- Step 6.3: Declare Paris-CE1 in AS 65031 as CE EBGP neighbor.
neighbor 31.3.3.2 remote-as 65031
neighbor 31.3.3.2 activate --- Step 6.4: Activate Paris-CE1.
no auto-summary
no synchronization
exit-address-family
 !
! --- Step 5.6: Enter the VPNv4 configuration mode to configure
! --- VPNv4 specific parameters on the respective MP-IBGP neighbors.
address-family vpnv4
! --- Step 5.7: Activate Brussels-PE for VPNv4 route exchange.
neighbor 223.0.0.1 activate
! --- Step 5.8: Activate London-PE for VPNv4 route exchange.
neighbor 223.0.0.2 activate

! --- Step 5.9: Disable the default IBGP next-hop processing
! --- for VPNv4 route exchange.
neighbor 223.0.0.1 next-hop-self
neighbor 223.0.0.2 next-hop-self

! --- Step 5.10: Propagate extended BGP communities
! --- to the respective MP-IBGP neighbors.
neighbor 223.0.0.1 send-community extended
neighbor 223.0.0.2 send-community extended
no auto-summary
exit-address-family
```

Code Listing 9.8 illustrates the MPLS VPN configuration for Paris-PE. In the configuration:

- Steps 1.1 to 1.5 configure VRF table vpn101 and assign the specified VRF to the corresponding PE-CE interfaces (FastEthernet0/0) that interconnects sites (Paris Site 1) belonging to vpn101.
- Steps 3.1 to 3.3 enable MPLS.
- Steps 4.1 to 4.7 configure the backbone IGP (Integrated IS-IS).
- Steps 5.1 to 5.10 configure MP-BGP.
- Steps 6.1 to 6.4 configure the per-VRF PE-CE routing protocol (EBGP) for vpn101.

9.5.5   Monitoring MPLS VPN operation

In this chapter, for simplicity we monitor the MPLS VPN operation by examining the PE router's as well as the CE router's respective routing tables and deduce whether the overall connectivity criteria are strictly adhered to.

9.5.5.1 Verifying the IP Connectivity between the PE Routers. Code Listings 9.9 to 9.11 illustrate the IPv4 forwarding tables for the three PE routers. These routing tables are used to verify the IP connectivity between the PE routers and also determine whether the configured backbone IGP (Integrated IS-IS) is fully operational.

The routing table in Code Listing 9.9 indicates that the loopback addresses of London-PE (223.0.0.2) and Paris-PE (223.0.0.3) are both reachable from Brussels-PE. This implies the IP connectivity from Brussels-PE to the other two PE routers is functioning.

Code Listing 9.9:   IPv4 forwarding table for Brussels-PE

```
Brussels-PE#show ip route
<Output Omitted>
      223.0.0.0/32 is subnetted, 3 subnets
i L1    223.0.0.3 [115/20] via 10.3.3.6, Serial1/0.2
i L1    223.0.0.2 [115/20] via 10.1.1.6, Serial1/0.1
C       223.0.0.1 is directly connected, Loopback0
      10.0.0.0/30 is subnetted, 3 subnets
i L1    10.2.2.4  [115/20] via 10.1.1.6, Serial1/0.1
                  [115/20] via 10.3.3.6, Serial1/0.2
C       10.3.3.4 is directly connected, Serial1/0.2
C       10.1.1.4 is directly connected, Serial1/0.1
```

The routing table in Code Listing 9.10 indicates that the loopback addresses of Brussels-PE (223.0.0.1) and Paris-PE (223.0.0.3) are both reachable from London-PE. This implies the IP connectivity from London-PE to the other two PE routers is functioning.

Code Listing 9.10:   IPv4 forwarding table for London-PE

```
London-PE#show ip route
<Output Omitted>
      223.0.0.0/32 is subnetted, 3 subnets
i L1    223.0.0.3 [115/20] via 10.2.2.6, Serial1/0.1
C       223.0.0.2 is directly connected, Loopback0
i L1    223.0.0.1 [115/20] via 10.1.1.5, Serial1/0.2
      10.0.0.0/30 is subnetted, 3 subnets
i L1    10.3.3.4  [115/20] via 10.2.2.6, Serial1/0.1
                  [115/20] via 10.1.1.5, Serial1/0.2
C       10.2.2.4 is directly connected, Serial1/0.1
C       10.1.1.4 is directly connected, Serial1/0.2
```

The routing table in Code Listing 9.11 indicates that the loopback addresses of Brussels-PE (223.0.0.1) and London-PE (223.0.0.2) are both reachable from Paris-PE. This implies the IP connectivity from Paris-PE to the other two PE routers is functioning.

Code Listing 9.11:   IPv4 forwarding table for Paris-PE

```
Paris-PE#show ip route
<Output Omitted>
      223.0.0.0/32 is subnetted, 3 subnets
C        223.0.0.3 is directly connected, Loopback0
i L1     223.0.0.2 [115/20] via 10.2.2.5, Serial1/0.2
i L1     223.0.0.1 [115/20] via 10.3.3.5, Serial1/0.1
      10.0.0.0/30 is subnetted, 3 subnets
i L1     10.1.1.4  [115/20] via 10.2.2.5, Serial1/0.2
                   [115/20] via 10.3.3.5, Serial1/0.1
C        10.2.2.4 is directly connected, Serial1/0.2
C        10.3.3.4 is directly connected, Serial1/0.1
```

9.5.5.2 Verifying VRFs Configured in the PE Routers.

Code Listings 9.12 to 9.14 display the set of defined VRFs and associated interfaces for the three PE routers. The "show ip vrf detail" command is used to verify the VRFs configured in the PE routers.

The "show ip vrf detail" command in Code Listing 9.12 displays detailed information on the configured VRFs and their associated interfaces for Brussels-PE.

Code Listing 9.12:   Detailed VRF configuration for Brussels-PE

```
Brussels-PE#show ip vrf detail
VRF vpn101; default RD 65000:101
  Interfaces:
    Serial1/7
  Connected addresses are not in global routing table
  Export VPN route-target communities
    RT:65000:101
  Import VPN route-target communities
    RT:65000:101
  No import route-map
  No export route-map
VRF vpn102; default RD 65000:102
  Interfaces:
```

```
FastEthernet0/0
Connected addresses are not in global routing table
Export VPN route-target communities
  RT:65000:102
Import VPN route-target communities
  RT:65000:102
No import route-map
No export route-map
```

The "show ip vrf detail" command in Code Listing 9.13 displays detailed information on the configured VRFs and their associated interfaces for London-PE.

Code Listing 9.13: Detailed VRF configuration for London-PE

```
London-PE#show ip vrf detail
VRF vpn101; default RD 65000:101
  Interfaces:
    Serial1/7
  Connected addresses are not in global routing table
  Export VPN route-target communities
    RT:65000:101
  Import VPN route-target communities
    RT:65000:101
  No import route-map
  No export route-map
VRF vpn102; default RD 65000:102
  Interfaces:
    FastEthernet0/0
  Connected addresses are not in global routing table
  Export VPN route-target communities
    RT:65000:102
  Import VPN route-target communities
    RT:65000:102
  No import route-map
  No export route-map
```

The "show ip vrf detail" command in Code Listing 9.14 displays detailed information on the configured VRFs and their associated interfaces for Paris-PE.

Code Listing 9.14:   Detailed VRF configuration for Paris-PE

```
Paris-PE#show ip vrf detail
VRF vpn101; default RD 65000:101
  Interfaces:
    FastEthernet0/0
  Connected addresses are not in global routing table
  Export VPN route-target communities
    RT:65000:101
  Import VPN route-target communities
    RT:65000:101
  No import route-map
  No export route-map
```

9.5.5.3 Per-VRF Label Forwarding Information Base of Brussels-PE. Code Listings 9.15 to 9.18 display Brussels-PE's label-forwarding information for locally and remotely advertised VRF routes. The "show mpls forwarding-table" command is used to monitor the VPN labels that are assigned by Brussels-PE for VRF routes advertised from the local CE routers, and the "show mpls forwarding-table vrf" command is used to monitor the VPN labels that are assigned by other PE routers for VRF routes advertised from remote CE routers.

Code Listing 9.15 indicates that:

- Destination prefix 9.0.0.0/8 received from FastEthernet0/0 (interconnecting Brussels-CE2 in vpn102) is assigned a local VPN label of value 20 by Brussels-PE. The outgoing label is "Untagged," meaning there is no label for the destination prefix from the next-hop.

- Destination prefix 8.8.8.0/24 received from Serial1/7 (interconnecting Brussels-CE1 in vpn101) is assigned a local VPN label of value 22 by Brussels-PE. The outgoing label is "Untagged."

- Local prefix 100.100.100.0/24 (from directly connected interface FastEthernet0/0) is assigned a local VPN label of value 19 by Brussels-PE. The outgoing label is "Aggregate," meaning Brussels-PE would need to perform an IP (FIB) lookup for this prefix besides the regular label lookup for determining the outgoing interface (see section 8.6.1).

- Local prefix 130.130.130.0/24 (from directly connected interface Serial1/7) is assigned a local VPN label of value 21 by Brussels-PE. The outgoing label is "Aggregate."

Code Listing 9.15:   Brussels-PE LFIB

```
Brussels-PE#show mpls forwarding-table

Local   Outgoing    Prefix         Bytes tag  Outgoing    Next Hop
tag     tag or VC   or Tunnel Id   switched   interface
<Output Omitted>
19      Aggregate   100.100.100.0/24[V]    4890
20      Untagged    9.0.0.0/8[V]   0                    Fa0/0       100.100.100.2
21      Aggregate   130.130.130.0/24[V]    0
22      Untagged    8.8.8.0/24[V]  2832                 Se1/7       point2point
```

Code Listing 9.16 indicates that the VPN label assigned to prefix 31.5.5.4/30 (in vpn101) by next-hop (Paris-PE) is 21.

Code Listing 9.16:   Brussels-PE Per-VRF LFIB for vpn101 with prefix 31.5.5.4/30

```
Brussels-PE#show mpls forwarding-table vrf vpn101 31.5.5.5 detail

Local   Outgoing    Prefix         Bytes tag  Outgoing    Next Hop
tag     tag or VC   or Tunnel Id   switched   interface
None    21          31.5.5.4/30    0          Se1/0.2     point2point
   <Output Omitted>
```

Code Listing 9.17 indicates that the VPN label assigned to prefix 9.9.9.9/32 (in vpn101) by next-hop (London-PE) is 22.

Code Listing 9.17:   Brussels-PE Per-VRF LFIB for vpn101 with prefix 9.9.9.9/32

```
Brussels-PE#show mpls forwarding-table vrf vpn101 9.9.9.9 detail

Local   Outgoing    Prefix      .  Bytes tag  Outgoing    Next Hop
tag     tag or VC   or Tunnel Id   switched   interface
None    22          9.9.9.9/32     0          Se1/0.1     point2point
   <Output Omitted>
```

Code Listing 9.18 indicates that the VPN label assigned to prefix 130.130.0.0/16 (in vpn102) by next-hop (London-PE) is 20.

Code Listing 9.18:   Brussels-PE Per-VRF LFIB for vpn102 with prefix 130.130.0.0/16

```
Brussels-PE#show mpls forwarding-table vrf vpn102 130.130.130.1 detail
Local   Outgoing    Prefix          Bytes tag   Outgoing    Next Hop
tag     tag or VC   or Tunnel Id    switched    interface
None    20          130.130.0.0/16  0           Se1/0.1     point2point
  <Output Omitted>
```

9.5.5.4 Per-VRF label forwarding information base of London-PE. Code Listings 9.19 to 9.22 display London-PE's label forwarding information for locally and remotely advertised VRF routes. The "show mpls forwarding-table" command is used to monitor the VPN labels that are assigned by London-PE for VRF routes advertised from the local CE routers, and the "show mpls forwarding-table vrf" command is used to monitor the VPN labels that are assigned by other PE routers for VRF routes advertised from remote CE routers.

Code Listing 9.19 indicates that:

- Destination prefix 130.130.0.0/16 received from FastEthernet0/0 (interconnecting London-CE2 in vpn102) is assigned a local VPN label of value 20 by London-PE. The outgoing label is "Untagged."
- Destination prefix 9.9.9.9/32 received from Serial1/7 (interconnecting London-CE1 in vpn101) is assigned a local VPN label of value 22 by London-PE. The outgoing label is "Untagged."
- Local prefix 202.202.202.0/24 (from directly connected interface FastEthernet0/0) is assigned a local VPN label of value 19 by London-PE. The outgoing label is "Aggregate."
- Local prefix 200.200.200.0/24 (from directly connected interface Serial1/7) is assigned a local VPN label of value 21 by London-PE. The outgoing label is "Aggregate."

Code Listing 9.19:   London-PE LFIB

```
London-PE#show mpls forwarding-table
Local   Outgoing    Prefix           Bytes tag   Outgoing    Next Hop
tag     tag or VC   or Tunnel Id     switched    interface
<Output Omitted>
19      Aggregate   202.202.202.0/24[V]   1272
20      Untagged    130.130.0.0/16[V] 4442    Fa0/0     202.202.202.2
21      Aggregate   200.200.200.0/24[V]   2544
22      Untagged    9.9.9.9/32[V]    3170     Se1/7     point2point
```

Code Listing 9.20 indicates that the VPN label assigned to prefix 8.8.8.0/24 (in vpn101) by next-hop (Brussels-PE) is 22.

Code Listing 9.20:   London-PE Per-VRF LFIB for vpn101 with prefix 8.8.8.0/24

```
London-PE#show mpls forwarding-table vrf vpn101 8.8.8.8 detail
Local   Outgoing    Prefix        Bytes tag   Outgoing      Next Hop
tag     tag or VC   or Tunnel Id  switched    interface
None    22          8.8.8.0/24    0           Se1/0.2       point2point
  <Output Omitted>
```

Code Listing 9.21 indicates that the VPN label assigned to prefix 31.5.5.4/30 (in vpn101) by next-hop (Paris-PE) is 21.

Code Listing 9.21:   London-PE Per-VRF LFIB for vpn101 with prefix 31.5.5.4/30

```
London-PE#show mpls forwarding-table vrf vpn101 31.5.5.5 detail
Local   Outgoing    Prefix        Bytes tag   Outgoing      Next Hop
tag     tag or VC   or Tunnel Id  switched    interface
None    21          31.5.5.4/30   0           Se1/0.1       point2point
  <Output Omitted>
```

Code Listing 9.22 indicates that the VPN label assigned to prefix 9.0.0.0/8 (in vpn102) by next-hop (Brussels-PE) is 20.

Code Listing 9.22:   London-PE Per-VRF LFIB for vpn102 with prefix 9.0.0.0/8

```
London-PE#show mpls forwarding-table vrf vpn102 9.9.9.9 detail
Local   Outgoing    Prefix        Bytes tag   Outgoing      Next Hop
tag     tag or VC   or Tunnel Id  switched    interface
None    20          9.0.0.0/8     0           Se1/0.2       point2point
  <Output Omitted>
```

9.5.5.5 Per-VRF label forwarding information base of Paris-PE. Code Listings 9.23 to 9.25 display Paris-PE's label-forwarding information for locally and remotely advertised VRF routes. The "show mpls forwarding-table" command is used to monitor the VPN labels that are assigned by Paris-PE for VRF routes advertised from the local CE routers, and the "show mpls forwarding-table vrf" command is used to monitor the VPN labels that are assigned by other PE routers for VRF routes advertised from remote CE routers.

Code Listing 9.23 indicates that:

- Destination prefix 31.5.5.4/30 received from FastEthernet0/0 (interconnecting Paris-CE1 in vpn101) is assigned a local VPN label of value 21 by Paris-PE. The outgoing label is "Untagged."

- Local prefix 31.3.3.0/24 (from directly connected interface FastEthernet0/0) is assigned a local VPN label of value 20 by Paris-PE. The outgoing label is "Aggregate."

Code Listing 9.23:   Paris-PE LFIB

```
Paris-PE#show mpls forwarding-table
Local    Outgoing     Prefix          Bytes tag   Outgoing    Next Hop
tag      tag or VC    or Tunnel Id    switched    interface
<Output Omitted>
20       Aggregate    31.3.3.0/24[V]  9176
21       Untagged     31.5.5.4/30[V]  0           Fa0/0       31.3.3.2
```

Code Listing 9.24 indicates that the VPN label assigned to prefix 8.8.8.0/24 (in vpn101) by next-hop (Brussels-PE) is 22.

Code Listing 9.24:   Paris-PE Per-VRF LFIB for vpn101 with prefix 8.8.8.0/24

```
Paris-PE#show mpls forwarding-table vrf vpn101 8.8.8.8 detail
Local    Outgoing     Prefix          Bytes tag   Outgoing    Next Hop
tag      tag or VC    or Tunnel Id    switched    interface
None     22           8.8.8.0/24      0           Se1/0.1     point2point
   <Output Omitted>
```

Code Listing 9.25 indicates that the VPN label assigned to prefix 9.9.9.9/32 (in vpn101) by next-hop (London-PE) is 22.

Code Listing 9.25:   Paris-PE Per-VRF LFIB for vpn101 with prefix 9.9.9.9/32

```
Paris-PE#show mpls forwarding-table vrf vpn101 9.9.9.9 detail
Local    Outgoing     Prefix          Bytes tag   Outgoing    Next Hop
tag      tag or VC    or Tunnel Id    switched    interface
None     22           9.9.9.9/32      0           Se1/0.2     point2point
   <Output Omitted>
```

9.5.5.6 Monitoring the PE routers for VRF vpn101. Code Listings 9.26 to 9.31 display the per-VRF BGP VPNv4 table and per-VRF routing table on the three PE routers for vpn101. The "show ip bgp vpnv4 vrf" command is used to display the per-VRF BGP VPNv4 table, and the "show ip route vrf" command is used to display the per-VRF routing table. Together these two commands are used to verify whether the configured MP-BGP is fully operational for propagating routes among sites belonging to vpn101 and also whether the PE-CE routing protocol (EBGP) configured between the respective PE and CE routers for vpn101 is functional.

The "show ip bgp vpnv4 vrf vpn101" command in Code Listing 9.26 displays Brussels-PE per-VRF BGP VPNv4 information associated with vpn101. Prefix 8.8.8.0/24 is propagated via EBGP from local CE router Brussels-CE1, prefix 9.9.9.9/32 along with 200.200.200.0/24 is propagated via MP-IBGP from London-PE, and prefix 31.5.5.4/30 along with 31.3.3.0/24 is propagated via MP-IBGP from Paris-PE. As such, these network prefixes meet the connectivity requirement defined in section 9.5.2.1 for vpn101.

Code Listing 9.26: Brussels-PE per-VRF BGP VPNv4 table for vpn101

```
Brussels-PE#show ip bgp vpnv4 vrf vpn101
BGP table version is 24, local router ID is 223.0.0.1
Status codes: s suppressed, d damped, h history, * valid, > best, i -
internal
Origin codes:  i - IGP, e - EGP, ? - incomplete

   Network          Next Hop          Metric LocPrf Weight Path
Route Distinguisher: 65000:101 (default for vrf vpn101)
*> 8.8.8.0/24       130.130.130.2     0            0 65011 i
*>i9.9.9.9/32       223.0.0.2         0      100   0 65021 i
*>i31.3.3.0/24      223.0.0.3         0      100   0 ?
*>i31.5.5.4/30      223.0.0.3         0      100   0 65031 i
*> 130.130.130.0/24 0.0.0.0           0            32768 ?
*>i200.200.200.0    223.0.0.2         0      100   0 ?
```

The "show ip route vrf vpn101" command in Code Listing 9.27 displays Brussels-PE per-VRF routing table associated with vpn101. All the BGP VPNv4 routes listed in Code Listing 9.26 are imported and installed into the VRF routing table so that they can be further distributed to the associated local CE router, in this case Brussels-CE1.

Code Listing 9.27: Brussels-PE per-VRF Routing table for vpn101

```
Brussels-PE#show ip route vrf vpn101
<Output Omitted>
```

```
B     200.200.200.0/24 [200/0] via 223.0.0.2, 02:25:24
      8.0.0.0/24 is subnetted, 1 subnets
B         8.8.8.0 [20/0] via 130.130.130.2, 02:28:38
      9.0.0.0/32 is subnetted, 1 subnets
B         9.9.9.9 [200/0] via 223.0.0.2, 02:24:54
      130.130.0.0/24 is subnetted, 1 subnets
C         130.130.130.0 is directly connected, Serial1/7
      31.0.0.0/8 is variably subnetted, 2 subnets, 2 masks
B         31.5.5.4/30 [200/0] via 223.0.0.3, 02:33:09
B         31.3.3.0/24 [200/0] via 223.0.0.3, 02:33:24
```

The "show ip bgp vpnv4 vrf vpn101" command in Code Listing 9.28 displays London-PE per-VRF BGP VPNv4 information associated with vpn101. Prefix 9.9.9.9/32 is propagated via EBGP from local CE router London-CE1, prefix 8.8.8.0/24 along with 130.130.130.0/24 is propagated via MP-IBGP from Brussels-PE, and prefix 31.5.5.4/30 along with 31.3.3.0/24 is propagated via MP-IBGP from Paris-PE. As such, these network prefixes meet the connectivity requirement defined in section 9.5.2.1 for vpn101.

Code Listing 9.28:   London-PE per-VRF BGP VPNv4 table for vpn101

```
London-PE#show ip bgp vpnv4 vrf vpn101
BGP table version is 24, local router ID is 223.0.0.2
Status codes: s suppressed, d damped, h history, * valid, > best, i -
internal
Origin codes: i - IGP, e - EGP, ? - incomplete

   Network          Next Hop          Metric LocPrf Weight Path
Route Distinguisher: 65000:101 (default for vrf vpn101)
*>i8.8.8.0/24        223.0.0.1           0      100      0 65011 i
*> 9.9.9.9/32        200.200.200.2       0               0 65021 i
*>i31.3.3.0/24       223.0.0.3           0      100      0 ?
*>i31.5.5.4/30       223.0.0.3           0      100      0 65031 i
*>i130.130.130.0/24 223.0.0.1           0      100      0 ?
*> 200.200.200.0     0.0.0.0             0           32768 ?
```

The "show ip route vrf vpn101" command in Code Listing 9.29 displays London-PE per-VRF routing table associated with vpn101. All the BGP VPNv4 routes listed in Code Listing 9.28 are imported and installed into the VRF routing table so that they can be further distributed to the associated local CE router, in this case London-CE1.

Code Listing 9.29:   London-PE per-VRF Routing table for vpn101

```
London-PE#show ip route vrf vpn101
<Output Omitted>

C     200.200.200.0/24 is directly connected, Serial1/7
      8.0.0.0/24 is subnetted, 1 subnets
B        8.8.8.0 [200/0] via 223.0.0.1, 02:49:18
      9.0.0.0/32 is subnetted, 1 subnets
B        9.9.9.9 [20/0] via 200.200.200.2, 02:45:53
      130.130.0.0/24 is subnetted, 1 subnets
B        130.130.130.0 [200/0] via 223.0.0.1, 02:49:48
      31.0.0.0/8 is variably subnetted, 2 subnets, 2 masks
B        31.5.5.4/30 [200/0] via 223.0.0.3, 02:54:04
B        31.3.3.0/24 [200/0] via 223.0.0.3, 02:54:19
```

The "show ip bgp vpnv4 vrf vpn101" command in Code Listing 9.30 displays Paris-PE per-VRF BGP VPNv4 information associated with vpn101. Prefix 31.5.5.4/30 is propagated via EBGP from local CE router Paris-CE1, prefix 8.8.8.0/24 along with 130.130.130.0/24 is propagated via MP-IBGP from Brussels-PE, and prefix 9.9.9.9/32 along with 200.200.200.0/24 is propagated via MP-IBGP from London-PE. As such, these network prefixes meet the connectivity requirement defined in section 9.5.2.1 for vpn101.

Code Listing 9.30:   Paris-PE per-VRF BGP VPNv4 table for vpn101

```
Paris-PE#show ip bgp vpnv4 vrf vpn101
BGP table version is 16, local router ID is 223.0.0.3
Status codes: s suppressed, d damped, h history, * valid, > best, i -
internal
Origin codes: i - IGP, e - EGP, ? - incomplete

   Network          Next Hop         Metric LocPrf Weight Path
Route Distinguisher: 65000:101 (default for vrf vpn101)
*>i8.8.8.0/24       223.0.0.1            0    100      0 65011 i
*>i9.9.9.9/32       223.0.0.2            0    100      0 65021 i
*>  31.3.3.0/24     0.0.0.0              0             32768 ?
*>  31.5.5.4/30     31.3.3.2            0             0 65031 i
*>i130.130.130.0/24 223.0.0.1            0    100      0 ?
*>i200.200.200.0    223.0.0.2            0    100      0 ?
```

The "show ip route vrf vpn101" command in Code Listing 9.31 displays Paris-PE per-VRF routing table associated with vpn101. All the BGP VPNv4 routes listed in Code Listing 9.30 are imported and installed into the VRF routing table so that they can be further distributed to the associated local CE router, in this case, Paris-CE1.

Code Listing 9.31: Paris-PE per-VRF Routing table for vpn101

```
Paris-PE#show ip route vrf vpn101
<Output Omitted>

B    200.200.200.0/24 [200/0] via 223.0.0.2, 01:39:31
     8.0.0.0/24 is subnetted, 1 subnets
B       8.8.8.0 [200/0] via 223.0.0.1, 01:42:31
     9.0.0.0/32 is subnetted, 1 subnets
B       9.9.9.9 [200/0] via 223.0.0.2, 01:39:01
     130.130.0.0/24 is subnetted, 1 subnets
B       130.130.130.0 [200/0] via 223.0.0.1, 01:43:01
     31.0.0.0/8 is variably subnetted, 2 subnets, 2 masks
B       31.5.5.4/30 [20/0] via 31.3.3.2, 01:47:27
C       31.3.3.0/24 is directly connected, FastEthernet0/0
```

9.5.5.7 Monitoring the CE routers in vpn101. Code Listings 9.32 to 9.37 display the BGP table and IPv4 routing table on the three CE routers belonging to vpn101. The "show ip bgp" command is used to display the regular BGP table, and the "show ip route" command is used to display the traditional IP routing table. Together these two commands are used to confirm that the PE-CE routing protocol (EBGP) configured between the respective PE and CE routers for vpn101 is truly functioning.

The "show ip bgp" command in Code Listing 9.32 displays all the BGP routes received from Brussels-PE. These routes belong to vpn101 and originate from AS 65021, AS 65031, or AS 65000.

Code Listing 9.32: Brussels-CE1 BGP table

```
Brussels-CE1#show ip bgp
BGP table version is 11, local router ID is 8.8.8.8
Status codes: s suppressed, d damped, h history, * valid, > best, i -
internal
Origin codes: i - IGP, e - EGP, ? - incomplete

   Network          Next Hop          Metric LocPrf Weight Path
<Output Omitted>
*> 9.9.9.9/32       130.130.130.1                   0 65000 65021 i
*> 31.3.3.0/24      130.130.130.1                   0 65000 ?
*> 31.5.5.4/30      130.130.130.1                   0 65000 65031 i
<Output Omitted>
*> 200.200.200.0    130.130.130.1                   0 65000 ?
```

First, routes that originate from AS 65021 (in this case only prefix 9.9.9.9/32) are advertised from London-CE1 to London-PE, which in turn propagate them across the MPLS VPN backbone to Brussels-PE. Next, routes that originate from AS 65031 (in this case only prefix 31.5.5.4/30) are advertised from Paris-CE1 to Paris-PE, which in turn propagate them across the MPLS VPN backbone to Brussels-PE.

Last but not least, routes that originate from AS 65000 (in this case prefixes 200.200.200.0/24 and 31.3.3.0/24) are derived from directly connected interfaces interconnecting London-CE1 and Paris-CE1. These routes are redistributed into their respective per-VRF BGP routing context and are propagated by MP-BGP to Brussels-PE. As such, the network prefixes belonging to London Site 1 and Paris Site 1 are all reachable from Brussels Site 1, satisfying the connectivity criteria for vpn101.

The "show ip route" command in Code Listing 9.33 displays the IP routing table for Brussels-CE1 belonging to vpn101. All the BGP routes listed in Code Listing 9.32 are installed into the IP routing table, attesting that the PE-CE routing protocol (EBGP) is fully operational.

Code Listing 9.33:   Brussels-CE1 IPv4 Forwarding Table

```
Brussels-CE1#show ip route
<Output Omitted>

B    200.200.200.0/24 [20/0] via 130.130.130.1, 05:10:08
     8.0.0.0/24 is subnetted, 1 subnets
C       8.8.8.0 is directly connected, Loopback0
     9.0.0.0/32 is subnetted, 1 subnets
B       9.9.9.9 [20/0] via 130.130.130.1, 05:09:38
     130.130.0.0/24 is subnetted, 1 subnets
C       130.130.130.0 is directly connected, Serial1/0
     31.0.0.0/8 is variably subnetted, 2 subnets, 2 masks
B       31.5.5.4/30 [20/0] via 130.130.130.1, 00:04:26
B       31.3.3.0/24 [20/0] via 130.130.130.1, 00:04:56
```

The "show ip bgp" command in Code Listing 9.34 displays all the BGP routes received from London-PE. These routes belong to vpn101 and originate from AS 65011, AS 65031, or AS 65000.

Code Listing 9.34:   London-CE1 BGP table

```
London-CE1#show ip bgp
BGP table version is 11, local router ID is 9.9.9.9
Status codes:  s suppressed, d damped, h history, * valid, > best, i - internal
Origin codes: i - IGP, e - EGP, ? - incomplete
```

```
        Network              Next Hop              Metric LocPrf Weight  Path
*>  8.8.8.0/24           200.200.200.1                     0 65000 65011 i
<Output Omitted>
*>  31.3.3.0/24          200.200.200.1                       0 65000 ?
*>  31.5.5.4/30          200.200.200.1                       0 65000 65031 i
*>  130.130.130.0/24  200.200.200.1                          0 65000 ?
<Output Omitted>
```

First, routes that originate from AS 65011 (in this case only prefix 8.8.8.0/24) are advertised from Brussels-CE1 to Brussels-PE, which in turn propagate them across the MPLS VPN backbone to London-PE. Next, routes that originate from AS 65031 (in this case only prefix 31.5.5.4/30) are advertised from Paris-CE1 to Paris-PE, which in turn propagate them across the MPLS VPN backbone to London-PE.

Last but not least, routes that originate from AS 65000 (in this case prefixes 130.130.130.0/24 and 31.3.3.0/24) are derived from directly connected interfaces interconnecting Brussels-CE1 and Paris-CE1. These routes are redistributed into their respective per-VRF BGP routing context and are propagated by MP-BGP to London-PE. As such, the network prefixes belonging to Brussels Site 1 and Paris Site 1 are all reachable from London Site 1, satisfying the connectivity criteria for vpn101.

The "show ip route" command in Code Listing 9.35 displays the IP routing table for London-CE1 belonging to vpn101. All the BGP routes listed in Code Listing 9.34 are installed into the IP routing table, attesting that the PE-CE routing protocol (EBGP) is fully operational.

Code Listing 9.35:   London-CE1 IPv4 Forwarding Table

```
London-CE1#show ip route
<Output Omitted>

C    200.200.200.0/24 is directly connected, Serial1/0
     8.0.0.0/24 is subnetted, 1 subnets
B       8.8.8.0 [20/0] via 200.200.200.1, 05:08:48
     9.0.0.0/32 is subnetted, 1 subnets
C       9.9.9.9 is directly connected, Loopback0
     130.130.0.0/24 is subnetted, 1 subnets
B       130.130.130.0 [20/0] via 200.200.200.1, 05:08:48
     31.0.0.0/8 is variably subnetted, 2 subnets, 2 masks
B       31.5.5.4/30 [20/0] via 200.200.200.1, 00:03:16
B       31.3.3.0/24 [20/0] via 200.200.200.1, 00:03:44
```

The "show ip bgp" command in Code Listing 9.36 displays all the BGP routes received from Paris-PE. These routes belong to vpn101 and originate from AS 65011, AS 65021, or AS 65000.

Code Listing 9.36:   Paris-CE1 BGP table

```
Paris-CE1#show ip bgp
BGP table version is 7, local router ID is 31.5.5.5
Status codes: s suppressed, d damped, h history, * valid, > best, i - internal
Origin codes: i - IGP, e - EGP, ? - incomplete

    Network            Next Hop          Metric LocPrf Weight Path
*>  8.8.8.0/24         31.3.3.1                        0 65000 65011 i
*>  9.9.9.9/32         31.3.3.1                        0 65000 65021 i
<Output Omitted>
*>  130.130.130.0/24 31.3.3.1                          0 65000 ?
*>  200.200.200.0      31.3.3.1                        0 65000 ?
```

First, routes that originate from AS 65011 (in this case only prefix 8.8.8.0/24) are advertised from Brussels-CE1 to Brussels-PE, which in turn propagate them across the MPLS VPN backbone to Paris-PE. Next, routes that originate from AS 65021 (in this case only prefix 9.9.9.9/32) are advertised from London-CE1 to London-PE, which in turn propagate them across the MPLS VPN backbone to Paris-PE.

Last but not least, routes that originate from AS 65000 (in this case prefixes 130.130.130.0/24 and 200.200.200.0/24) are derived from directly connected interfaces interconnecting Brussels-CE1 and London-CE1. These routes are redistributed into their respective per-VRF BGP routing context and are propagated by MP-BGP to Paris-PE. As such, the network prefixes belonging to Brussels Site 1 and London Site 1 are all reachable from Paris Site 1, satisfying the connectivity criteria for vpn101.

The "show ip route" command in Code Listing 9.37 displays the IP routing table for Paris-CE1 belonging to vpn101. All the BGP routes listed in Code Listing 9.36 are installed into the IP routing table, attesting that the PE-CE routing protocol (EBGP) is fully operational.

Code Listing 9.37:   Paris-CE1 IPv4 forwarding table

```
Paris-CE1#show ip route
<Output Omitted>

B    200.200.200.0/24 [20/0] via 31.3.3.1, 03:38:17
     8.0.0.0/24 is subnetted, 1 subnets
B       8.8.8.0 [20/0] via 31.3.3.1, 03:41:15
     9.0.0.0/32 is subnetted, 1 subnets
B       9.9.9.9 [20/0] via 31.3.3.1, 03:37:49
     130.130.0.0/24 is subnetted, 1 subnets
```

```
B        130.130.130.0 [20/0] via 31.3.3.1, 03:41:42
         31.0.0.0/8 is variably subnetted, 2 subnets, 2 masks
C        31.5.5.4/30 is directly connected, Loopback0
C        31.3.3.0/24 is directly connected, Ethernet0/0
```

9.5.5.8 Monitoring the PE routers for VRF vpn102. Code Listings 9.38 to 9.41 display the per-VRF BGP VPNv4 table and per-VRF routing table on the three PE routers for vpn102. The "show ip bgp vpnv4 vrf" command is used to display the per-VRF BGP VPNv4 table, and the "show ip route vrf" command is used to display the per-VRF routing table. Together these two commands are used to verify whether the configured MP-BGP is fully operational for propagating routes among sites belonging to vpn102 and also whether the PE-CE routing protocol (EBGP) configured between the respective PE and CE routers for vpn102 is functional.

The "show ip bgp vpnv4 vrf vpn102" command in Code Listing 9.38 displays Brussels-PE per-VRF BGP VPNv4 information associated with vpn102. Prefix 9.0.0.0/8 is propagated via EBGP from local CE router Brussels-CE2, and prefix 130.130.0.0/16 along with 202.202.202.0/24 is propagated via MP-IBGP from London-PE. As such, these network prefixes meet the connectivity requirement defined in section 9.5.2.1 for vpn102.

Code Listing 9.38:   Brussels-PE Per-VRF BGP VPNv4 table for vpn102

```
Brussels-PE#show ip bgp vpnv4 vrf vpn102
BGP table version is 24, local router ID is 223.0.0.1
Status codes: s suppressed, d damped, h history, * valid, > best, i -
internal
Origin codes: i - IGP, e - EGP, ? - incomplete


   Network          Next Hop          Metric LocPrf Weight Path
Route Distinguisher: 65000:102 (default for vrf vpn102)
*> 9.0.0.0          100.100.100.2     0              0 65012 i
*> 100.100.100.0/24 0.0.0.0           0          32768 ?
*>i130.130.0.0      223.0.0.2         0      100     0 65022 i
*>i202.202.202.0    223.0.0.2         0      100     0 ?
```

The "show ip route vrf vpn102" command in Code Listing 9.39 displays Brussels-PE per-VRF routing table associated with vpn102. All the BGP VPNv4 routes listed in Code Listing 9.38 are imported and installed into the VRF routing table so that they can be further distributed to the associated local CE router, in this case Brussels-CE2.

Code Listing 9.39:   Brussels-PE per-VRF routing table for vpn102

```
Brussels-PE#show ip route vrf vpn102
<Output Omitted>

B       202.202.202.0/24 [200/0] via 223.0.0.2, 02:27:03
        100.0.0.0/24 is subnetted, 1 subnets
C          100.100.100.0 is directly connected, FastEthernet0/0
B       9.0.0.0/8 [20/0] via 100.100.100.2, 02:30:17
B       130.130.0.0/16 [200/0] via 223.0.0.2, 02:26:33
```

The "show ip bgp vpnv4 vrf vpn102" command in Code Listing 9.40 displays London-PE per-VRF BGP VPNv4 information associated with vpn102. Prefix 130.130.0.0/24 is propagated via EBGP from local CE router London-CE2, and prefix 9.0.0.0/8 along with 100.100.100.0/24 is propagated via MP-IBGP from Brussels-PE. As such, these network prefixes meet the connectivity requirement defined in section 9.5.2.1 for vpn102.

Code Listing 9.40:   London-PE per-VRF BGP VPNv4 table for vpn102

```
London-PE#show ip bgp vpnv4 vrf vpn102
BGP table version is 24, local router ID is 223.0.0.2
Status codes: s suppressed, d damped, h history, * valid, > best, i -
internal
Origin codes: i - IGP, e - EGP, ? - incomplete

    Network          Next Hop         Metric LocPrf Weight Path
Route Distinguisher: 65000:102 (default for vrf vpn102)
*>i9.0.0.0          223.0.0.1            0    100      0 65012 i
*>i100.100.100.0/24  223.0.0.1           0    100      0 ?
*> 130.130.0.0       202.202.202.2        0             0 65022 i
*> 202.202.202.0     0.0.0.0              0         32768 ?
```

The "show ip route vrf vpn102" command in Code Listing 9.41 displays London-PE per-VRF routing table associated with vpn102. All the BGP VPNv4 routes listed in Code Listing 9.40 are imported and installed into the VRF routing table so that they can be further distributed to the associated local CE router, in this case London-CE2.

Code Listing 9.41:   London-PE per-VRF routing table for vpn102

```
London-PE#show ip route vrf vpn102
<Output Omitted>


C     202.202.202.0/24 is directly connected, FastEthernet0/0
      100.0.0.0/24 is subnetted, 1 subnets
B        100.100.100.0 [200/0] via 223.0.0.1, 02:51:40
B        9.0.0.0/8 [200/0] via 223.0.0.1, 02:50:55
B        130.130.0.0/16 [20/0] via 202.202.202.2, 02:47:36
```

9.5.5.9   Monitoring the CE routers in vpn102.   Code Listings 9.42 to 9.45 display the BGP table and IPv4 routing table on the two CE routers belonging to vpn102. The "show ip bgp" command is used to display the regular BGP table, and the "show ip route" command is used to display the traditional IP routing table. Together these two commands are used to confirm that the PE-CE routing protocol (EBGP) configured between the respective PE and CE routers for vpn102 is truly functioning.

The "show ip bgp" command in Code Listing 9.42 displays all the BGP routes received from Brussels-PE. These routes belong to vpn102 and originate either from AS 65022 or AS 65000. Routes that originate from AS 65022 (in this case only prefix 130.130.0.0/16) are advertised from London-CE2 to London-PE, which in turn propagate them across the MPLS VPN backbone to Brussels-PE. The route that originates from AS 65000 (in this case only prefix 202.202.202.0/24) is derived from the directly connected interface interconnecting London-CE2. This route is redistributed into its respective per-VRF BGP routing context and is propagated by MP-BGP to Brussels-PE. As such, the network prefixes belonging to London Site 2 are all reachable from Brussels Site 2, satisfying the connectivity criteria for vpn102.

Code Listing 9.42:   Brussels-CE2 BGP table

```
Brussels-CE2#show ip bgp
BGP table version is 5, local router ID is 9.9.9.9
Status codes: s suppressed, d damped, h history, * valid, > best, i -
internal
Origin codes: i - IGP, e - EGP, ? - incomplete

    Network          Next Hop          Metric LocPrf Weight Path
*> 9.0.0.0           0.0.0.0                0         32768 i
*> 100.100.100.0/24  100.100.100.1          0             0 65000 ?
*> 130.130.0.0       100.100.100.1          0               65000 65022 i
*> 202.202.202.0     100.100.100.1          0               65000 ?
```

The "show ip route" command in Code Listing 9.43 displays the IP routing table for Brussels-CE2 belonging to vpn102. All the BGP routes listed in Code Listing 9.42 are installed into the IP routing table, attesting that the PE-CE routing protocol (EBGP) is fully operational.

Code Listing 9.43:   Brussels-CE2 IPv4 forwarding table

```
Brussels-CE2#show ip route
<Output Omitted>


B    202.202.202.0/24 [20/0] via 100.100.100.1, 04:55:54
     100.0.0.0/24 is subnetted, 1 subnets
C        100.100.100.0 is directly connected, Ethernet0/0
C    9.0.0.0/8 is directly connected, Loopback0
B    130.130.0.0/16 [20/0] via 100.100.100.1, 04:55:28
```

The "show ip bgp" command in Code Listing 9.44 displays all the BGP routes received from London-PE. These routes belong to vpn102 and originate either from AS 65012, or AS 65000. Routes that originate from AS 65012 (in this case only prefix 9.0.0.0/8) are advertised from Brussels-CE2 to Brussels-PE, which in turn propagate them across the MPLS VPN backbone to London-PE. The route that originates from AS 65000 (in this case only prefix 100.100.100.0/24) is derived from the directly connected interface interconnecting Brussels-CE2. This route is redistributed into its respective per-VRF BGP routing context and is propagated by MP-BGP to London-PE. As such, the network prefixes belonging to Brussels Site 2 are all reachable from London Site 2, satisfying the connectivity criteria for vpn102.

Code Listing 9.44:   London-CE2 BGP Table

```
London-CE2#show ip bgp
BGP table version is 5, local router ID is 130.130.130.1
Status codes: s suppressed, d damped, h history, * valid, > best, i -
internal
Origin codes:  i - IGP, e - EGP, ? - incomplete

   Network          Next Hop          Metric LocPrf Weight Path
*> 9.0.0.0          202.202.202.1                    0 65000 65012 i
*> 100.100.100.0/24 202.202.202.1                    0 65000 ?
*> 130.130.0.0      0.0.0.0                0  32768 i
*> 202.202.202.0    202.202.202.1          0        0 65000 ?
```

The "show ip route" command in Code Listing 9.45 displays the IP routing table for London-CE2 belonging to vpn102. All the BGP routes listed in Code

Listing 9.44 are installed into the IP routing table, attesting that the PE-CE routing protocol (EBGP) is fully operational.

Code Listing 9.45:   London-CE2 IPv4 forwarding table

```
London-CE2#show ip route
<Output Omitted>

C    202.202.202.0/24 is directly connected, Ethernet0/0
     100.0.0.0/24 is subnetted, 1 subnets
B       100.100.100.0 [20/0] via 202.202.202.1, 04:54:39
B    9.0.0.0/8 [20/0] via 202.202.202.1, 04:54:39
C    130.130.0.0/16 is directly connected, Loopback0
```

9.5.5.10 Verifying the coexistent of overlapping address spaces. As mentioned in section 9.5.2.1, there is an occurrence of overlapping address space at London Site 1 (in vpn101) and Brussels Site 2 (in vpn102) for host route 9.9.9.9/32 and network 9.0.0.0/8, respectively. The tests performed in Code Listings 9.46 to 9.48 are used to verify that the overlapping address prefixes can coexist without any clash.

As illustrated in Code Listing 9.46, the ping test executed from Paris-CE1 to destination address 9.9.9.9 is successful.

Code Listing 9.46:   Ping test at Paris-CE1

```
Paris-CE1#ping 9.9.9.9

Type escape sequence to abort.
Sending 5, 100-byte ICMP Echos to 9.9.9.9, timeout is 2 seconds:
!!!!!
Success rate is 100 percent (5/5), round-trip min/avg/max = 92/101/125 ms
```

As illustrated in Code Listing 9.47, London-CE1 has turned on the "debug ip icmp" command. When London-CE1 receives the five ICMP echoes (pings) from Paris-CE1 in Code Listing 9.46, it responds back to Paris-CE1 with five ICMP echo replies.

Code Listing 9.47:   Enable debug IP ICMP at London-CE1

```
London-CE1#debug ip icmp
ICMP packet debugging is on
```

```
London-CE1#
1d00h: ICMP: echo reply sent, src 9.9.9.9, dst 31.3.3.2
1d00h: ICMP: echo reply sent, src 9.9.9.9, dst 31.3.3.2
1d00h: ICMP: echo reply sent, src 9.9.9.9, dst 31.3.3.2
1d00h: ICMP: echo reply sent, src 9.9.9.9, dst 31.3.3.2
1d00h: ICMP: echo reply sent, src 9.9.9.9, dst 31.3.3.2
```

Code Listing 9.48 performs a VRF-based traceroute on destination address 9.9.9.9 (belonging to vpn101) that leads straight to London-CE1.

Code Listing 9.48:   Traceroute test at Paris-PE

```
Paris-PE#traceroute vrf vpn101 ip 9.9.9.9

Type escape sequence to abort.
Tracing the route to 9.9.9.9

  1 200.200.200.1 [MPLS: Label 22 Exp 0] 104 msec 104 msec 104 msec
  2 200.200.200.2 44 msec * 332 msec
```

Based on the results obtained from the tests conducted in Code Listings 9.46 to 9.48, we can conclude that the overlapping address prefixes at London Site 1 (9.9.9.9/32) and Brussels Site 2 (9.0.0.0/8) can coexist with no conflict.

9.6   Case Study 9.2: Static Routes and RIPv2 as PE-CE Routing Protocol

9.6.1   Case overview and network topology

Case Study 9.2 is adapted directly from Case Study 9.1. The main objective of this case study is to demonstrate to the reader how static routing and RIPv2 (dynamic routing) can be used as the per-VRF PE-CE routing protocol. As illustrated in Figure 9.4, the variations from the previous case study include:

- Static routing is used as the per-VRF PE-CE routing protocol for vpn101 with London-CE1 at London Site 1.

- RIPv2 is used as the per-VRF PE-CE routing protocol for vpn102 with Brussels-CE2 at Brussels Site 2.

- To ease monitoring and verification in the later sections, no overlapping address spaces are used:
  — London-CE1 advertises prefix 6.6.6.6/32 instead of 9.9.9.9/32.
  — London-CE2 advertises prefix 5.5.0.0/16 instead of 130.130.0.0/16.

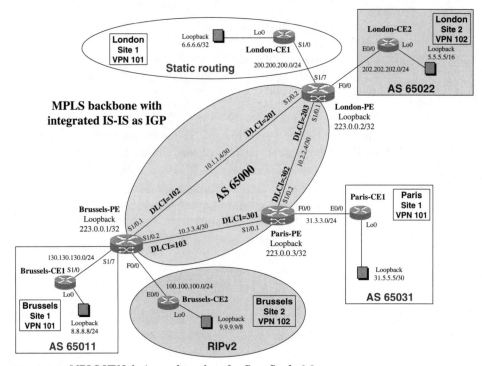

Figure 9.4: MPLS VPN design and topology for Case Study 9.2.

9.6.2 CE configurations

Code Listings 9.49 to 9.51 illustrate the configurations for the affected CE routers in Case Study 9.2.

Code Listing 9.49 illustrates the default route configuration for London-CE1. Note that the loopback interface address is 6.6.6.6/32.

Code Listing 9.49: Default route configuration for London-CE1

```
hostname London-CE1
!
interface Loopback0
! --- Note the loopback address is now 6.6.6.6/32.
 ip address 6.6.6.6 255.255.255.255
!
interface Serial1/0
 bandwidth 64
 ip address 200.200.200.2 255.255.255.0
!
! --- Configure default route pointing to Serial1/0.
ip route 0.0.0.0 0.0.0.0 Serial1/0
```

Code Listing 9.50 illustrates the regular RIPv2 configuration for Brussels-CE2.

Code Listing 9.50: RIPv2 configuration for Brussels-CE2

```
hostname Brussels-CE2
!
interface Loopback0
 ip address 9.9.9.9 255.0.0.0
!
interface Ethernet0/0
 ip address 100.100.100.2 255.255.255.0
!
! --- Configure regular RIPv2 for directly connected
! --- networks 9.0.0.0 and 100.0.0.0.
router rip
 version 2
 network 9.0.0.0
 network 100.0.0.0
 no auto-summary
```

Code Listing 9.51 illustrates the EBGP configuration for London-CE2. Note that the loopback interface address is 5.5.5.5/16.

Code Listing 9.51: EBGP configuration for London-CE2

```
hostname London-CE2
!
interface Loopback0
 ip address 5.5.5.5 255.255.0.0 ! --- Note the loopback address is now
5.5.5.5/16.
!
interface Ethernet0/0
 ip address 202.202.202.2 255.255.255.0
!
router bgp 65022
 no synchronization
 network 5.5.0.0 mask 255.255.0.0 ! --- Advertise prefix 5.5.0.0/16.
 neighbor 202.202.202.1 remote-as 65000
 no auto-summary
```

9.6.3 Configuring RIPv2 as PE-CE routing protocol

Code listing 9.52 illustrates the per-VRF RIPv2 configuration for Brussels-PE. In the configuration, self-explanatory comments (in *italics*) are embedded in between the configuration lines to provide clarification to the configuration commands and procedures.

Code Listing 9.52: Per-VRF RIPv2 configuration for Brussels-PE

```
hostname Brussels-PE
!
<Output Omitted>
!
! --- Configure RIPv2 as per-VRF PE-CE routing protocol for
! --- Brussels Site 2 belonging to vpn102.
router rip

! --- RIP parameters defined in global RIP configuration are inherited by all
! --- address families for this routing process and can be overridden for each
! --- individual address family.

 version 2 ! --- Configure RIP version 2 as a global RIP parameter.
 !
! --- Enter the per-VRF RIP address family (routing context) for VRF vpn102.
 address-family ipv4 vrf vpn102

! --- Configure all per-VRF RIP parameters here

! --- Redistribute BGP into RIP with metric propagation.
! --- With metric transparent option, the value of the MED attribute,
! --- containing the original RIP hop count, is copied into the RIP hop
count
! --- if the BGP route is redistributed back in RIP.
 redistribute bgp 65000 metric transparent

! --- Specify which directly connected network (configured on FastEthernet0/0
! --- in VRF vpn102) to run RIP.
 network 100.0.0.0
 no auto-summary ! --- Disable auto-summarization.
 exit-address-family
 !
router bgp 65000
<Output Omitted>
 address-family ipv4 vrf vpn102
```

```
! --- The BGP address family (routing context) is configured for every VRF,
! --- in this case vpn102. Therefore, redistribution of RIP routes into
BGP has
! --- to be configured for every VRF (in this case vpn102) for which the
! --- RIP routing context is configured.
! --- The RIP hop count is copied into the BGP MED attribute when
! --- the RIP route is redistributed into BGP (enabled by default).
 redistribute rip
 no auto-summary
 no synchronization
 exit-address-family
 !
<Output Omitted>
```

9.6.4 Configuring per-VRF static routes

Code listing 9.53 illustrates the per-VRF static routes configurations for London-PE. In the configuration, self-explanatory comments (in *italics*) are embedded in between the configuration lines to provide clarification to the configuration commands and procedures.

Code Listing 9.53:   Per-VRF static routes configuration for London-PE

```
hostname London-PE
!
<Output Omitted>
!
router bgp 65000
<Output Omitted>
 address-family ipv4 vrf vpn101
 redistribute connected

! --- Static routes must be redistributed into MP-BGP to inform
! --- other PE routers of the remote networks belonging to London Site 1.
 redistribute static
 no auto-summary
 no synchronization
 exit-address-family
 !
<Output Omitted>
! --- This command configures the per-VRF static routes for vpn101.
ip route vrf vpn101 6.6.6.6 255.255.255.255 Serial1/7
! --- The route is entered into the VRF table.
! --- Always specify the outgoing interface even if the next-hop is known.
```

9.6.5 Verifying the intersite connectivity for vpn101

The CE routers' routing tables and ping test results are illustrated in Code Listings 9.54 to 9.60, and are used to verify the intersite connectivity for vpn101. For simplicity the main focus in this section is to verify the intersite reachability of prefixes 6.6.6.6/32, 8.8.8.0/24, and 31.5.5.4/30.

Code Listing 9.54 displays the IP routing table of London-CE1, indicating that a default route has been configured on London-CE1 pointing to Serial1/0.

Code Listing 9.54:   London-CE1 IPv4 forwarding table

```
London-CE1#show ip route
<Output Omitted>


C    200.200.200.0/24 is directly connected, Serial1/0
     6.0.0.0/32 is subnetted, 1 subnets
C       6.6.6.6 is directly connected, Loopback0
S*   0.0.0.0/0 is directly connected, Serial1/0
```

Code Listing 9.55 displays the per-VRF routing table associated with vpn101 for London-PE. After configuring the per-VRF static route to prefix 6.6.6.6/32 (see Code Listing 9.53), the static route is entered into the VRF routing table.

Code Listing 9.55:   London-PE Per-VRF routing table for vpn101

```
London-PE#show ip route vrf vpn101
<Output Omitted>


C    200.200.200.0/24 is directly connected, Serial1/7
     6.0.0.0/32 is subnetted, 1 subnets
S       6.6.6.6 is directly connected, Serial1/7
     8.0.0.0/24 is subnetted, 1 subnets
B       8.8.8.0 [200/0] via 223.0.0.1, 00:36:19
     130.130.0.0/24 is subnetted, 1 subnets
B       130.130.130.0 [200/0] via 223.0.0.1, 00:36:19
     31.0.0.0/8 is variably subnetted, 2 subnets, 2 masks
B       31.5.5.4/30 [200/0] via 223.0.0.3, 00:36:19
B       31.3.3.0/24 [200/0] via 223.0.0.3, 00:36:19
```

Code Listing 9.56 displays the IP routing table of Brussels-CE1, indicating that Brussels-CE1 has successfully received the prefix 6.6.6.6/32 via EBGP.

Code Listing 9.56: Brussels-CE1 IPv4 forwarding table

```
Brussels-CE1#show ip route
<Output Omitted>

B    200.200.200.0/24 [20/0] via 130.130.130.1, 00:20:51
     6.0.0.0/32 is subnetted, 1 subnets
B       6.6.6.6 [20/0] via 130.130.130.1, 00:20:51
     8.0.0.0/24 is subnetted, 1 subnets
C       8.8.8.0 is directly connected, Loopback0
     130.130.0.0/24 is subnetted, 1 subnets
C       130.130.130.0 is directly connected, Serial1/0
     31.0.0.0/8 is variably subnetted, 2 subnets, 2 masks
B       31.5.5.4/30 [20/0] via 130.130.130.1, 00:45:10
B       31.3.3.0/24 [20/0] via 130.130.130.1, 02:01:31
```

Code Listing 9.57 displays the IP routing table of Paris-CE1, indicating that Paris-CE1 has successfully received the prefix 6.6.6.6/32 via EBGP.

Code Listing 9.57: Paris-CE1 IPv4 forwarding table

```
Paris-CE1#show ip route
<Output Omitted>

B    200.200.200.0/24 [20/0] via 31.3.3.1, 00:31:04
     6.0.0.0/32 is subnetted, 1 subnets
B       6.6.6.6 [20/0] via 31.3.3.1, 00:31:04
     8.0.0.0/24 is subnetted, 1 subnets
B       8.8.8.0 [20/0] via 31.3.3.1, 00:55:20
     130.130.0.0/24 is subnetted, 1 subnets
B       130.130.130.0 [20/0] via 31.3.3.1, 00:55:20
     31.0.0.0/8 is variably subnetted, 2 subnets, 2 masks
C       31.5.5.4/30 is directly connected, Loopback0
C       31.3.3.0/24 is directly connected, Ethernet0/0
```

Code Listing 9.58 displays the results for the successful ping test conducted from Brussels-CE1 to destination address 6.6.6.6 located in London Site 1.

Code Listing 9.58:   Ping test at Brussels-CE1

```
Brussels-CE1#ping 6.6.6.6

Type escape sequence to abort.
Sending 5, 100-byte ICMP Echos to 6.6.6.6, timeout is 2 seconds:
!!!!!
Success rate is 100 percent (5/5), round-trip min/avg/max = 120/161/320 ms
```

Code Listing 9.59 displays the results for the successful ping test conducted from Paris-CE1 to destination address 6.6.6.6 located in London Site 1.

Code Listing 9.59:   Ping test at Paris-CE1

```
Paris-CE1#ping 6.6.6.6

Type escape sequence to abort.
Sending 5, 100-byte ICMP Echos to 6.6.6.6, timeout is 2 seconds:
!!!!!
Success rate is 100 percent (5/5), round-trip min/avg/max = 88/125/272 ms
```

Code Listing 9.60 displays the results for two successful ping tests. One is conducted from London-CE1 to destination address 8.8.8.8 located in Brussels Site 1, and the other is conducted from London-CE1 to destination address 31.5.5.5 located in Paris Site 1.

Code Listing 9.60:   Ping tests at London-CE1

```
London-CE1#ping 8.8.8.8

Type escape sequence to abort.
Sending 5, 100-byte ICMP Echos to 8.8.8.8, timeout is 2 seconds:
!!!!!
Success rate is 100 percent (5/5), round-trip min/avg/max = 120/127/144 ms

London-CE1#ping 31.5.5.5

Type escape sequence to abort.
Sending 5, 100-byte ICMP Echos to 31.5.5.5, timeout is 2 seconds:
!!!!!
Success rate is 100 percent (5/5), round-trip min/avg/max = 92/98/112 ms
```

Together Code Listings 9.54 to 9.60 indicate that all three sites belonging to vpn101 have full intersite connectivity with each other.

9.6.6 Verifying the intersite connectivity for vpn102

The CE routers' routing tables and ping test results are illustrated in Code Listings 9.61 to 9.65 and are used to verify the intersite connectivity for vpn102. For simplicity, the main focus in this section is to verify the intersite reachability of prefixes 5.5.0.0/16 and 9.0.0.0/8.

Code Listing 9.61 displays the IP routing table of Brussels-CE2, indicating that Brussels-CE2 has successfully received the prefix 5.5.0.0/16 via RIP.

Code Listing 9.61:   Brussels-CE2 IPv4 forwarding table

```
Brussels-CE2#show ip route
<Output Omitted>

R    202.202.202.0/24 [120/1] via 100.100.100.1, 00:00:14, Ethernet0/0
     100.0.0.0/24 is subnetted, 1 subnets
C       100.100.100.0 is directly connected, Ethernet0/0
     5.0.0.0/16 is subnetted, 1 subnets
R       5.5.0.0 [120/1] via 100.100.100.1, 00:00:14, Ethernet0/0
C    9.0.0.0/8 is directly connected, Loopback0
```

Code Listing 9.62 displays the per-VRF routing table associated with vpn102 for Brussels-PE. Notice that prefix 9.0.0.0/8, originating from Brussels-CE2, has been installed into the VRF routing table.

Code Listing 9.62:   Brussels-PE Per-VRF routing table for vpn102

```
Brussels-PE#show ip route vrf vpn102
<Output Omitted>

B    202.202.202.0/24 [200/0] via 223.0.0.2, 00:18:52
     100.0.0.0/24 is subnetted, 1 subnets
C       100.100.100.0 is directly connected, FastEthernet0/0
     5.0.0.0/16 is subnetted, 1 subnets
B       5.5.0.0 [200/0] via 223.0.0.2, 00:18:52
R    9.0.0.0/8 [120/1] via 100.100.100.2, 00:00:05, FastEthernet0/0
```

Code Listing 9.63 displays the IP routing table of London-CE2, indicating that London-CE2 has successfully received the prefix 9.0.0.0/8 via EBGP.

Code Listing 9.63:   London-CE2 IPv4 forwarding table

```
London-CE2#show ip route
<Output Omitted>

C    202.202.202.0/24 is directly connected, Ethernet0/0
     100.0.0.0/24 is subnetted, 1 subnets
B       100.100.100.0 [20/0] via 202.202.202.1, 00:41:00
     5.0.0.0/16 is subnetted, 1 subnets
C       5.5.0.0 is directly connected, Loopback0
B    9.0.0.0/8 [20/0] via 202.202.202.1, 00:41:00
```

Code Listing 9.64 displays the results for the successful ping test conducted from Brussels-CE2 to destination address 5.5.5.5 located in London Site 2.

Code Listing 9.64:   Ping test at Brussels-CE2

```
Brussels-CE2#ping 5.5.5.5

Type escape sequence to abort.
Sending 5, 100-byte ICMP Echos to 5.5.5.5, timeout is 2 seconds:
!!!!!
Success rate is 100 percent (5/5), round-trip min/avg/max = 60/61/64 ms
```

Code Listing 9.65 displays the results for the successful ping test conducted from London-CE2 to destination address 9.9.9.9 located in Brussels Site 2.

Code Listing 9.65:   Ping test at London-CE2

```
London-CE2#ping 9.9.9.9

Type escape sequence to abort.
Sending 5, 100-byte ICMP Echos to 9.9.9.9, timeout is 2 seconds:
!!!!!
Success rate is 100 percent (5/5), round-trip min/avg/max = 60/61/65 ms
```

Together Code Listings 9.61 to 9.65 indicate that the two sites belonging to vpn102 have full intersite connectivity with each other.

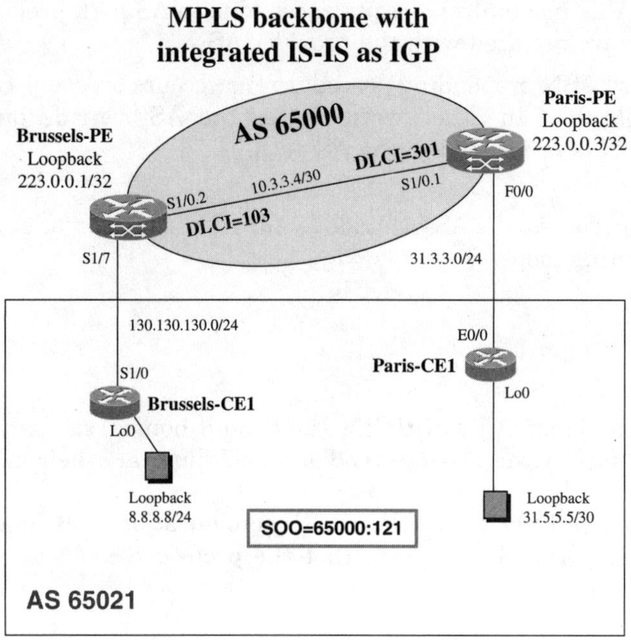

**MPLS backbone with
integrated IS-IS as IGP**

Figure 9.5: Applying site-of-origin in a multihomed environment.

9.7   Case Study 9.3: Site-of-Origin Application

9.7.1   Case overview

Case Study 9.3 uses a multihomed site portrayed in Figure 9.5 to illustrate how the Site-of-Origin (SOO) extended BGP community attribute can be used as an additional loop prevention mechanism when the regular AS path–based BGP loop prevention is bypassed with the AS-override feature.

9.7.2   AS-override

To support customer topologies where identical customer ASN is applied at more than one site, the AS path update procedure in BGP has been adjusted to overcome the loop prevention rules of BGP. The new AS path update procedure, also known as AS-override, supports the use of one ASN at multiple sites (even among various overlapping VPNs). The procedure allows the use of private as well as public ASNs.

With AS-override configured, the AS path update procedure on the PE router is as follows:

- If the first ASN in the AS path is equal to the neighboring one, it is replaced with the provider ASN.

- If the first ASN has multiple occurrences (due to AS path prepend), all such occurrences are replaced with the provider ASN.

- The standard ASN prepending procedure that occurs on every EBGP update is still applicable. In other words, after the AS-override operation, the provider ASN is prepended to the AS path.

In multihomed sites, the AS-override feature is used in conjunction with SOO to prevent routing loops.

9.7.3 CE configurations

Brussels-CE1 and Paris-CE1 are in the same (multihomed) site and autonomous system (AS 65021). Code Listings 9.66 and 9.67 illustrate their EBGP configurations, respectively.

Code Listing 9.66 illustrates the EBGP configuration of Brussels-CE1 with EBGP neighbor Brussels-PE. Note that the prefix 8.8.8.0/24 is advertised to Brussels-PE.

Code Listing 9.66:   EBGP configuration for Brussels-CE1

```
hostname Brussels-CE1
!
interface Loopback0
 ip address 8.8.8.8 255.255.255.0
!
interface Serial1/0
 bandwidth 64
 ip address 130.130.130.2 255.255.255.0
!
! --- Regular BGP Configuration for AS 65021.
router bgp 65021
 no synchronization
 network 8.8.8.0 mask 255.255.255.0 ! --- Advertise prefix 8.8.8.0/24.
 neighbor 130.130.130.1 remote-as 65000 ! --- Declare EBGP neighbor in AS
65000.
 no auto-summary
```

Code Listing 9.67 illustrates the EBGP configuration of Paris-CE1 with EBGP neighbor Paris-PE. Note that the prefix 31.5.5.4/30 is advertised to Paris-PE.

Code Listing 9.67:   EBGP configuration for Paris-CE1

```
hostname Paris-CE1
!
interface Loopback0
  ip address 31.5.5.5 255.255.255.252
!
interface Ethernet0/0
  ip address 31.3.3.2 255.255.255.0
!
! --- Regular BGP Configuration for AS 65021.
router bgp 65021
 no synchronization
 network 31.5.5.4 mask 255.255.255.252 ! --- Advertise prefix 31.5.5.4/30.
 neighbor 31.3.3.1 remote-as 65000 ! --- Declare EBGP neighbor in AS 65000.
 no auto-summary
```

9.7.4   PE configurations

Since AS-override and SOO are the main highlights in this case study, the core focus of this section will also be on the configuration of these BGP mechanisms as illustrated in Code Listing 9.68 for Brussels-PE and Code Listing 9.69 for Paris-PE.

To demonstrate the functionality of SOO in the later sections, we first disable the regular AS path update procedure with AS-override in Brussels-PE. AS-override is configured for Brussels-PE's EBGP neighbor Brussels-CE1 in the VRF address family of the BGP process as illustrated in Code Listing 9.68. Thereafter, we set the SOO attribute to 65000:121 for routes announced from Brussels-CE1.

Code Listing 9.68:   AS-override and SOO configuration for Brussels-PE

```
hostname Brussels-PE
!
ip vrf site2a ! --- Create VRF site2a.
 rd 65000:121 ! --- Assign route distinguisher 65000:121 to VRF site2a.

! --- Specify RT 65000:121 to be attached
! --- to every route from this VRF to MP-BGP.
 route-target export 65000:121

! --- Specify RT 65000:121 to be used as an import filter.
! --- Only routes matching this RT are imported into VRF site2a.
```

```
   route-target import 65000:121
 !
 ip cef
 !
 interface Loopback0
  ip address 223.0.0.1 255.255.255.255
  ip router isis
 !
 interface Serial1/0
  encapsulation frame-relay
  frame-relay lmi-type ansi
 !
 interface Serial1/0.2 point-to-point
  bandwidth 64
  ip address 10.3.3.5 255.255.255.252
  ip router isis
  tag-switching ip
  frame-relay interface-dlci 103
 !
 interface Serial1/7
  bandwidth 64
  ip vrf forwarding site2a ! --- Associate Serial1/7 with VRF site2a.
  ip address 130.130.130.1 255.255.255.0
  clockrate 64000
 !
 router isis
  net 49.0001.0000.0000.0001.00
  is-type level-1
  metric-style wide
 !
 ! --- Configure MP-BGP for Brussels-PE in AS 65000.
 router bgp 65000
  no synchronization
  neighbor 223.0.0.2 remote-as 65000
  neighbor 223.0.0.2 update-source Loopback0
  neighbor 223.0.0.3 remote-as 65000
  neighbor 223.0.0.3 update-source Loopback0
  no auto-summary
  !
 ! --- Enter the per-VRF BGP routing context for VRF site2a.
  address-family ipv4 vrf site2a

 ! --- Redistribute directly connected interface Serial1/7
 ! --- into the per-VRF BGP routing context for VRF site2a.
  redistribute connected
```

```
! --- Declare Brussels-CE1 in AS 65021 as CE EBGP neighbor.
 neighbor 130.130.130.2 remote-as 65021
 neighbor 130.130.130.2 activate


! --- AS-override is configured for CE EBGP neighbors
! --- in the VRF address family of the BGP process.
! --- Configure the AS-override feature for Brussels-CE1 in AS 65021.
! --- In this case, Brussels-PE will override AS 65021 with AS 65000.
 neighbor 130.130.130.2 as-override


! --- Apply inbound route-map setsoo to EBGP neighbor Brussels-CE1.
 neighbor 130.130.130.2 route-map setsoo in
 no auto-summary
 no synchronization
 exit-address-family
 !
 address-family vpnv4
 neighbor 223.0.0.2 activate
 neighbor 223.0.0.2 send-community extended
 neighbor 223.0.0.3 activate
 neighbor 223.0.0.3 send-community extended
 no auto-summary
 exit-address-family
!
route-map setsoo permit 10 ! --- Create route-map setsoo.
 set extcommunity soo 65000:121 ! --- Set SOO attribute to 65000:121.
```

To demonstrate the functionality of SOO in the later sections, we first disable the regular AS path update procedure with AS-override in Paris-PE. AS-override is configured for Paris-PE's EBGP neighbor Paris-CE1 in the VRF address family of the BGP process as illustrated in Code Listing 9.69. Thereafter, we set the SOO attribute to 65000:121 for routes announced from Paris-CE1.

Code Listing 9.69:   AS-override and SOO configuration for Paris-PE

```
hostname Paris-PE
!
ip vrf site2b ! --- Create VRF site2b.
 rd 65000:121 ! --- Assign route distinguisher 65000:121 to VRF site2b.


! --- Specify RT 65000:121 to be attached
! --- to every route from this VRF to MP-BGP.
 route-target export 65000:121
```

```
! --- Specify RT 65000:121 to be used as an import filter.
! --- Only routes matching this RT are imported into VRF site2b.
 route-target import 65000:121
!
ip cef
!
interface Loopback0
 ip address 223.0.0.3 255.255.255.255
 ip router isis
!
interface FastEthernet0/0
  ip vrf forwarding site2b ! --- Associate FastEthernet0/0 with VRF
site2b.
 ip address 31.3.3.1 255.255.255.0
!
interface Serial1/0
 encapsulation frame-relay
 frame-relay lmi-type ansi
!
interface Serial1/0.1 point-to-point
 ip address 10.3.3.6 255.255.255.252
 ip router isis
 tag-switching ip
 frame-relay interface-dlci 301
!
router isis
 net 49.0001.0000.0000.0003.00
 is-type level-1
 metric-style wide
!
! --- Configure MP-BGP for Brussels-PE in AS 65000.
router bgp 65000
 no synchronization
 neighbor 223.0.0.1 remote-as 65000
 neighbor 223.0.0.1 update-source Loopback0
 neighbor 223.0.0.2 remote-as 65000
 neighbor 223.0.0.2 update-source Loopback0
 no auto-summary
 !
! --- Enter the per-VRF BGP routing context for VRF site2b.
 address-family ipv4 vrf site2b

! --- Redistribute directly connected interface FastEthernet0/0
! --- into the per-VRF BGP routing context for VRF site2b.
 redistribute connected
```

```
! --- Declare Paris-CE1 in AS 65021 as CE EBGP neighbor.
 neighbor 31.3.3.2 remote-as 65021
 neighbor 31.3.3.2 activate


! --- AS-override is configured for CE EBGP neighbors
! --- in the VRF address family of the BGP process.
! --- Configure the AS-override feature for Paris-CE1 in AS 65021.
! --- In this case, Paris-PE will override AS 65021 with AS 65000.
 neighbor 31.3.3.2 as-override

! --- Apply inbound route-map setsoo to EBGP neighbor Paris-CE1.
 neighbor 31.3.3.2 route-map setsoo in
 no auto-summary
 no synchronization
 exit-address-family
 !
 address-family vpnv4
 neighbor 223.0.0.1 activate
 neighbor 223.0.0.1 send-community extended
 neighbor 223.0.0.2 activate
 neighbor 223.0.0.2 send-community extended
 no auto-summary
 exit-address-family
 !
route-map setsoo permit 10 ! --- Create route-map setsoo.
 set extcommunity soo 65000:121 ! --- Set SOO attribute to 65000:121.
```

9.7.5 Before implementing AS-override

Code Listings 9.70 and 9.71 illustrate the BGP tables of the two CE routers before implementing AS-override. For simplicity we shall verify only the visibility of prefixes 8.8.8.0/24 (originating from Brussels-CE1) and 31.5.5.4/30 (originating from Paris-CE1) in all the subsequent displays.

As illustrated in Code Listing 9.70, the prefix 31.5.5.4/30 originating from Paris-CE1 is not displayed on Brussels-CE1's BGP table. This is because the standard AS path update procedure is still in effect before AS-override is implemented.

Code Listing 9.70:   Brussels-CE1 BGP table before AS-override

---

```
Brussels-CE1#show ip bgp
BGP table version is 32, local router ID is 8.8.8.8
Status codes: s suppressed, d damped, h history, * valid, > best, i - internal
Origin codes: i - IGP, e - EGP, ? - incomplete
```

```
    Network           Next Hop            Metric LocPrf Weight Path
*>  8.8.8.0/24        0.0.0.0                        0         32768 i
<Output Omitted>
```

As illustrated in Code Listing 9.71, the prefix 8.8.8.0/24 originating from Brussels-CE1 is not displayed on Paris-CE1's BGP table. This is because the standard AS path update procedure is still in effect before AS-override is implemented.

Code Listing 9.71:   Paris-CE1 BGP table before AS-override

```
Paris-CE1#show ip bgp
BGP table version is 6, local router ID is 31.5.5.5
Status codes: s suppressed, d damped, h history, * valid, > best, i - internal
Origin codes: i - IGP, e - EGP, ? - incomplete

    Network           Next Hop            Metric LocPrf Weight Path
<Output Omitted>
*>  31.5.5.4/30       0.0.0.0                        0         32768 i
<Output Omitted>
```

9.7.6   After implementing AS-override

Code Listings 9.72 to 9.75 illustrate the BGP tables and IP routing tables of the two CE routers after implementing AS-override.

Since the AS-override feature has been enabled, prefix 31.5.5.4/30 originating from Paris-CE1 is now displayed on Brussels-CE1's BGP table in Code Listing 9.72. The AS-override mechanism, configured on Brussels-PE, replaces the customer ASN 65021 (associated with prefix 31.5.5.4/30) with the provider ASN 65000 before sending the prefix to Brussels-CE1. Another copy of the provider ASN (65000) is prepended to the AS path when the prefix traverses from provider AS to the customer AS. As a result, the AS-path appears as "65000 65000" in Brussels-CE1.

Code Listing 9.72:   Brussels-CE1 BGP table after AS-override

---

```
Brussels-CE1#show ip bgp
BGP table version is 13, local router ID is 8.8.8.8
Status codes: s suppressed, d damped, h history, * valid, > best, i - internal
Origin codes: i - IGP, e - EGP, ? - incomplete

    Network           Next Hop            Metric LocPrf Weight Path
*>  8.8.8.0/24        0.0.0.0                        0    32768 i
<Output Omitted>
*>  31.5.5.4/30       130.130.130.1                       0 65000 65000 i
<Output Omitted>
```

As illustrated in Code Listing 9.73, the prefix 31.5.5.4/30 is further installed into the Brussels-CE1's IP routing table as a destination subnet reachable via EBGP. This would produce a routing loop if the prefix were also reachable locally within the customer site.

Code Listing 9.73:   Brussels-CE1 IPv4 forwarding table after AS-override

---

```
Brussels-CE1#show ip route
<Output Omitted>


     8.0.0.0/24 is subnetted, 1 subnets
C        8.8.8.0 is directly connected, Loopback0
<Output Omitted>
B        31.5.5.4/30 [20/0] via 130.130.130.1, 00:01:17
<Output Omitted>
```

Since the AS-override feature has been enabled, prefix 8.8.8.0/24 originating from Brussels-CE1 is now displayed on Paris-CE1 BGP table in Code Listing 9.74. The AS-override mechanism, configured on Paris-PE, replaces the customer ASN 65021 (associated with prefix 8.8.8.0/24) with the provider ASN 65000 before sending the prefix to Paris-CE1. Another copy of the provider ASN (65000) is prepended to the AS-path when the prefix traverses from provider AS to the customer AS. As a result, the AS-path appears as "65000 65000" in Paris-CE1.

Code Listing 9.74:   Paris-CE1 BGP table after AS-override

```
Paris-CE1#show ip bgp
BGP table version is 21, local router ID is 31.5.5.5
Status codes: s suppressed, d damped, h history, * valid, > best, i - internal
Origin codes: i - IGP, e - EGP, ? - incomplete

   Network          Next Hop         Metric LocPrf Weight Path
*> 8.8.8.0/24       31.3.3.1                         0 65000 65000 i
<Output Omitted>
*> 31.5.5.4/30      0.0.0.0               0        32768 i
<Output Omitted>
```

As illustrated in Code Listing 9.75, the prefix 8.8.8.0/24 is further installed into the Paris-CE1's IP routing table as a destination subnet reachable via EBGP. This would produce a routing loop if the prefix were also reachable locally within the customer site.

Code Listing 9.75:   Paris-CE1 IPv4 forwarding table after AS-override

---

```
Paris-CE1#show ip route
<Output Omitted>


     8.0.0.0/24 is subnetted, 1 subnets
B        8.8.8.0 [20/0] via 31.3.3.1, 00:05:04
<Output Omitted>
C        31.5.5.4/30 is directly connected, Loopback0
<Output Omitted>
```

9.7.7   After setting SOO

Code Listings 9.76 to 9.79 are used to verify the SOO attribute value associated with prefixes 8.8.8.0/24 and 31.5.5.4/30, and Code Listings 9.80 and 9.81 are used to verify the visibility of these two prefixes in Brussels-CE1 and Paris-CE1.

The SOO value 65000:121 is assigned to prefix 8.8.8.0/24 (originating from Brussels-CE1) at Brussels-PE as illustrated in Code Listing 9.76.

Code Listing 9.76:   Brussels-PE BGP VPNv4 table with subnet 8.8.8.0/24

```
Brussels-PE#show ip bgp vpnv4 all 8.8.8.8
BGP routing table entry for 65000:121:8.8.8.0/24, version 9
Paths: (1 available, best #1, table site2a)
  Advertised to non peer-group peers:
  223.0.0.3
  65021
    130.130.130.2 from 130.130.130.2 (8.8.8.8)
      Origin IGP, metric 0, localpref 100, valid, external, best
      Extended Community: SoO:65000:121 RT:65000:121
```

The SOO value 65000:121 is associated with prefix 8.8.8.0/24 at Paris-PE as illustrated in Code Listing 9.77. As prefix 8.8.8.0/24 originates from Brussels-CE1 and is propagated to Paris-PE via Brussels-PE, Paris-PE will not propagate the prefix to Paris-CE1 since the prefix's SOO is equal to the one configured for the site.

Code Listing 9.77:   Paris-PE BGP VPNv4 table with subnet 8.8.8.0/24

```
Paris-PE#show ip bgp vpnv4 all 8.8.8.8
BGP routing table entry for 65000:121:8.8.8.0/24, version 24
Paths: (1 available, best #1, table site2b)
```

```
Not advertised to any peer
65021
   223.0.0.1 (metric 20) from 223.0.0.1 (223.0.0.1)
      Origin IGP, metric 0, localpref 100, valid, internal, best
      Extended Community: SoO:65000:121 RT:65000:121
```

The SOO value 65000:121 is assigned to prefix 31.5.5.4/30 (originating from Paris-CE1) at Paris-PE as illustrated in Code Listing 9.78.

Code Listing 9.78:   Paris-PE BGP VPNv4 table with subnet 31.5.5.4/30

```
Paris-PE#show ip bgp vpnv4 all 31.5.5.5
BGP routing table entry for 65000:121:31.5.5.4/30, version 9
Paths: (1 available, best #1, table site2b)
  Advertised to non peer-group peers:
  223.0.0.1
  65021
    31.3.3.2 from 31.3.3.2 (31.5.5.5)
      Origin IGP, metric 0, localpref 100, valid, external, best
      Extended Community: SoO:65000:121 RT:65000:121
```

The SOO value 65000:121 is associated with prefix 31.5.5.4/30 at Brussels-PE as illustrated in Code Listing 9.79. As prefix 31.5.5.4/30 originates from Paris-CE1 and is propagated to Brussels-PE via Paris-PE, Brussels-PE will not propagate the prefix to Brussels-CE1 since the prefix's SOO is equal to the one configured for the site.

Code Listing 9.79:   Brussels-PE BGP VPNv4 table with subnet 31.5.5.4/30

```
Brussels-PE#show ip bgp vpnv4 all 31.5.5.5
BGP routing table entry for 65000:121:31.5.5.4/30, version 7
Paths: (1 available, best #1, table site2a)
  Not advertised to any peer
  65021
    223.0.0.3 (metric 20) from 223.0.0.3 (223.0.0.3)
      Origin IGP, metric 0, localpref 100, valid, internal, best
      Extended Community: SoO:65000:121 RT:65000:121
```

As a result of Code Listing 9.77, Paris-CE1's BGP table in Code Listing 9.80 no longer reflects the prefix 8.8.8.0/24 originating from Brussels-CE1, thus preventing the occurrence of a routing loop.

Code Listing 9.80:   Paris-CE1 BGP table after setting SOO

```
Paris-CE1#show ip bgp
BGP table version is 50, local router ID is 31.5.5.5
Status codes: s suppressed, d damped, h history, * valid, > best, i - internal
Origin codes: i - IGP, e - EGP, ? - incomplete

   Network          Next Hop          Metric LocPrf Weight Path
<Output Omitted>
*> 31.5.5.4/30      0.0.0.0                0           32768 i
<Output Omitted>
```

As a result of Code Listing 9.79, Brussels-CE1's BGP table in Code Listing 9.81 no longer reflects the prefix 31.5.5.4/30 originating from Paris-CE1, thus preventing the occurrence of a routing loop.

Code Listing 9.81:   Brussels-CE1 BGP table after setting SOO

```
Brussels-CE1#show ip bgp
BGP table version is 48, local router ID is 8.8.8.8
Status codes: s suppressed, d damped, h history, * valid, > best, i -
internal
Origin codes: i - IGP, e - EGP, ? - incomplete

   Network          Next Hop          Metric LocPrf Weight Path
*> 8.8.8.0/24       0.0.0.0                0           32768 i
<Output Omitted>
```

9.8 Summary

Chapter 9 addressed the implementation of MPLS VPN from the practical perspective, which includes the design considerations and implementation procedures that are required when building the MPLS VPN. The actual deployment of MPLS VPN was illustrated with three different case studies.

Case Study 9.1 gave an intensive coverage on how to design and implement MPLS VPN practically from scratch. EBGP is the PE-CE routing protocol adopted by this first case study. Case Study 9.2 is a variant of Case Study 9.1 in which it also deploys static routes and RIPv2 as the PE-CE routing protocols, instead of just EBGP. Finally, Case Study 9.3 illustrated how to prevent potential BGP loops by using the Site-of-Origin (SOO) extended BGP community attribute when the BGP AS-override feature has circumvented the standard AS-path–based BGP loop prevention.

MPLS VPN Topologies

10.1 Introduction

Recall from Chapter 9 that a VRF table can be used for a single VPN site or for multiple VPN sites connected to the same PE router, provided these sites share identical connectivity requirements in simple VPN topologies. This chapter goes a step further by venturing into VPN topologies that do not share connectivity requirements. In such complex VPN topologies, multiple VRF tables per VPN are required for dissimilar connectivity requirements. The following VPN topologies are discussed in the subsequent sections, and supporting case studies address all these topology settings in detail:

- *Overlapping VPN topology*: This topology is applicable to either a company where central sites participate in corporate network and also in an Extranet (see section 1.2.2) or a company with several security-aware departments that require restricted visibility when exchanging data between their servers.
- *Central services topology*: This topology is applicable to clients who need access to central servers, servers that can communicate with each other, and clients that can communicate with all servers but not with each other.
- *Hybrid VPN topology*: This topology uses a combination of the rules implemented in overlapping VPN and central services VPN.
- *Hub-and-spoke topology*: In this topology, one central (hub) site has full routing knowledge of all other sites of the same VPN, whereas other (spoke) sites send traffic to the central site for any destination. In other words, the hub site becomes the central transit point between the spoke sites. Security services (filters), traffic logging, traffic accounting, and intrusion-detection systems can be deployed in the hub site. Moreover, in a hub-and-spoke topology, traffic from one spoke to another will need to travel across the hub site. Thus, the BGP Allowas-in feature has to be enabled at the appropriate PE routers if we are using BGP as the PE-CE routing protocol. Besides, if the

spoke sites are using the same AS number for reuse purposes, the BGP AS-override mechanism (see section 9.7.2) will need to be configured.

Complex VPN topologies need to rely on some kind of identifiers that can be attached to a VPNv4 route to specify its VPN association. Route targets (see section 8.4.2) fit into this requirement snugly. Route targets (RTs) are BGP extended community attributes that can be attached to a VPNv4 route (carried in an MP-BGP update) to indicate its VPN membership, thus fulfilling the requirements of complex VPN topologies. In short, RTs allow the implementation of complex VPN topologies that were hard to implement with other VPN architectures. We shall use case studies to demonstrate how RTs are used to resolve the different connectivity requirements of the various complex VPN topologies. In addition, two other important aspects to consider when deploying MPLS VPN—Internet access and performance optimization—will also be discussed.

10.2   Case Study 10.1:Overlapping VPN Topology

10.2.1   Case overview

This case study provides a sample configuration of an overlapping VPN topology. The MPLS VPN deployment uses Cisco routers and spans four different customer sites: Brussels Client-1, Brussels Central-1, London Client-2, and London Central-2. The MPLS VPN design and physical topology for Case Study 10.1 are shown in Figure 10.1.

10.2.2   Overlapping VPN topology

Figure 10.2 illustrates the VPN connectivity requirements (logical VPN topology) and the VPN memberships of the four different sites. As shown in the figure, we have two customers, in VPN 1 and VPN 2, who want to share some information through their central sites. To achieve this a third VPN, VPN 12, is created that partially overlaps with both VPN 1 and VPN 2 and connects only their central sites: Brussels Central-1 and London Central-2. The central sites can talk to each other but not to other sites belonging to the third VPN. The addresses used in the central sites, however, have to be unique in both VPNs.

In this example, we have four sites with different VPN connectivity requirements and memberships. This means we have to have at least four VRFs:

1. Client-1 is a member of VPN 1 only.

2. Client-2 is a member of VPN 2 only.

3. Central-1 is a member of VPN 1 and VPN 12. Networks originating in Central-1 are exported with two RTs—one for its VPN (VPN 1) and one for the overlapping VPN (VPN 12).

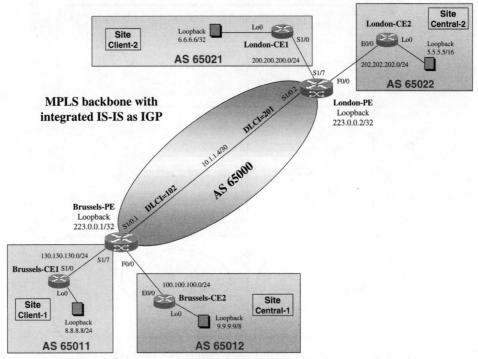

Figure 10.1: MPLS VPN design and topology for Case Study 10.1.

4. Central-2 is a member of VPN 2 and VPN 12. Networks originating in Central-2 are also exported with two RTs—one for its VPN (VPN 2) and one for the overlapping VPN (VPN 12).

Tables 10.1 and 10.2 illustrate the RD and RT numbering scheme for Brussels-PE and London-PE, respectively.

Tables 10.1 and 10.2 are illustrated schematically in Figure 10.3 for which:

- Central-1 and Client-1 will import and export only routes from each other tagged with RT 65000:101.

- Central-1 and Central-2 will import and export only routes from each other tagged with RT 65000:102.

- Central-2 and Client-2 will import and export only routes from each other tagged with RT 65000:202.

In addition, according to the four VRFs defined in Tables 10.1 and 10.2, we can construe the following intersite connectivity:

- VRF Client-1 contains routes from VRF Central-1: 9.0.0.0/8 and 100.100.100.0/24.

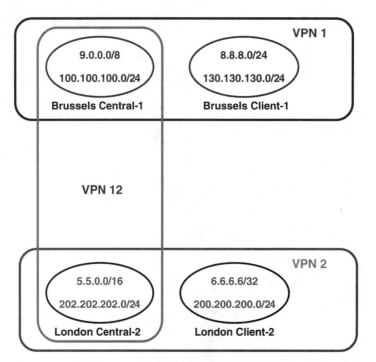

Figure 10.2: VPN connectivity requirements for Case Study 10.1.

TABLE 10.1   Per-VRF RD and RT numbering scheme for Brussels-PE.

VRF	Client-1	Central-1
Export RT	65000:101	65000:101 65000:102
Import RT	65000:101	65000:101 65000:102
RD	65000:101	65000:666

- VRF Central-1 contains routes from VRF Client-1: 8.8.8.0/24 and 130.130.130.0/24.
- VRF Central-1 contains routes from VRF Central-2: 5.5.0.0/16 and 202.202.202.0/24.
- VRF Central-2 contains routes from VRF Central-1: 9.0.0.0/8 and 100.100.100.0/24.
- VRF Central-2 contains routes from VRF Client-2: 6.6.6.6/32 and 200.200.200.0/24.
- VRF Client-2 contains routes from VRF Central-2: 5.5.0.0/16 and 202.202.202.0/24.

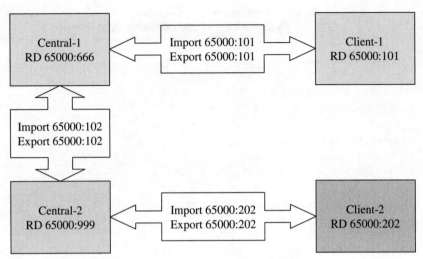

Figure 10.3: RD and RT numbering scheme for Case Study 10.1.

TABLE 10.2 Per-VRF RD and RT numbering scheme for London-PE.

VRF	Client-2	Central-2
Export RT	65000:202	65000:202 65000:102
Import RT	65000:202	65000:202 65000:102
RD	65000:202	65000:999

Note that VRF Client-1 does not contain routes from VRF Client-2 and vice versa, implying that Client-1 and Client-2 cannot communicate with each other. These intersite connectivity deductions will be verified in the later sections.

10.2.3 CE configurations

Code Listings 10.1 to 10.4 illustrates the EBGP configurations for the CE routers in the case study.

Code Listing 10.1 illustrates the EBGP configuration for Brussels-CE1. Brussels-CE1 advertises subnet 8.8.8.0/24 to Brussels-PE and belongs to site Client-1.

Code Listing 10.1:   EBGP configuration for Brussels-CE1

```
hostname Brussels-CE1
!
interface Loopback0
 ip address 8.8.8.8 255.255.255.0
!
interface Serial1/0
 bandwidth 64
 ip address 130.130.130.2 255.255.255.0
!
router bgp 65011
 no synchronization
 network 8.8.8.0 mask 255.255.255.0
 neighbor 130.130.130.1 remote-as 65000
 no auto-summary
```

Code Listing 10.2 illustrates the EBGP configuration for Brussels-CE2. Brussels-CE2 advertises network 9.0.0.0/8 to Brussels-PE and belongs to site Central-1.

Code Listing 10.2:   EBGP configuration for Brussels-CE2

```
hostname Brussels-CE2
!
interface Loopback0
 ip address 9.9.9.9 255.0.0.0
!
interface Ethernet0/0
 ip address 100.100.100.2 255.255.255.0
!
router bgp 65012
 no synchronization
 network 9.0.0.0
 neighbor 100.100.100.1 remote-as 65000
 no auto-summary
```

Code Listing 10.3 illustrates the EBGP configuration for London-CE1. London-CE1 advertises host route 6.6.6.6/32 to London-PE and belongs to site Client-2.

Code Listing 10.3:   EBGP configuration for London-CE1

```
hostname London-CE1
!
interface Loopback0
 ip address 6.6.6.6 255.255.255.255
!
interface Serial1/0
 bandwidth 64
 ip address 200.200.200.2 255.255.255.0
!
router bgp 65021
 no synchronization
 network 6.6.6.6 mask 255.255.255.255
 neighbor 200.200.200.1 remote-as 65000
 no auto-summary
```

Code Listing 10.4 illustrates the EBGP configuration for London-CE2. London-CE2 advertises subnet 5.5.0.0/16 to London-PE and belongs to site Central-2.

Code Listing 10.4:   EBGP configuration for London-CE2

```
hostname London-CE2
!
interface Loopback0
 ip address 5.5.5.5 255.255.0.0
!
interface Ethernet0/0
 ip address 202.202.202.2 255.255.255.0
!
router bgp 65022
 no synchronization
 network 5.5.0.0 mask 255.255.0.0
 neighbor 202.202.202.1 remote-as 65000
 no auto-summary
```

10.2.4   PE configurations

Code Listings 10.5 and 10.6 illustrate the MPLS VPN configurations for Brussels-PE and London-PE, respectively. The configurations reflect the RT and RD numbering scheme defined in Tables 10.1 and 10.2.

The VRF configuration for Brussels-PE highlighted in Code Listing 10.5 is based on the RD and RT numbering scheme defined in Table 10.1.

Code Listing 10.5: MPLS VPN configuration for Brussels-PE

```
hostname Brussels-PE
!
ip vrf central-1 ! --- Create VRF central-1.
 rd 65000:666 ! --- Assign route distinguisher 65000:666 to VRF Central-1.

! --- Specify RT 65000:101 to be attached
! --- to every route from this VRF to MP-BGP.
 route-target export 65000:101

! --- Specify RT 65000:102 to be attached
! --- to every route from this VRF to MP-BGP.
 route-target export 65000:102

! --- Specify RT 65000:101 to be used as an import filter.
 route-target import 65000:101

! --- Specify RT 65000:102 to be used as an import filter.
 route-target import 65000:102
!
ip vrf client-1 ! --- Create VRF Client-1.
 rd 65000:101 ! --- Assign route distinguisher 65000:101 to VRF Client-1.

! --- Specify RT 65000:101 to be attached
! --- to every route from this VRF to MP-BGP.
 route-target export 65000:101

! --- Specify RT 65000:101 to be used as an import filter.
 route-target import 65000:101
!
ip cef
!
interface Loopback0
 ip address 223.0.0.1 255.255.255.255
 ip router isis
!
interface FastEthernet0/0
 ip vrf forwarding central-1 ! --- Associate FastEthernet0/0 with VRF Central-1.
 ip address 100.100.100.1 255.255.255.0
!
```

```
interface Serial1/0
 encapsulation frame-relay
 frame-relay lmi-type ansi
!
interface Serial1/0.1 point-to-point
 bandwidth 64
 ip address 10.1.1.5 255.255.255.252
 ip router isis
 tag-switching ip
 frame-relay interface-dlci 102
!
interface Serial1/7
 bandwidth 64
 ip vrf forwarding client-1 ! --- Associate Serial1/7 with VRF Client-1.
 ip address 130.130.130.1 255.255.255.0
 clockrate 64000
!
router isis
 net 49.0001.0000.0000.0001.00
 is-type level-1
 metric-style wide
!
! --- Configure MP-BGP for Brussels-PE in AS 65000.
router bgp 65000
 no synchronization
 neighbor 223.0.0.2 remote-as 65000
 neighbor 223.0.0.2 update-source Loopback0
 neighbor 223.0.0.3 remote-as 65000
 neighbor 223.0.0.3 update-source Loopback0
 no auto-summary
 !
! --- Enter the per-VRF BGP routing context for VRF Client-1.
 address-family ipv4 vrf client-1
 redistribute connected
 neighbor 130.130.130.2 remote-as 65011
 neighbor 130.130.130.2 activate
 no auto-summary
 no synchronization
 exit-address-family
 !
! --- Enter the per-VRF BGP routing context for VRF Central-1.
 address-family ipv4 vrf central-1
 redistribute connected
 neighbor 100.100.100.2 remote-as 65012
 neighbor 100.100.100.2 activate
```

```
no auto-summary
no synchronization
exit-address-family
!
address-family vpnv4
neighbor 223.0.0.2 activate
neighbor 223.0.0.2 send-community extended
neighbor 223.0.0.3 activate
neighbor 223.0.0.3 send-community extended
no auto-summary
exit-address-family
```

The VRF configuration for London-PE highlighted in Code Listing 10.6 is based on the RD and RT numbering scheme defined in Table 10.2.

Code Listing 10.6:   MPLS VPN configuration for London-PE

```
hostname London-PE
!
ip vrf central-2 ! --- Create VRF Central-2.
 rd 65000:999 ! --- Assign route distinguisher 65000:999 to VRF Central-2.

! --- Specify RT 65000:202 to be attached
! --- to every route from this VRF to MP-BGP.
 route-target export 65000:202

! --- Specify RT 65000:102 to be attached
! --- to every route from this VRF to MP-BGP.
 route-target export 65000:102

! --- Specify RT 65000:202 to be used as an import filter.
 route-target import 65000:202

! --- Specify RT 65000:102 to be used as an import filter.
 route-target import 65000:102
!
ip vrf client-2 ! --- Create VRF client-2.
 rd 65000:202 ! --- Assign route distinguisher 65000:202 to VRF Client-2.

! --- Specify RT 65000:202 to be attached
! --- to every route from this VRF to MP-BGP.
 route-target export 65000:202
```

```
! --- Specify RT 65000:202 to be used as an import filter.
 route-target import 65000:202
!
ip cef
!
interface Loopback0
 ip address 223.0.0.2 255.255.255.255
 ip router isis
!
interface FastEthernet0/0
 ip vrf forwarding central-2 ! --- Associate FastEthernet0/0 with VRF central-2.
 ip address 202.202.202.1 255.255.255.0
!
interface Serial1/0
 encapsulation frame-relay
 frame-relay lmi-type ansi
!
interface Serial1/0.2 point-to-point
 ip address 10.1.1.6 255.255.255.252
 ip router isis
 tag-switching ip
 frame-relay interface-dlci 201
!
interface Serial1/7
 bandwidth 64
 ip vrf forwarding client-2 ! --- Associate Serial1/7 with VRF Client-2.
 ip address 200.200.200.1 255.255.255.0
 clockrate 64000
!
router isis
 net 49.0001.0000.0000.0002.00
 is-type level-1
 metric-style wide
!
! --- Configure MP-BGP for London-PE in AS 65000.
router bgp 65000
 no synchronization
 neighbor 223.0.0.1 remote-as 65000
 neighbor 223.0.0.1 update-source Loopback0
 neighbor 223.0.0.3 remote-as 65000
 neighbor 223.0.0.3 update-source Loopback0
 no auto-summary
 !
! --- Enter the per-VRF BGP routing context for VRF Client-2.
 address-family ipv4 vrf client-2
```

```
redistribute connected
neighbor 200.200.200.2 remote-as 65021
neighbor 200.200.200.2 activate
no auto-summary
no synchronization
exit-address-family
!
! --- Enter the per-VRF BGP routing context for VRF Central-2.
address-family ipv4 vrf central-2
redistribute connected
neighbor 202.202.202.2 remote-as 65022
neighbor 202.202.202.2 activate
no auto-summary
no synchronization
exit-address-family
!
address-family vpnv4
neighbor 223.0.0.1 activate
neighbor 223.0.0.1 send-community extended
neighbor 223.0.0.3 activate
neighbor 223.0.0.3 send-community extended
no auto-summary
exit-address-family
```

10.2.5 Verifying VRFs configured in the PE routers

Code Listings 10.7 and 10.8 are used to verify the detailed VRF configurations for the two PE routers.

The "show ip vrf detail" command can be used to verify the VRF configuration for Brussels-PE, and the command output in Code Listing 10.7 corresponds with the RD and RT numbering scheme configured previously in Code Listing 10.5 for Brussels-PE.

Code Listing 10.7:   Detailed VRF configuration for Brussels-PE

```
Brussels-PE#show ip vrf detail
VRF central-1; default RD 65000:666
  Interfaces:
    FastEthernet0/0
  Connected addresses are not in global routing table
  Export VPN route-target communities
    RT:65000:101              RT:65000:102
  Import VPN route-target communities
    RT:65000:101              RT:65000:102
```

```
No import route-map
No export route-map
```
VRF client-1; default RD 65000:101
```
  Interfaces:
    Serial1/7
  Connected addresses are not in global routing table
```
  Export VPN route-target communities
    RT:65000:101
  Import VPN route-target communities
    RT:65000:101
```
No import route-map
No export route-map
```

The "show ip vrf detail" command can be used to verify the VRF configuration for London-PE, and the command output in Code Listing 10.8 corresponds with the RD and RT numbering scheme configured previously in Code Listing 10.6 for London-PE.

Code Listing 10.8:   Detailed VRF configuration for London-PE

```
London-PE#show ip vrf detail
```
VRF central-2; default RD 65000:999
```
  Interfaces:
    FastEthernet0/0
  Connected addresses are not in global routing table
```
  Export VPN route-target communities
    RT:65000:202   RT:65000:102
  Import VPN route-target communities
    RT:65000:202   RT:65000:102
```
No import route-map
No export route-map
```
VRF client-2; default RD 65000:202
```
  Interfaces:
    Serial1/7
  Connected addresses are not in global routing table
```
  Export VPN route-target communities
    RT:65000:202
  Import VPN route-target communities
    RT:65000:202
```
No import route-map
No export route-map
```

10.2.6 Monitoring BGP VPNv4 tables of the PE routers

Code Listings 10.9 and 10.10 illustrate the entire content of the BGP VPNv4 table for the two PE routers.

The "show ip bgp vpnv4 all" command in Code Listing 10.9 displays the entire content of the BGP VPNv4 table for Brussels-PE. From the table, we can see the per-VRF BGP VPNv4 information associated with VRF Client-1, which includes prefix 9.0.0.0/8 as well as 100.100.100.0/24 from Brussels-CE2 (VRF Central-1); and VRF Central-1, which includes prefixes 8.8.8.0/24 and 130.130.130.0/24 from Brussels-CE1 (VRF Client-1) as well as prefixes 5.5.0.0/16 and 202.202.202.0/24 from London-CE2 (VRF Central-2). This command output concurs with the intersite connectivity requirements defined in section 10.2.2.

Code Listing 10.9: Brussels-PE BGP VPNv4 table

```
Brussels-PE#show ip bgp vpnv4 all
BGP table version is 17, local router ID is 223.0.0.1
Status codes:  s suppressed, d damped, h history, * valid, > best, i - internal
Origin codes: i - IGP, e - EGP, ? - incomplete

    Network            Next Hop           Metric LocPrf Weight Path
Route Distinguisher: 65000:101 (default for vrf client-1)
*> 8.8.8.0/24         130.130.130.2           0             0 65011 i
*> 9.0.0.0            100.100.100.2           0             0 65012 i
*> 100.100.100.0/24 0.0.0.0                   0         32768 ?
*> 130.130.130.0/24 0.0.0.0                   0         32768 ?
Route Distinguisher: 65000:666 (default for vrf central-1)
*>i5.5.0.0/16         223.0.0.2               0    100      0 65022 i
*> 8.8.8.0/24         130.130.130.2           0             0 65011 i
*> 9.0.0.0            100.100.100.2           0             0 65012 i
*> 100.100.100.0/24 0.0.0.0                   0         32768 ?
*> 130.130.130.0/24 0.0.0.0                   0         32768 ?
*>i202.202.202.0     223.0.0.2               0    100      0 ?
<Output Omitted>
```

The "show ip bgp vpnv4 all" command in Code Listing 10.10 displays the entire content of the BGP VPNv4 table for London-PE. From the table, we can see the per-VRF BGP VPNv4 information associated with VRF Client-2, which includes prefix 5.5.0.0/16 as well as 202.202.202.0/24 from London-CE2 (VRF Central-2); and VRF Central-2, which includes prefixes 6.6.6.6/32 and 200.200.200.0/24 from London-CE1 (VRF Client-2) as well as prefixes 9.0.0.0/8 and 100.100.100.0/24 from Brussels-CE2 (VRF Central-1). This command utput concurs with the intersite connectivity requirements defined in section 10.2.2.

Code Listing 10.10:   London-PE BGP VPNv4 table

```
London-PE#show ip bgp vpnv4 all
BGP table version is 17, local router ID is 223.0.0.2
Status codes: s suppressed, d damped, h history, * valid, > best, i -
internal
Origin codes: i - IGP, e - EGP, ? - incomplete
      Network            Next Hop            Metric LocPrf Weight Path
Route Distinguisher: 65000:202 (default for vrf client-2)
*> 5.5.0.0/16         202.202.202.2            0            0 65022 i
*> 6.6.6.6/32         200.200.200.2            0            0 65021 i
*> 200.200.200.0      0.0.0.0                  0        32768 ?
*> 202.202.202.0      0.0.0.0                  0        32768 ?
<Output Omitted>
Route Distinguisher: 65000:999 (default for vrf central-2)
*> 5.5.0.0/16         202.202.202.2            0            0 65022 i
*> 6.6.6.6/32         200.200.200.2            0            0 65021 i
*>i9.0.0.0            223.0.0.1                0    100     0 65012 i
*>i100.100.100.0/24 223.0.0.1                  0    100     0 ?
*> 200.200.200.0      0.0.0.0                  0        32768 ?
*> 202.202.202.0      0.0.0.0                  0        32768 ?
```

10.2.7 Verifying the intersite connectivity

Code Listings 10.11 to 10.14 illustrate the IPv4 table for the four CE routers at different locations.

Code Listing 10.11 displays the IP routing table for Brussels-CE1 (Client-1) belonging to VPN 1. The BGP routes listed in Code Listing 10.9 for VRF Client-1 are installed into the IP routing table, thus fulfilling the intersite connectivity requirements defined in section 10.2.2.

Code Listing 10.11:   Brussels-CE1 IPv4 forwarding table

```
Brussels-CE1#show ip route
<Output Omitted>

     100.0.0.0/24 is subnetted, 1 subnets
B       100.100.100.0 [20/0] via 130.130.130.1, 00:09:20
     8.0.0.0/24 is subnetted, 1 subnets
C       8.8.8.0 is directly connected, Loopback0
B    9.0.0.0/8 [20/0] via 130.130.130.1, 00:08:51
     130.130.0.0/24 is subnetted, 1 subnets
C       130.130.130.0 is directly connected, Serial1/0
```

Code Listing 10.12 displays the IP routing table for Brussels-CE2 (Central-1) overlapping VPN 1 and VPN 12. The BGP routes listed in Code Listing 10.9 for VRF Central-1 are installed into the IP routing table, thus fulfilling the intersite connectivity requirements defined in section 10.2.2.

Code Listing 10.12:   Brussels-CE2 IPv4 forwarding table

```
Brussels-CE2#show ip route
<Output Omitted>
B    202.202.202.0/24 [20/0] via 100.100.100.1, 00:11:24
       100.0.0.0/24 is subnetted, 1 subnets
C       100.100.100.0 is directly connected, Ethernet0/0
       5.0.0.0/16 is subnetted, 1 subnets
B        5.5.0.0 [20/0] via 100.100.100.1, 00:10:53
       8.0.0.0/24 is subnetted, 1 subnets
B        8.8.8.0 [20/0] via 100.100.100.1, 00:10:53
C     9.0.0.0/8 is directly connected, Loopback0
       130.130.0.0/24 is subnetted, 1 subnets
B        130.130.130.0 [20/0] via 100.100.100.1, 00:10:53
```

Code Listing 10.13 displays the IP routing table for London-CE1 (Client-2) belonging to VPN 2. The BGP routes listed in Code Listing 10.10 for VRF Client-2 are installed into the IP routing table, thus fulfilling the intersite connectivity requirements defined in section 10.2.2.

Code Listing 10.13:   London-CE1 IPv4 forwarding table

```
London-CE1#show ip route
<Output Omitted>
C    200.200.200.0/24 is directly connected, Serial1/0
B    202.202.202.0/24 [20/0] via 200.200.200.1, 00:14:29
       5.0.0.0/16 is subnetted, 1 subnets
B        5.5.0.0 [20/0] via 200.200.200.1, 00:14:01
       6.0.0.0/32 is subnetted, 1 subnets
C        6.6.6.6 is directly connected, Loopback0
```

Code Listing 10.14 displays the IP routing table for London-CE2 (Central-2) overlapping VPN 2 and VPN 12. The BGP routes listed in Code Listing 10.10 for VRF Central-2 are installed into the IP routing table, thus fulfilling the intersite connectivity requirements defined in section 10.2.2.

Code Listing 10.14:   London-CE2 IPv4 forwarding table

```
London-CE2#show ip route
<Output Omitted>
B    200.200.200.0/24 [20/0] via 202.202.202.1, 00:17:07
C    202.202.202.0/24 is directly connected, Ethernet0/0
     100.0.0.0/24 is subnetted, 1 subnets
B       100.100.100.0 [20/0] via 202.202.202.1, 00:16:40
     5.0.0.0/16 is subnetted, 1 subnets
C       5.5.0.0 is directly connected, Loopback0
     6.0.0.0/32 is subnetted, 1 subnets
B       6.6.6.6 [20/0] via 202.202.202.1, 00:16:41
B    9.0.0.0/8 [20/0] via 202.202.202.1, 00:16:40
```

10.3   Case Study 10.2: Central Services VPN Topology

10.3.1   Case overview

This case study is an extension of Case Study 10.1. It provides a sample configuration of a central services VPN topology. This time the MPLS VPN deployment spans five different customer sites: Brussels Client-1, Brussels Server-1, London Client-2, London Server-2, and Paris Client-3. The MPLS VPN design and physical topology for Case Study 10.2 are shown in Figure 10.4.

10.3.2   Central services VPN topology

Figure 10.5 illustrates the VPN connectivity requirements (logical VPN topology) and the VPN memberships of the five different sites. As shown in Figure 10.5, the server sites—Brussels Server-1 and London Server-2—can communicate with all other sites (Server-1 with Client-1 as well as Client-3; and Server-2 with Client-1 as well as Client-2), while the client sites—Brussels Client-1, London Client-2, and Paris Client-3—can communicate only with the server sites but not with each other. In this case, Paris Client-3 can communicate only with Brussels Server-1; London Client-2 can communicate only with London Server-2; and Brussels Client-1 can communicate with both Brussels Server-1 and London Server-2.

In this example, we have five sites with different VPN connectivity requirements and memberships. This means we have to have at least five VRFs:

1. Client-1 is a member of VPN 1 and VPN 112.

2. Client-2 is a member of VPN 2 and VPN 22.

3. Client-3 is a member of VPN 3 and VPN 13.

4. Server-1 is a member of VPN 112 and VPN 13.

5. Server-2 is a member of VPN 112 and VPN 22.

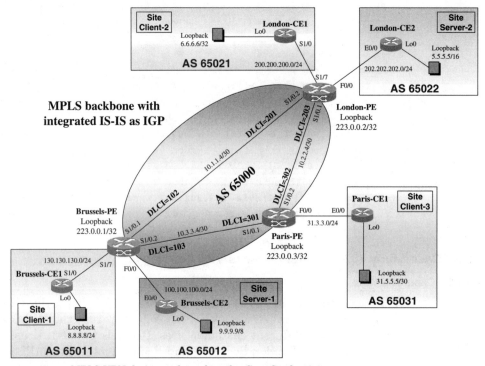

Figure 10.4: MPLS VPN design and topology for Case Study 10.2.

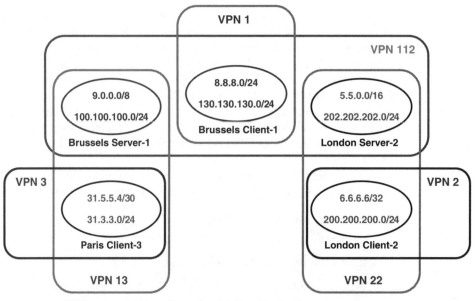

Figure 10.5: VPN connectivity requirements for Case Study 10.2.

TABLE 10.3   Per-VRF RD and RT numbering scheme for Brussels-PE.

VRF	Client-1	Server-1
Export RT	65000:101 65000:182 65000:192	65000:181 65000:222
Import RT	65000:101 65000:181 65000:191	65000:181 65000:182 65000:222
RD	65000:101	65000:180

TABLE 10.4   Per-VRF RD and RT numbering scheme for London-PE.

VRF	Client-2	Server-2
Export RT	65000:202 65000:192	65000:191 65000:222
Import RT	65000:202 65000:191	65000:191 65000:192 65000:222
RD	65000:202	65000:190

TABLE 10.5   Per-VRF RD and RT numbering scheme for Paris-PE.

VRF	Client-3
Export RT	65000:103 65000:182
Import RT	65000:103 65000:181
RD	65000:103

Tables 10.3 to 10.5 illustrate the RD and RT numbering scheme for Brussels-PE, London-PE and Paris-PE, respectively.

Tables 10.3 to 10.5 are illustrated schematically in Figure 10.6 for which:

- Client-1 imports all networks that carry RT 65000:101 or 65000:181 (from Server-1) or 65000:191 (from Server-2) and exports all networks with RTs 65000:101, 65000:182 (to Server-1), and 65000:192 (to Server-2).

- Client-2 imports all networks that carry RT 65000:202 or 65000:191 (from Server-2) and exports all networks with RTs 65000:202 and 65000:192 (to Server-2).

- Client-3 imports all networks that carry RT 65000:103 or 65000:181 (from Server-1) and exports all networks with RTs 65000:103 and 65000:182 (to Server-1).

- Server-1 imports all networks that carry RT 65000:181 or 65000:182 (from Client-1 and Client-3) or 65000:222 (from Server-2) and exports all networks with RTs 65000:181 (to Client-1 and Client-3) and 65000:222 (to Server-2).

- Server-2 imports all networks that carry RT 65000:191 or 65000:192 (from Client-1 and Client-2) or 65000:222 (from Server-1) and exports all networks with RTs 65000:191 (to Client-1 and Client-2) and 65000:222 (to Server-1).

In addition, according to the five VRFs defined in Tables 10.3 to 10.5, we can construe the following intersite connectivity:

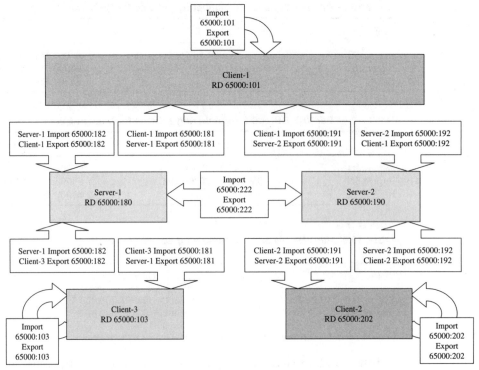

Figure 10.6: RD and RT numbering scheme for Case Study 10.2.

- VRF Client-1 contains routes from VRF Server-1: 9.0.0.0/8 and 100.100.100.0/24.

- VRF Client-1 contains routes from VRF Server-2: 5.5.0.0/16 and 202.202.202.0/24.

- VRF Client-2 contains routes from VRF Server-2: 5.5.0.0/16 and 202.202.202.0/24.

- VRF Client-3 contains routes from VRF Server-1: 9.0.0.0/8 and 100.100.100.0/24.

- VRF Server-1 contains routes from VRF Client-1: 8.8.8.0/24 and 130.130.130.0/24.

- VRF Server-1 contains routes from VRF Client-3: 31.5.5.4/30 and 31.3.3.0/24.

- VRF Server-1 contains routes from VRF Server-2: 5.5.0.0/16 and 202.202.202.0/24.

- VRF Server-2 contains routes from VRF Client-1: 8.8.8.0/24 and 130.130.130.0/24.

- VRF Server-2 contains routes from VRF Client-2: 6.6.6.6/32 and 200.200.200.0/24.

- VRF Server-2 contains routes from VRF Server-1: 9.0.0.0/8 and 100.100.100.0/24.

Note that the client-VRFs do not contain routes from other clients, implying that all the clients cannot communicate with each other. These intersite connectivity deductions will be verified in the later sections.

10.3.3 CE configurations

In this section, we shall only illustrate the EBGP configuration for Paris-CE1. The other four CE routers share the same configurations as those illustrated in Code Listing 10.1 to 10.4.

Code Listing 10.15 illustrates the EBGP configuration for Paris-CE1. Paris-CE1 advertises subnet 31.5.5.4/30 to Paris-PE and belongs to site Client-3.

Code Listing 10.15:   EBGP configuration for Paris-CE1

```
hostname Paris-CE1
!
interface Loopback0
  ip address 31.5.5.5 255.255.255.252
!
interface Ethernet0/0
  ip address 31.3.3.2 255.255.255.0
!
router bgp 65031
  no synchronization
  network 31.5.5.4 mask 255.255.255.252
  neighbor 31.3.3.1 remote-as 65000
  no auto-summary
```

10.3.4 PE configurations

Code Listings 10.16 to 10.18 illustrate the MPLS VPN configurations for Brussels-PE, London-PE and Paris-PE, respectively. The configurations reflect the RT and RD numbering scheme defined in Tables 10.3 to 10.5.

The VRF configuration for Brussels-PE highlighted in Code Listing 10.16 is based on the RD and RT numbering scheme defined in Table 10.3.

Code Listing 10.16:   MPLS VPN configuration for Brussels-PE

```
hostname Brussels-PE
!
ip vrf client-1 ! --- Create VRF Client-1.
 rd 65000:101 ! --- Assign route distinguisher 65000:101 to VRF Client-1.

! --- Specify RT 65000:101 to be attached
! --- to every route from this VRF to MP-BGP.
 route-target export 65000:101

! --- Specify RT 65000:182 to be attached
! --- to every route from this VRF to MP-BGP.
 route-target export 65000:182

! --- Specify RT 65000:192 to be attached
! --- to every route from this VRF to MP-BGP.
 route-target export 65000:192

! --- Specify RT 65000:101 to be used as an import filter.
 route-target import 65000:101

! --- Specify RT 65000:181 to be used as an import filter.
 route-target import 65000:181

! --- Specify RT 65000:191 to be used as an import filter.
 route-target import 65000:191
!
ip vrf server-1 ! --- Create VRF Server-1.
 rd 65000:180 ! --- Assign route distinguisher 65000:180 to VRF Server-1.

! --- Specify RT 65000:181 to be attached
! --- to every route from this VRF to MP-BGP.
 route-target export 65000:181

! --- Specify RT 65000:222 to be attached
! --- to every route from this VRF to MP-BGP.
 route-target export 65000:222

! --- Specify RT 65000:181 to be used as an import filter.
 route-target import 65000:181

! --- Specify RT 65000:182 to be used as an import filter.
 route-target import 65000:182
```

```
! --- Specify RT 65000:222 to be used as an import filter.
 route-target import 65000:222
!
ip cef
!
interface Loopback0
 ip address 223.0.0.1 255.255.255.255
 ip router isis
!
interface FastEthernet0/0
 ip vrf forwarding server-1 ! --- Associate FastEthernet0/0 with VRF Server-
 1.
 ip address 100.100.100.1 255.255.255.0
!
interface Serial1/0
 encapsulation frame-relay
 frame-relay lmi-type ansi
!
interface Serial1/0.1 point-to-point
 bandwidth 64
 ip address 10.1.1.5 255.255.255.252
 ip router isis
 tag-switching ip
 frame-relay interface-dlci 102
!
interface Serial1/0.2 point-to-point
 bandwidth 64
 ip address 10.3.3.5 255.255.255.252
 ip router isis
 tag-switching ip
 frame-relay interface-dlci 103
!
interface Serial1/7
 bandwidth 64
 ip vrf forwarding client-1 ! --- Associate Serial1/7 with VRF Client-1.
 ip address 130.130.130.1 255.255.255.0
 clockrate 64000
!
router isis
 net 49.0001.0000.0000.0001.00
 is-type level-1
 metric-style wide
!
! --- Configure MP-BGP for Brussels-PE in AS 65000.
router bgp 65000
```

```
no synchronization
neighbor 223.0.0.2 remote-as 65000
neighbor 223.0.0.2 update-source Loopback0
neighbor 223.0.0.3 remote-as 65000
neighbor 223.0.0.3 update-source Loopback0
no auto-summary
!
! --- Enter the per-VRF BGP routing context for VRF Server-1.
address-family ipv4 vrf server-1
redistribute connected
neighbor 100.100.100.2 remote-as 65012
neighbor 100.100.100.2 activate
no auto-summary
no synchronization
exit-address-family
!
! --- Enter the per-VRF BGP routing context for VRF Client-1.
address-family ipv4 vrf client-1
redistribute connected
neighbor 130.130.130.2 remote-as 65011
neighbor 130.130.130.2 activate
no auto-summary
no synchronization
exit-address-family
!
address-family vpnv4
neighbor 223.0.0.2 activate
neighbor 223.0.0.2 send-community extended
neighbor 223.0.0.3 activate
neighbor 223.0.0.3 send-community extended
no auto-summary
exit-address-family
```

The VRF configuration for London-PE highlighted in Code Listing 10.17 is based on the RD and RT numbering scheme defined in Table 10.4.

Code Listing 10.17:   MPLS VPN configuration for London-PE

```
hostname London-PE
!
ip vrf client-2 ! --- Create VRF Client-2.
  rd 65000:202 ! --- Assign route distinguisher 65000:202 to VRF Client-2.
```

```
! --- Specify RT 65000:202 to be attached
! --- to every route from this VRF to MP-BGP.
 route-target export 65000:202

! --- Specify RT 65000:192 to be attached
! --- to every route from this VRF to MP-BGP.
 route-target export 65000:192

! --- Specify RT 65000:202 to be used as an import filter.
 route-target import 65000:202

! --- Specify RT 65000:191 to be used as an import filter.
 route-target import 65000:191
!
ip vrf server-2 ! --- Create VRF Server-2.
 rd 65000:190 ! --- Assign route distinguisher 65000:190 to VRF Server-2.

! --- Specify RT 65000:191 to be attached
! --- to every route from this VRF to MP-BGP.
 route-target export 65000:191

! --- Specify RT 65000:222 to be attached
! --- to every route from this VRF to MP-BGP.
 route-target export 65000:222

! --- Specify RT 65000:191 to be used as an import filter.
 route-target import 65000:191

! --- Specify RT 65000:192 to be used as an import filter.
 route-target import 65000:192

! --- Specify RT 65000:222 to be used as an import filter.
 route-target import 65000:222
!
ip cef
!
interface Loopback0
 ip address 223.0.0.2 255.255.255.255
 ip router isis
!
interface FastEthernet0/0
 ip vrf forwarding server-2 ! --- Associate FastEthernet0/0 with VRF Server-
 2.
 ip address 202.202.202.1 255.255.255.0
!
```

```
interface Serial1/0
 encapsulation frame-relay
 frame-relay lmi-type ansi
!
interface Serial1/0.1 point-to-point
 ip address 10.2.2.5 255.255.255.252
 ip router isis
 tag-switching ip
 frame-relay interface-dlci 203
!
interface Serial1/0.2 point-to-point
 ip address 10.1.1.6 255.255.255.252
 ip router isis
 tag-switching ip
 frame-relay interface-dlci 201
!
interface Serial1/7
 bandwidth 64
 ip vrf forwarding client-2 ! --- Associate Serial1/7 with VRF Client-2.
 ip address 200.200.200.1 255.255.255.0
 clockrate 64000
!
router isis
 net 49.0001.0000.0000.0002.00
 is-type level-1
 metric-style wide
!
! --- Configure MP-BGP for London-PE in AS 65000.
router bgp 65000
 no synchronization
 neighbor 223.0.0.1 remote-as 65000
 neighbor 223.0.0.1 update-source Loopback0
 neighbor 223.0.0.3 remote-as 65000
 neighbor 223.0.0.3 update-source Loopback0
 no auto-summary
 !
! --- Enter the per-VRF BGP routing context for VRF Server-2.
 address-family ipv4 vrf server-2
 redistribute connected
 neighbor 202.202.202.2 remote-as 65022
 neighbor 202.202.202.2 activate
 no auto-summary
 no synchronization
 exit-address-family
 !
```

```
! --- Enter the per-VRF BGP routing context for VRF Client-2.
 address-family ipv4 vrf client-2
 redistribute connected
 neighbor 200.200.200.2 remote-as 65021
 neighbor 200.200.200.2 activate
 no auto-summary
 no synchronization
 exit-address-family
 !
 address-family vpnv4
 neighbor 223.0.0.1 activate
 neighbor 223.0.0.1 send-community extended
 neighbor 223.0.0.3 activate
 neighbor 223.0.0.3 send-community extended
 no auto-summary
 exit-address-family
```

The VRF configuration for Paris-PE highlighted in Code Listing 10.18 is based on the RD and RT numbering scheme defined in Table 10.5.

Code Listing 10.18: MPLS VPN configuration for Paris-PE

```
hostname Paris-PE
!
ip vrf client-3 ! --- Create VRF Client-3.
 rd 65000:103 ! --- Assign route distinguisher 65000:103 to VRF Client-3.

! --- Specify RT 65000:103 to be attached
! --- to every route from this VRF to MP-BGP.
 route-target export 65000:103

! --- Specify RT 65000:182 to be attached
! --- to every route from this VRF to MP-BGP.
 route-target export 65000:182

! --- Specify RT 65000:103 to be used as an import filter.
 route-target import 65000:103

! --- Specify RT 65000:181 to be used as an import filter.
 route-target import 65000:181
!
ip cef
!
```

```
interface Loopback0
 ip address 223.0.0.3 255.255.255.255
 ip router isis
!
interface FastEthernet0/0
 ip vrf forwarding client-3 ! --- Associate FastEthernet0/0 with VRF Client-
 3.
 ip address 31.3.3.1 255.255.255.0
!
interface Serial1/0
 encapsulation frame-relay
 frame-relay lmi-type ansi
!
interface Serial1/0.1 point-to-point
 ip address 10.3.3.6 255.255.255.252
 ip router isis
 tag-switching ip
 frame-relay interface-dlci 301
!
interface Serial1/0.2 point-to-point
 ip address 10.2.2.6 255.255.255.252
 ip router isis
 tag-switching ip
 frame-relay interface-dlci 302
!
router isis
 net 49.0001.0000.0000.0003.00
 is-type level-1
 metric-style wide
!
! --- Configure MP-BGP for Paris-PE in AS 65000.
router bgp 65000
 no synchronization
 neighbor 223.0.0.1 remote-as 65000
 neighbor 223.0.0.1 update-source Loopback0
 neighbor 223.0.0.2 remote-as 65000
 neighbor 223.0.0.2 update-source Loopback0
 no auto-summary
 !
! --- Enter the per-VRF BGP routing context for VRF Client-3.
 address-family ipv4 vrf client-3
 redistribute connected
 neighbor 31.3.3.2 remote-as 65031
 neighbor 31.3.3.2 activate
 no auto-summary
```

```
no synchronization
exit-address-family
!
address-family vpnv4
neighbor 223.0.0.1 activate
neighbor 223.0.0.1 send-community extended
neighbor 223.0.0.2 activate
neighbor 223.0.0.2 send-community extended
no auto-summary
exit-address-family
```

10.3.5 Verifying VRFs configured in the PE routers

Code Listings 10.19 to 10.21 are used to verify the detailed VRF configurations for the three PE routers.

The "show ip vrf detail" command can be used to verify the VRF configuration for Brussels-PE, and the command output in Code Listing 10.19 corresponds with the RD and RT numbering scheme configured previously in Code Listing 10.16 for Brussels-PE.

Code Listing 10.19: Detailed VRF configuration for Brussels-PE

```
Brussels-PE#show ip vrf detail
VRF client-1; default RD 65000:101
  Interfaces:
    Serial1/7
  Connected addresses are not in global routing table
  Export VPN route-target communities
    RT:65000:101            RT:65000:182            RT:65000:192
  Import VPN route-target communities
    RT:65000:101            RT:65000:181            RT:65000:191
  No import route-map
  No export route-map
VRF server-1; default RD 65000:180
  Interfaces:
    FastEthernet0/0
  Connected addresses are not in global routing table
  Export VPN route-target communities
    RT:65000:181            RT:65000:222
  Import VPN route-target communities
    RT:65000:181            RT:65000:182            RT:65000:222
  No import route-map
  No export route-map
```

The "show ip vrf detail" command can be used to verify the VRF configuration for London-PE, and the command output in Code Listing 10.20 corresponds with the RD and RT numbering scheme configured previously in Code Listing 10.17 for London-PE.

Code Listing 10.20:   Detailed VRF configuration for London-PE

```
London-PE#show ip vrf detail
VRF client-2; default RD 65000:202
  Interfaces:
    Serial1/7
  Connected addresses are not in global routing table
  Export VPN route-target communities
    RT:65000:202              RT:65000:192
  Import VPN route-target communities
    RT:65000:202              RT:65000:191
  No import route-map
  No export route-map
VRF server-2; default RD 65000:190
  Interfaces:
    FastEthernet0/0
  Connected addresses are not in global routing table
  Export VPN route-target communities
    RT:65000:191              RT:65000:222
  Import VPN route-target communities
    RT:65000:191              RT:65000:192          RT:65000:222
  No import route-map
  No export route-map
```

The "show ip vrf detail" command can be used to verify the VRF configuration for Paris-PE, and the command output in Code Listing 10.21 corresponds with the RD and RT numbering scheme configured previously in Code Listing 10.18 for Paris-PE.

Code Listing 10.21:   Detailed VRF configuration for Paris-PE

```
Paris-PE#show ip vrf detail
VRF client-3; default RD 65000:103
  Interfaces:
    FastEthernet0/0
  Connected addresses are not in global routing table
  Export VPN route-target communities
    RT:65000:103              RT:65000:182
```

```
Import VPN route-target communities
   RT:65000:103                RT:65000:181
No import route-map
No export route-map
```

10.3.6 Monitoring BGP VPNv4 tables of the PE routers

Code Listings 10.22 to 10.24 illustrate the entire content of the BGP VPNv4 Table for the three PE routers.

The "show ip bgp vpnv4 all" command in Code Listing 10.22 displays the entire content of the BGP VPNv4 table for Brussels-PE. From the table, the per-VRF BGP VPNv4 information associated with VRF Client-1 includes prefix 9.0.0.0/8 as well as 100.100.100.0/24 from Brussels-CE2 (VRF Server-1), and prefix 5.5.0.0/16 as well as 202.202.202.0/24 from London-CE2 (VRF Server-2). Also, the per-VRF BGP VPNv4 information associated with VRF Server-1 includes prefix 8.8.8.0/24 as well as 130.130.130.0/24 from Brussels-CE1 (VRF Client-1), prefix 31.5.5.4/30 as well as 31.3.3.0/24 from Paris-CE1 (VRF Client-3), and prefix 5.5.0.0/16 as well as 202.202.202.0/24 from London-CE2 (VRF Server-2). This command output concurs with the intersite connectivity requirements defined in section 10.3.2.

Code Listing 10.22:   Brussels-PE BGP VPNv4 table

```
Brussels-PE#show ip bgp vpnv4 all
BGP table version is 62, local router ID is 223.0.0.1
Status codes: s suppressed, d damped, h history, * valid, > best, i - internal
Origin codes: i - IGP, e - EGP, ? - incomplete


    Network          Next Hop          Metric LocPrf Weight Path
Route Distinguisher: 65000:101 (default for vrf client-1)
*>i5.5.0.0/16       223.0.0.2              0    100      0 65022 i
*> 8.8.8.0/24       130.130.130.2         0              0 65011 i
*> 9.0.0.0          100.100.100.2         0              0 65012 i
*> 100.100.100.0/24 0.0.0.0               0          32768 ?
*> 130.130.130.0/24 0.0.0.0               0          32768 ?
*>i202.202.202.0    223.0.0.2              0    100      0 ?
<Output Omitted>
Route Distinguisher: 65000:180 (default for vrf server-1)
*>i5.5.0.0/16       223.0.0.2              0    100      0 65022 i
*> 8.8.8.0/24       130.130.130.2         0              0 65011 i
*> 9.0.0.0          100.100.100.2         0              0 65012 i
*>i31.3.3.0/24      223.0.0.3              0    100      0 ?
*>i31.5.5.4/30      223.0.0.3              0    100      0 65031 i
*> 100.100.100.0/24 0.0.0.0               0          32768 ?
```

```
*>  130.130.130.0/24 0.0.0.0                  0            32768 ?
*>i202.202.202.0    223.0.0.2                 0      100      0 ?
<Output Omitted>
```

The "show ip bgp vpnv4 all" command in Code Listing 10.23 displays the entire content of the BGP VPNv4 table for London-PE. From the table, the per-VRF BGP VPNv4 information associated with VRF Client-2 includes prefix 5.5.0.0/16 as well as 202.202.202.0/24 from London-CE2 (VRF Server-2). Also, the per-VRF BGP VPNv4 information associated with VRF Server-2 includes prefix 8.8.8.0/24 as well as 130.130.130.0/24 from Brussels-CE1 (VRF Client-1), prefix 6.6.6.6/32 as well as 200.200.200.0/24 from London-CE1 (VRF Client-2), and prefix 9.0.0.0/8 as well as 100.100.100.0/24 from Brussels-CE2 (VRF Server-1). This command output concurs with the intersite connectivity requirements defined in section 10.3.2.

Code Listing 10.23:   London-PE BGP VPNv4 table

```
London-PE#show ip bgp vpnv4 all
BGP table version is 21, local router ID is 223.0.0.2
Status codes: s suppressed, d damped, h history, * valid, > best, i - internal
Origin codes: i - IGP, e - EGP, ? - incomplete

    Network          Next Hop            Metric LocPrf Weight Path
<Output Omitted>
Route Distinguisher: 65000:190 (default for vrf server-2)
*>  5.5.0.0/16       202.202.202.2            0              0 65022 i
*>  6.6.6.6/32       200.200.200.2            0              0 65021 i
*>i8.8.8.0/24        223.0.0.1                0      100      0 65011 i
*>i9.0.0.0           223.0.0.1                0      100      0 65012 i
*>i100.100.100.0/24 223.0.0.1                 0      100      0 ?
*>i130.130.130.0/24 223.0.0.1                 0      100      0 ?
*>  200.200.200.0    0.0.0.0                  0            32768 ?
*>  202.202.202.0    0.0.0.0                  0            32768 ?
Route Distinguisher: 65000:202 (default for vrf client-2)
*>  5.5.0.0/16       202.202.202.2            0              0 65022 i
*>  6.6.6.6/32       200.200.200.2            0              0 65021 i
*>  200.200.200.0    0.0.0.0                  0            32768 ?
*>  202.202.202.0    0.0.0.0                  0            32768 ?
```

The "show ip bgp vpnv4 all" command in Code Listing 10.24 displays the entire content of the BGP VPNv4 table for Paris-PE. From the table, the per-VRF BGP VPNv4 information associated with VRF Client-3 includes prefix 9.0.0.0/8 as well as 100.100.100.0/24 from Brussels-CE2 (VRF Server-1). This command output concurs with the intersite connectivity requirements defined in section 10.3.2.

Code Listing 10.24: Paris-PE BGP VPNv4 table

```
Paris-PE#show ip bgp vpnv4 all
BGP table version is 73, local router ID is 223.0.0.3
Status codes: s suppressed, d damped, h history, * valid, > best, i -
internal
Origin codes: i - IGP, e - EGP, ? - incomplete

   Network          Next Hop          Metric LocPrf Weight Path
Route Distinguisher: 65000:103 (default for vrf client-3)
*>i9.0.0.0          223.0.0.1              0    100       0 65012 i
*> 31.3.3.0/24      0.0.0.0               0          32768 ?
*> 31.5.5.4/30      31.3.3.2             0              0 65031 i
*>i100.100.100.0/24 223.0.0.1             0    100       0 ?
<Output Omitted>
```

10.3.7 Verifying the intersite connectivity

Code Listings 10.25 to 10.29 illustrate the IPv4 table for the five CE routers in different locations.

Code Listing 10.25 displays the IP routing table for Brussels-CE1 (Client-1) overlapping VPN 1 and VPN 112. The BGP routes listed in Code Listing 10.22 for VRF Client-1 are installed into the IP routing table, thus fulfilling the intersite connectivity requirements defined in section 10.3.2.

Code Listing 10.25: Brussels-CE1 IPv4 Forwarding Table

```
Brussels-CE1#show ip route
<Output Omitted>

B    202.202.202.0/24 [20/0] via 130.130.130.1, 00:10:06
     100.0.0.0/24 is subnetted, 1 subnets
B       100.100.100.0 [20/0] via 130.130.130.1, 00:21:48
     5.0.0.0/16 is subnetted, 1 subnets
B       5.5.0.0 [20/0] via 130.130.130.1, 00:10:05
     8.0.0.0/24 is subnetted, 1 subnets
C       8.8.8.0 is directly connected, Loopback0
B    9.0.0.0/8 [20/0] via 130.130.130.1, 00:21:48
     130.130.0.0/24 is subnetted, 1 subnets
C       130.130.130.0 is directly connected, Serial1/0
```

Code Listing 10.26 displays the IP routing table for Brussels-CE2 (Server-1) overlapping VPN 13 and VPN 112. The BGP routes listed in Code Listing 10.22 for VRF Server-1 are installed into the IP routing table, thus fulfilling the inter-site connectivity requirements defined in section 10.3.2.

Code Listing 10.26:   Brussels-CE2 IPv4 forwarding table

```
Brussels-CE2#show ip route
<Output Omitted>


B      202.202.202.0/24 [20/0] via 100.100.100.1, 00:12:29
       100.0.0.0/24 is subnetted, 1 subnets
C         100.100.100.0 is directly connected, Ethernet0/0
       5.0.0.0/16 is subnetted, 1 subnets
B         5.5.0.0 [20/0] via 100.100.100.1, 00:12:29
       8.0.0.0/24 is subnetted, 1 subnets
B         8.8.8.0 [20/0] via 100.100.100.1, 00:23:46
C      9.0.0.0/8 is directly connected, Loopback0
       130.130.0.0/24 is subnetted, 1 subnets
B         130.130.130.0 [20/0] via 100.100.100.1, 00:27:51
       31.0.0.0/8 is variably subnetted, 2 subnets, 2 masks
B         31.5.5.4/30 [20/0] via 100.100.100.1, 00:27:51
B         31.3.3.0/24 [20/0] via 100.100.100.1, 00:27:52
```

Code Listing 10.27 displays the IP routing table for London-CE1 (Client-2) overlapping VPN 2 and VPN 22. The BGP routes listed in Code Listing 10.23 for VRF Client-2 are installed into the IP routing table, thus fulfilling the inter-site connectivity requirements defined in section 10.3.2.

Code Listing 10.27:   London-CE1 IPv4 forwarding table

```
London-CE1#show ip route
<Output Omitted>


C      200.200.200.0/24 is directly connected, Serial1/0
B      202.202.202.0/24 [20/0] via 200.200.200.1, 00:01:19
       5.0.0.0/16 is subnetted, 1 subnets
B         5.5.0.0 [20/0] via 200.200.200.1, 00:00:53
       6.0.0.0/32 is subnetted, 1 subnets
C         6.6.6.6 is directly connected, Loopback0
```

Code Listing 10.28 displays the IP routing table for London-CE2 (Server-2) overlapping VPN 22 and VPN 112. The BGP routes listed in Code Listing 10.23

for VRF Server-2 are installed into the IP routing table, thus fulfilling the inter-site connectivity requirements defined in section 10.3.2.

Code Listing 10.28:   London-CE2 IPv4 forwarding table

```
London-CE2#show ip route
<Output Omitted>


B     200.200.200.0/24 [20/0] via 202.202.202.1, 00:02:20
C     202.202.202.0/24 is directly connected, Ethernet0/0
      100.0.0.0/24 is subnetted, 1 subnets
B        100.100.100.0 [20/0] via 202.202.202.1, 00:02:20
      5.0.0.0/16 is subnetted, 1 subnets
C        5.5.0.0 is directly connected, Loopback0
      6.0.0.0/32 is subnetted, 1 subnets
B        6.6.6.6 [20/0] via 202.202.202.1, 00:01:53
      8.0.0.0/24 is subnetted, 1 subnets
B        8.8.8.0 [20/0] via 202.202.202.1, 00:02:20
B     9.0.0.0/8 [20/0] via 202.202.202.1, 00:02:20
      130.130.0.0/24 is subnetted, 1 subnets
B        130.130.130.0 [20/0] via 202.202.202.1, 00:02:20
```

Code Listing 10.29 displays the IP routing table for Paris-CE1 (Client-3) over-lapping VPN 3 and VPN 13. The BGP routes listed in Code Listing 10.24 for VRF Client-3 are installed into the IP routing table, thus fulfilling the intersite connectivity requirements defined in section 10.3.2.

Code Listing 10.29:   Paris-CE1 IPv4 forwarding table

```
Paris-CE1#show ip route
<Output Omitted>


      100.0.0.0/24 is subnetted, 1 subnets
B        100.100.100.0 [20/0] via 31.3.3.1, 00:32:50
B     9.0.0.0/8 [20/0] via 31.3.3.1, 00:32:50
      31.0.0.0/8 is variably subnetted, 2 subnets, 2 masks
C        31.5.5.4/30 is directly connected, Loopback0
C        31.3.3.0/24 is directly connected, Ethernet0/0
```

10.4   Case Study 10.3: Hybrid VPN Topology

10.4.1   Case overview

This case study is a variant of Case Study 10.2. It provides a sample configuration of a central services with simple VPN topology—a hybrid VPN topology. Similar to Case Study 10.2, the MPLS VPN deployment spans five different customer sites: Brussels Client-1, Brussels Central-1, London Client-2, London Central-2, and Paris Server. The MPLS VPN design and physical topology for Case Study 10.3 are shown in Figure 10.7.

10.4.2   Hybrid VPN topology

Figure 10.8 illustrates the VPN connectivity requirements (logical VPN topology) and the VPN memberships of the five different sites. As shown in Figure 10.8, some of the customer sites (Central-1 and Central-2) need access to the central server (Server), and all other sites (Client-1 and Client-2) just need intra-VPN access. In other words, Brussels Central-1, London-Central-2, and Paris Server can communicate with each other, whereas Brussels Client-1 can communicate only with Brussels Central-1 and London Client-2 can communi-

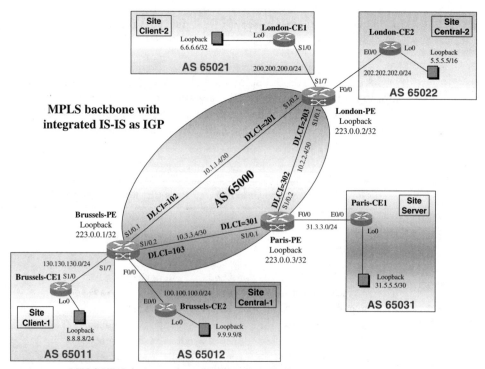

Figure 10.7: MPLS VPN design and topology for Case Study 10.3.

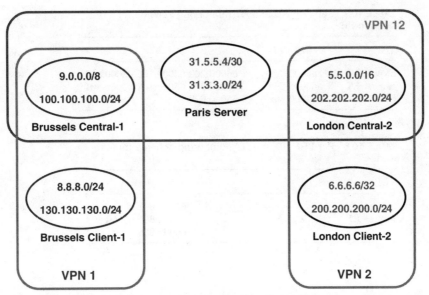

Figure 10.8: VPN connectivity requirements for Case Study 10.3.

TABLE 10.6 Per-VRF RD and RT numbering scheme for Brussels-PE.

VRF	Client-1	Central-1
Export RT	65000:101	65000:101 65000:102 65000:802
Import RT	65000:101	65000:101 65000:102 65000:801
RD	65000:101	65000:666

cate only with London Central-2. As a result, this design is a mix of central services plus simple VPN topology.

In this example, we have five sites with different VPN connectivity requirements and memberships. This means we have to have at least five VRFs:

1. Client-1 is a member of VPN 1.

2. Client-2 is a member of VPN 2.

3. Server is a member of VPN 12.

4. Central-1 is a member of VPN 1 and VPN 12.

5. Central-2 is a member of VPN 2 and VPN 12.

Tables 10.6 to 10.8 illustrate the RD and RT numbering scheme for Brussels-PE, London-PE and Paris-PE, respectively.

Tables 10.6 to 10.8 are illustrated schematically in Figure 10.9 for which:

TABLE 10.7   Per-VRF RD and RT numbering scheme for
London-PE.

VRF	Client-2	Central-2
Export RT	65000:202	65000:202 65000:102 65000:802
Import RT	65000:202	65000:202 65000:102 65000:801
RD	65000:202	65000:999

TABLE 10.8   Per-VRF RD and RT numbering scheme for
Paris-PE.

VRF	Server
Export RT	65000:801
Import RT	65000:801 65000:802
RD	65000:888

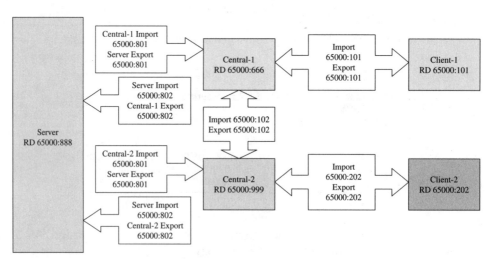

Figure 10.9:   RD and RT numbering scheme for Case Study 10.3.

- Client-1 imports all networks that carry RT 65000:101 and exports all networks with RT 65000:101.

- Client-2 imports all networks that carry RT 65000:202 and exports all networks with RT 65000:202.

- Server imports all networks that carry RT 65000:801 or 65000:802 (from Central-1 and Central-2); and exports all networks with RT 65000:801 (to Central-1 and Central-2).

- Central-1 imports all networks that carry RT 65000:101 (from Client-1) or 65000:102 (from Central-2) or 65000:801 (from Server); and exports all networks with RTs 65000:101 (to Client-1), 65000:102 (to Central-2) and 65000:802 (to Server).

- Central-2 imports all networks that carry RT 65000:202 (from Client-2) or 65000:102 (from Central-1) or 65000:801 (from Server); and exports all networks with RTs 65000:202 (to Client-2), 65000:102 (to Central-1) and 65000:802 (to Server).

In addition, according to the five VRFs defined in Tables 10.6 to 10.8, we can construe the following inter-site connectivity:

- VRF Client-1 contains routes from VRF Central-1: 9.0.0.0/8 and 100.100.100.0/24.

- VRF Client-2 contains routes from VRF Central-2: 5.5.0.0/16 and 202.202.202.0/24.

- VRF Server contains routes from VRF Central-1: 9.0.0.0/8 and 100.100.100.0/24.

- VRF Server contains routes from VRF Central-2: 5.5.0.0/16 and 202.202.202.0/24.

- VRF Central-1 contains routes from VRF Client-1: 8.8.8.0/24 and 130.130.130.0/24.

- VRF Central-1 contains routes from VRF Server: 31.5.5.4/30 and 31.3.3.0/24.

- VRF Central-1 contains routes from VRF Central-2: 5.5.0.0/16 and 202.202.202.0/24.

- VRF Central-2 contains routes from VRF Client-2: 6.6.6.6/32 and 200.200.200.0/24.

- VRF Central-2 contains routes from VRF Server: 31.5.5.4/30 and 31.3.3.0/24.

- VRF Central-2 contains routes from VRF Central-1: 9.0.0.0/8 and 100.100.100.0/24.

Note that VRF Client-1 does not contain routes from VRF Client-2 and vice versa, implying that Client-1 and Client-2 cannot communicate with each other. These intersite connectivity deductions will be verified in the later sections.

10.4.3   CE configurations

The five CE routers in this case study reuse the same configuration listings as those illustrated in Code Listings 10.1 to 10.4 and Code Listing 10.15.

10.4.4   PE configurations

Code Listings 10.30 to 10.32 illustrate the MPLS VPN configurations for Brussels-PE, London-PE, and Paris-PE, respectively. The configurations reflect the RT and RD numbering scheme defined in Tables 10.6 to 10.8.

The VRF configuration for Brussels-PE highlighted in Code Listing 10.30 is based on the RD and RT numbering scheme defined in Table 10.6.

Code Listing 10.30:   MPLS VPN configuration for Brussels-PE

```
hostname Brussels-PE
!
ip vrf central-1
 rd 65000:666
 route-target export 65000:101
 route-target export 65000:102
 route-target export 65000:802
 route-target import 65000:101
 route-target import 65000:102
 route-target import 65000:801
!
ip vrf client-1
 rd 65000:101
 route-target export 65000:101
 route-target import 65000:101
!
ip cef
!
interface Loopback0
 ip address 223.0.0.1 255.255.255.255
 ip router isis
!
interface FastEthernet0/0
 ip vrf forwarding central-1
 ip address 100.100.100.1 255.255.255.0
!
interface Serial1/0
 encapsulation frame-relay
 frame-relay lmi-type ansi
!
interface Serial1/0.1 point-to-point
 bandwidth 64
 ip address 10.1.1.5 255.255.255.252
 ip router isis
 tag-switching ip
 frame-relay interface-dlci 102
!
interface Serial1/0.2 point-to-point
```

```
  bandwidth 64
  ip address 10.3.3.5 255.255.255.252
  ip router isis
  tag-switching ip
  frame-relay interface-dlci 103
!
interface Serial1/7
  bandwidth 64
  ip vrf forwarding client-1
  ip address 130.130.130.1 255.255.255.0
  clockrate 64000
!
router isis
  net 49.0001.0000.0000.0001.00
  is-type level-1
  metric-style wide
!
router bgp 65000
  no synchronization
  neighbor 223.0.0.2 remote-as 65000
  neighbor 223.0.0.2 update-source Loopback0
  neighbor 223.0.0.3 remote-as 65000
  neighbor 223.0.0.3 update-source Loopback0
  no auto-summary
  !
  address-family ipv4 vrf client-1
  redistribute connected
  neighbor 130.130.130.2 remote-as 65011
  neighbor 130.130.130.2 activate
  no auto-summary
  no synchronization
  exit-address-family
  !
  address-family ipv4 vrf central-1
  redistribute connected
  neighbor 100.100.100.2 remote-as 65012
  neighbor 100.100.100.2 activate
  no auto-summary
  no synchronization
  exit-address-family
  !
  address-family vpnv4
  neighbor 223.0.0.2 activate
  neighbor 223.0.0.2 send-community extended
  neighbor 223.0.0.3 activate
```

```
neighbor 223.0.0.3 send-community extended
no auto-summary
exit-address-family
```

The VRF configuration for London-PE highlighted in Code Listing 10.31 is based on the RD and RT numbering scheme defined in Table 10.7.

Code Listing 10.31:   MPLS VPN configuration for London-PE

```
hostname London-PE
!
ip vrf central-2
 rd 65000:999
 route-target export 65000:202
 route-target export 65000:102
 route-target export 65000:802
 route-target import 65000:202
 route-target import 65000:102
 route-target import 65000:801
!
ip vrf client-2
 rd 65000:202
 route-target export 65000:202
 route-target import 65000:202
!
ip cef
!
interface Loopback0
 ip address 223.0.0.2 255.255.255.255
 ip router isis
!
interface FastEthernet0/0
 ip vrf forwarding central-2
 ip address 202.202.202.1 255.255.255.0
!
interface Serial1/0
 encapsulation frame-relay
 frame-relay lmi-type ansi
!
interface Serial1/0.1 point-to-point
 ip address 10.2.2.5 255.255.255.252
 ip router isis
 tag-switching ip
```

```
   frame-relay interface-dlci 203
!
interface Serial1/0.2 point-to-point
 ip address 10.1.1.6 255.255.255.252
 ip router isis
 tag-switching ip
 frame-relay interface-dlci 201
!
interface Serial1/7
 bandwidth 64
 ip vrf forwarding client-2
 ip address 200.200.200.1 255.255.255.0
 clockrate 64000
!
router isis
 net 49.0001.0000.0000.0002.00
 is-type level-1
 metric-style wide
!
router bgp 65000
 no synchronization
 neighbor 223.0.0.1 remote-as 65000
 neighbor 223.0.0.1 update-source Loopback0
 neighbor 223.0.0.3 remote-as 65000
 neighbor 223.0.0.3 update-source Loopback0
 no auto-summary
 !
 address-family ipv4 vrf client-2
 redistribute connected
 neighbor 200.200.200.2 remote-as 65021
 neighbor 200.200.200.2 activate
 no auto-summary
 no synchronization
 exit-address-family
 !
 address-family ipv4 vrf central-2
 redistribute connected
 neighbor 202.202.202.2 remote-as 65022
 neighbor 202.202.202.2 activate
 no auto-summary
 no synchronization
 exit-address-family
 !
 address-family vpnv4
 neighbor 223.0.0.1 activate
```

```
neighbor 223.0.0.1 send-community extended
neighbor 223.0.0.3 activate
neighbor 223.0.0.3 send-community extended
no auto-summary
exit-address-family
```

The VRF configuration for Paris-PE highlighted in Code Listing 10.32 is based on the RD and RT numbering scheme defined in Table 10.8.

Code Listing 10.32:   MPLS VPN configuration for Paris-PE

```
hostname Paris-PE
!
ip vrf server
 rd 65000:888
 route-target export 65000:801
 route-target import 65000:801
 route-target import 65000:802
!
ip cef
!
interface Loopback0
 ip address 223.0.0.3 255.255.255.255
 ip router isis
!
interface FastEthernet0/0
 ip vrf forwarding server
 ip address 31.3.3.1 255.255.255.0
!
interface Serial1/0
 encapsulation frame-relay
 frame-relay lmi-type ansi
!
interface Serial1/0.1 point-to-point
 ip address 10.3.3.6 255.255.255.252
 ip router isis
 tag-switching ip
 frame-relay interface-dlci 301
!
interface Serial1/0.2 point-to-point
 ip address 10.2.2.6 255.255.255.252
 ip router isis
 tag-switching ip
```

```
  frame-relay interface-dlci 302
!
router isis
 net 49.0001.0000.0000.0003.00
 is-type level-1
 metric-style wide
!
router bgp 65000
 no synchronization
 neighbor 223.0.0.1 remote-as 65000
 neighbor 223.0.0.1 update-source Loopback0
 neighbor 223.0.0.2 remote-as 65000
 neighbor 223.0.0.2 update-source Loopback0
 no auto-summary
 !
 address-family ipv4 vrf server
 redistribute connected
 neighbor 31.3.3.2 remote-as 65031
 neighbor 31.3.3.2 activate
 no auto-summary
 no synchronization
 exit-address-family
 !
 address-family vpnv4
 neighbor 223.0.0.1 activate
 neighbor 223.0.0.1 send-community extended
 neighbor 223.0.0.2 activate
 neighbor 223.0.0.2 send-community extended
 no auto-summary
 exit-address-family
```

10.4.5 Verifying VRFs configured in the PE routers

Code Listings 10.33 to 10.35 are used to verify the detailed VRF configurations for the three PE routers.

The "show ip vrf detail" command can be used to verify the VRF configuration for Brussels-PE, and the command output in Code Listing 10.33 corresponds with the RD and RT numbering scheme configured previously in Code Listing 10.30 for Brussels-PE.

Code Listing 10.33:   Detailed VRF configuration for Brussels-PE

```
Brussels-PE#show ip vrf detail
VRF central-1; default RD 65000:666
  Interfaces:
```

```
    FastEthernet0/0
    Connected addresses are not in global routing table
    Export VPN route-target communities
      RT:65000:101              RT:65000:102              RT:65000:802
    Import VPN route-target communities
      RT:65000:101              RT:65000:102              RT:65000:801
    No import route-map
    No export route-map
VRF client-1; default RD 65000:101
    Interfaces:
      Serial1/7
    Connected addresses are not in global routing table
    Export VPN route-target communities
      RT:65000:101
    Import VPN route-target communities
      RT:65000:101
    No import route-map
    No export route-map
```

The "show ip vrf detail" command can be used to verify the VRF configuration for London-PE, and the command output in Code Listing 10.34 corresponds with the RD and RT numbering scheme configured previously in Code Listing 10.31 for London-PE.

Code Listing 10.34:   Detailed VRF configuration for London-PE

```
London-PE#show ip vrf detail
VRF central-2; default RD 65000:999
    Interfaces:
      FastEthernet0/0
    Connected addresses are not in global routing table
    Export VPN route-target communities
      RT:65000:202              RT:65000:102              RT:65000:802
    Import VPN route-target communities
      RT:65000:202              RT:65000:102              RT:65000:801
    No import route-map
    No export route-map
VRF client-2; default RD 65000:202
    Interfaces:
      Serial1/7
    Connected addresses are not in global routing table
    Export VPN route-target communities
      RT:65000:202
    Import VPN route-target communities
```

```
    RT:65000:202
No import route-map
No export route-map
```

The "show ip vrf detail" command can be used to verify the VRF configuration for Paris-PE, and the command output in Code Listing 10.35 corresponds with the RD and RT numbering scheme configured previously in Code Listing 10.32 for Paris-PE.

Code Listing 10.35:   Detailed VRF configuration for Paris-PE

```
Paris-PE#show ip vrf detail
VRF server; default RD 65000:888
  Interfaces:
    FastEthernet0/0
  Connected addresses are not in global routing table
  Export VPN route-target communities
    RT:65000:801
  Import VPN route-target communities
    RT:65000:801              RT:65000:802
  No import route-map
  No export route-map
```

10.4.6   Monitoring BGP VPNv4 tables of the PE routers

Code Listings 10.36 to 10.38 illustrate the entire content of the BGP VPNv4 table for the three PE routers.

The "show ip bgp vpnv4 all" command in Code Listing 10.36 displays the entire content of the BGP VPNv4 table for Brussels-PE. From the table, the per-VRF BGP VPNv4 information associated with VRF Client-1 includes prefix 9.0.0.0/8 as well as 100.100.100.0/24 from Brussels-CE2 (VRF Central-1). Also, the per-VRF BGP VPNv4 information associated with VRF Central-1 includes prefix 8.8.8.0/24 as well as 130.130.130.0/24 from Brussels-CE1 (VRF Client-1), prefix 31.5.5.4/30 as well as 31.3.3.0/24 from Paris-CE1 (VRF Server), and prefix 5.5.0.0/16 as well as 202.202.202.0/24 from London-CE2 (VRF Central-2). This command output concurs with the intersite connectivity requirements defined in section 10.4.2.

Code Listing 10.36:   Brussels-PE BGP VPNv4 table

```
Brussels-PE#show ip bgp vpnv4 all
BGP table version is 29, local router ID is 223.0.0.1
Status codes: s suppressed, d damped, h history, * valid, > best, i - internal
Origin codes: i - IGP, e - EGP, ? - incomplete
```

```
      Network            Next Hop             Metric LocPrf Weight Path
Route Distinguisher: 65000:101 (default for vrf client-1)
*> 8.8.8.0/24          130.130.130.2             0               0 65011 i
*> 9.0.0.0             100.100.100.2             0               0 65012 i
*> 100.100.100.0/24 0.0.0.0                      0           32768 ?
*> 130.130.130.0/24 0.0.0.0                      0           32768 ?
Route Distinguisher: 65000:666 (default for vrf central-1)
*>i5.5.0.0/16          223.0.0.2                 0    100        0 65022 i
*> 8.8.8.0/24          130.130.130.2             0               0 65011 i
*> 9.0.0.0             100.100.100.2             0               0 65012 i
*>i31.3.3.0/24         223.0.0.3                 0    100        0 ?
*>i31.5.5.4/30         223.0.0.3                 0    100        0 65031 i
*> 100.100.100.0/24 0.0.0.0                      0           32768 ?
*> 130.130.130.0/24 0.0.0.0                      0           32768 ?
*>i202.202.202.0      223.0.0.2                 0    100        0 ?
<Output Omitted>
```

The "show ip bgp vpnv4 all" command in Code Listing 10.37 displays the entire content of the BGP VPNv4 table for London-PE. From the table, the per-VRF BGP VPNv4 information associated with VRF Client-2 includes prefix 5.5.0.0/16 as well as 202.202.202.0/24 from London-CE2 (VRF Central-2). Also, the per-VRF BGP VPNv4 information associated with VRF Central-2 includes prefix 31.5.5.4/30 as well as 31.3.3.0/24 from Paris-CE1 (VRF Server), prefix 6.6.6.6/32 as well as 200.200.200.0/24 from London-CE1 (VRF Client-2), and prefix 9.0.0.0/8 as well as 100.100.100.0/24 from Brussels-CE2 (VRF Central-1). This command output concurs with the intersite connectivity requirements defined in section 10.4.2.

Code Listing 10.37:   London-PE BGP VPNv4 table

```
London-PE#show ip bgp vpnv4 all
BGP table version is 81, local router ID is 223.0.0.2
Status codes: s suppressed, d damped, h history, * valid, > best, i - internal
Origin codes: i - IGP, e - EGP, ? - incomplete

      Network            Next Hop             Metric LocPrf Weight Path
Route Distinguisher: 65000:202 (default for vrf client-2)
*> 5.5.0.0/16          202.202.202.2             0               0 65022 i
*> 6.6.6.6/32          200.200.200.2             0               0 65021 i
*> 200.200.200.0       0.0.0.0                   0           32768 ?
*> 202.202.202.0       0.0.0.0                   0           32768 ?
<Output Omitted>
Route Distinguisher: 65000:999 (default for vrf central-2)
```

```
*> 5.5.0.0/16        202.202.202.2              0                    0 65022 i
*> 6.6.6.6/32        200.200.200.2              0                    0 65021 i
*>i9.0.0.0           223.0.0.1                  0      100           0 65012 i
*>i31.3.3.0/24       223.0.0.3                  0      100           0 ?
*>i31.5.5.4/30       223.0.0.3                  0      100           0 65031 i
*>i100.100.100.0/24  223.0.0.1                  0      100           0 ?
*> 200.200.200.0     0.0.0.0                    0             32768  ?
*> 202.202.202.0     0.0.0.0                    0             32768  ?
```

The "show ip bgp vpnv4 all" command in Code Listing 10.38 displays the entire content of the BGP VPNv4 table for Paris-PE. From the table, the per-VRF BGP VPNv4 information associated with VRF Server includes prefix 9.0.0.0/8 as well as 100.100.100.0/24 from Brussels-CE2 (VRF Central-1), and prefix 5.5.0.0/16 as well as 202.202.202.0/24 from London-CE2 (VRF Central-2). This command output concurs with the intersite connectivity requirements defined in section 10.4.2.

Code Listing 10.38:   Paris-PE BGP VPNv4 table

```
Paris-PE#show ip bgp vpnv4 all
BGP table version is 13, local router ID is 223.0.0.3
Status codes: s suppressed, d damped, h history, * valid, > best, i - internal
Origin codes: i - IGP, e - EGP, ? - incomplete

    Network            Next Hop            Metric LocPrf Weight Path
<Output Omitted>
Route Distinguisher: 65000:888 (default for vrf server)
*>i5.5.0.0/16         223.0.0.2                  0      100           0 65022 i
*>i9.0.0.0            223.0.0.1                  0      100           0 65012 i
*> 31.3.3.0/24        0.0.0.0                    0             32768  ?
*> 31.5.5.4/30        31.3.3.2                   0                    0 65031 i
*>i100.100.100.0/24   223.0.0.1                  0      100           0 ?
*>i202.202.202.0      223.0.0.2                  0      100           0 ?
<Output Omitted>
```

10.4.7   Verifying the intersite connectivity

Code Listings 10.39 to 10.43 illustrate the IPv4 table for the five CE routers at different locations.

Code Listing 10.39 displays the IP routing table for Brussels-CE1 (Client-1) belonging to VPN 1. The BGP routes listed in Code Listing 10.36 for VRF Client-1 are installed into the IP routing table, thus fulfilling the intersite connectivity requirements defined in section 10.4.2.

Code Listing 10.39:   Brussels-CE1 IPv4 forwarding table

```
Brussels-CE1#show ip route
<Output Omitted>

     100.0.0.0/24 is subnetted, 1 subnets
B       100.100.100.0 [20/0] via 130.130.130.1, 00:19:43
     8.0.0.0/24 is subnetted, 1 subnets
C       8.8.8.0 is directly connected, Loopback0
B    9.0.0.0/8 [20/0] via 130.130.130.1, 00:19:43
     130.130.0.0/24 is subnetted, 1 subnets
C       130.130.130.0 is directly connected, Serial1/0
```

Code Listing 10.40 displays the IP routing table for Brussels-CE2 (Central-1) overlapping VPN 1 and VPN 12. The BGP routes listed in Code Listing 10.36 for VRF Central-1 are installed into the IP routing table, thus fulfilling the intersite connectivity requirements defined in section 10.4.2.

Code Listing 10.40:   Brussels-CE2 IPv4 forwarding table

```
Brussels-CE2#show ip route
<Output Omitted>

B    202.202.202.0/24 [20/0] via 100.100.100.1, 00:18:33
     100.0.0.0/24 is subnetted, 1 subnets
C       100.100.100.0 is directly connected, Ethernet0/0
     5.0.0.0/16 is subnetted, 1 subnets
B       5.5.0.0 [20/0] via 100.100.100.1, 00:18:33
     8.0.0.0/24 is subnetted, 1 subnets
B       8.8.8.0 [20/0] via 100.100.100.1, 00:18:04
C    9.0.0.0/8 is directly connected, Loopback0
     130.130.0.0/24 is subnetted, 1 subnets
B       130.130.130.0 [20/0] via 100.100.100.1, 00:19:00
     31.0.0.0/8 is variably subnetted, 2 subnets, 2 masks
B       31.5.5.4/30 [20/0] via 100.100.100.1, 00:08:17
B       31.3.3.0/24 [20/0] via 100.100.100.1, 00:08:17
```

Code Listing 10.41 displays the IP routing table for London-CE1 (Client-2) belonging to VPN 2. The BGP routes listed in Code Listing 10.37 for VRF Client-2 are installed into the IP routing table, thus fulfilling the intersite connectivity requirements defined in section 10.4.2.

Code Listing 10.41:   London-CE1 IPv4 forwarding table

```
London-CE1#show ip route
<Output Omitted>

C    200.200.200.0/24 is directly connected, Serial1/0
B    202.202.202.0/24 [20/0] via 200.200.200.1, 00:42:45
     5.0.0.0/16 is subnetted, 1 subnets
B       5.5.0.0 [20/0] via 200.200.200.1, 00:18:33
     6.0.0.0/32 is subnetted, 1 subnets
C       6.6.6.6 is directly connected, Loopback0
```

Code Listing 10.42 displays the IP routing table for London-CE2 (Central-2) overlapping VPN 2 and VPN 12. The BGP routes listed in Code Listing 10.37 for VRF Central-2 are installed into the IP routing table, thus fulfilling the intersite connectivity requirements defined in section 10.4.2.

Code Listing 10.42:   London-CE2 IPv4 forwarding table

```
London-CE2#show ip route
<Output Omitted>

B    200.200.200.0/24 [20/0] via 202.202.202.1, 00:17:38
C    202.202.202.0/24 is directly connected, Ethernet0/0
     100.0.0.0/24 is subnetted, 1 subnets
B       100.100.100.0 [20/0] via 202.202.202.1, 00:15:50
     5.0.0.0/16 is subnetted, 1 subnets
C       5.5.0.0 is directly connected, Loopback0
     6.0.0.0/32 is subnetted, 1 subnets
B       6.6.6.6 [20/0] via 202.202.202.1, 00:17:38
B    9.0.0.0/8 [20/0] via 202.202.202.1, 00:15:50
     31.0.0.0/8 is variably subnetted, 2 subnets, 2 masks
B       31.5.5.4/30 [20/0] via 202.202.202.1, 00:05:36
B       31.3.3.0/24 [20/0] via 202.202.202.1, 00:05:36
```

Code Listing 10.43 displays the IP routing table for Paris-CE1 (server) belonging to VPN 12. The BGP routes listed in Code Listing 10.38 for VRF Server are installed into the IP routing table, thus fulfilling the intersite connectivity requirements defined in section 10.4.2.

Code Listing 10.43:   Paris-CE1 IPv4 forwarding table

```
Paris-CE1#show ip route
<Output Omitted>


B    202.202.202.0/24 [20/0] via 31.3.3.1, 00:03:54
     100.0.0.0/24 is subnetted, 1 subnets
B       100.100.100.0 [20/0] via 31.3.3.1, 00:03:54
     5.0.0.0/16 is subnetted, 1 subnets
B       5.5.0.0 [20/0] via 31.3.3.1, 00:03:54
B    9.0.0.0/8 [20/0] via 31.3.3.1, 00:03:54
     31.0.0.0/8 is variably subnetted, 2 subnets, 2 masks
C       31.5.5.4/30 is directly connected, Loopback0
C       31.3.3.0/24 is directly connected, Ethernet0/0
```

10.5   Case Study 10.4: Hub-and-Spoke VPN Topology

10.5.1   Case overview

This case study is adapted from Case Studies 10.2 and 10.3. It provides a sample configuration of a hub-and-spoke VPN topology. In this case, the MPLS VPN deployment spans four different customer sites: Paris Site 1 (Spoke), Brussels Site 2 (Spoke), London Hub (Central Site), and London Spoke (Central Site). Notice that London Hub and the London Spoke are interconnected in AS 65021. The MPLS VPN design and physical topology for Case Study 10.4 are shown in Figure 10.10.

10.5.2   Hub-and-Spoke VPN topology

MPLS VPNs impeccably create a fully meshed topology between sites belonging to the same VPN, with optimal routing in the service provider (SP) MPLS VPN core network. However, some customers might want to preserve the centralized control inherent with overlay VPN hub-and-spoke topology, in which all traffic is exchanged via a central site (or sites). The central site is commonly referred to as a hub site and has full routing knowledge of all other sites belonging to the same VPN. These other sites are commonly referred to as spoke sites, and they send traffic to the hub site for any destinations. The reader should take note that the hub-and-spoke VPN topology defeats the scalability of MPLS VPN, and the optimum intersite routing provided by the MPLS VPN routing mechanisms. In a hub-and-spoke topology, an extra hop is always incurred during routing because all traffic needs to be sent to the hub site for any destinations.

From the routing perspective, the hub-and-spoke VPN topology results in asymmetrical routing. In other words, traffic going from Paris Site 1 to Brussels Site 2 will enter the hub site through interface hub (Serial1/7) and

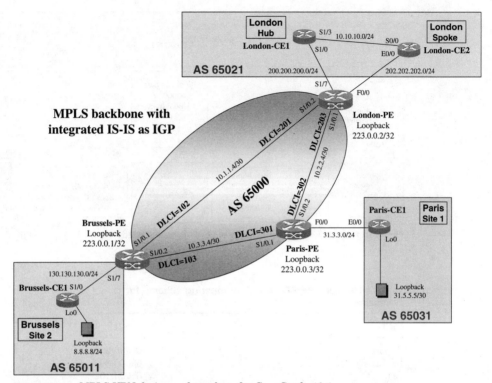

Figure 10.10: MPLS VPN design and topology for Case Study 10.4.

exit through interface spoke (FastEthernet0/0); the returning traffic will also enter the hub site through interface hub (Serial1/7) and exit through interface spoke (FastEthernet0/0). As a result, the customer may be unable to deploy stateful filters or similar mechanisms that monitor and verify the direction of traffic.

The design criteria for implementing the hub-and-spoke topology over an MPLS VPN backbone ensure that the spoke sites can exchange routing information only with the hub site, and not with each other. To achieve this, two interfaces, Serial1/7 and FastEthernet0/0 (see Fig. 10.10), in two separate VRFs, VRF Hub and VRF Spoke (see Table 10.10), are required for the hub site. Interface FastEthernet0/0 interconnecting London-CE2 (London Spoke) is used to receive packets from the hub site and send them to the spoke sites; interface Serial1/7 interconnecting London-CE1 (London Hub) is used to receive packets from the spoke sites and send them to the hub site. Furthermore, a separate VRF (VRF Site 1 for Paris Site 1 [see Table 10.11], and VRF Site 2 for Brussels Site 2 [see Table 10.9]) is required for every spoke site to prevent the spoke sites from exchanging routing information with each other directly.

In Figure 10.10, Paris Site 1 and Brussels Site 2 tag all routing information sent to London-PE with a Hub RT 65000:203 that can be imported only into

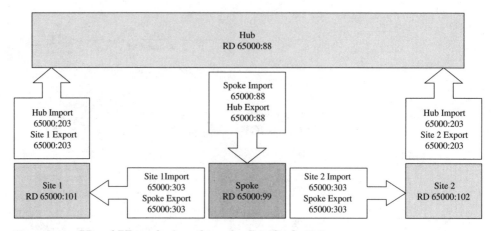

Figure 10.11:   RD and RT numbering scheme for Case Study 10.4.

TABLE 10.9   Per-VRF RD and RT numbering scheme for Brussels-PE.

VRF	Site2
Export RT	65000:203
Import RT	65000:303
RD	65000:102

TABLE 10.10   Per-VRF RD and RT numbering scheme for London-PE.

VRF	Hub	Spoke
Export RT	65000:88	65000:303
Import RT	65000:203	65000:88
RD	65000:88	65000:99

VRF Hub. London-PE then distributes the same information out through interface Serial1/7 to London-CE1, which in turn forwards the routing information to London-CE2 and back to London-PE again through its interface FastEthernet0/0. London-PE tags the information with a Spoke RT 65000:303 that can be imported in all other spoke VRFs (in this case, VRF Site 1 and VRF Site 2), and propagates them to Paris Site 1 and Brussels Site 2.

Tables 10.9 to 10.11 illustrate the RD and RT numbering scheme for Brussels-PE, London-PE and Paris-PE, respectively.

Tables 10.9 to 10.11 are illustrated schematically in Figure 10.11 for which:

TABLE 10.11   Per-VRF RD and RT numbering scheme for Paris-PE.

VRF	Site1
Export RT	65000:203
Import RT	65000:303
RD	65000:101

- Site 1 imports all networks that carry RT 65000:303 (from Spoke) and exports all networks with RT 65000:203 (to Hub) that can be imported only into VRF Hub.

- Site 2 imports all networks that carry RT 65000:303 (from Spoke) and exports all networks with RT 65000:203 (to Hub) that can be imported only into VRF Hub.

- Hub imports all networks that carry RT 65000:203 (from Site 1 and Site 2) and exports all networks with RT 65000:88 (to Spoke), which can be imported only into VRF Spoke.

- Spoke imports all networks that carry RT 65000:88 (from Hub) and exports all networks with RT 65000:303 (to Site 1 and Site 2).

In addition, according to the four VRFs defined in Tables 10.9 to 10.11, we can construe the following intersite connectivity:

- VRF Site 1 contains routes from VRF Site 2: 8.8.8.0/24 and 130.130.130.0/24.

- VRF Site 2 contains routes from VRF Site 1: 31.5.5.4/30 and 31.3.3.0/24.

- VRF Hub contains routes from VRF Site 1: 31.5.5.4/30 and 31.3.3.0/24.

- VRF Hub contains routes from VRF Site 2: 8.8.8.0/24 and 130.130.130.0/24.

- VRF Spoke contains routes from VRF Site 1: 31.5.5.4/30 and 31.3.3.0/24.

- VRF Spoke contains routes from VRF Site 2: 8.8.8.0/24 and 130.130.130.0/24.

The above-mentioned intersite connectivity deductions will be verified in the later sections. Since the BGP VPNv4 table (see section 10.5.7) for all spoke VRFs (Site 1 and Site 2) is pointing to the Hub site for all destinations in the VPN, traffic from Paris Site 1 to Brussels Site 2, and vice versa, will flow through the Hub site.

10.5.3   Allowas-in

When the customer is using BGP with the hub-and-spoke VPN topology, the SP's PE router will not accept routing updates coming from the hub site if the updates have previously been sent to the hub site via other BGP session. In such cases, the PE router would receive routes with its own AS number (ASN) already in the AS-path, regard it as a routing loop, and drop the routes.

With the Allowas-in BGP feature configured on the PE router for its BGP neighbor in the hub site, the PE router would not drop incoming BGP updates with its ASN in the AS-path if the updates are received from that neighbor. Nevertheless, to avoid real BGP routing loops, the maximum number of occurrences of the MPLS VPN backbone ASN can be specified, and incoming updates exceeding the maximum limit will be dropped.

10.5.4   CE configurations

In this section, we illustrate only the BGP configuration for London-CE1 and London-CE2. Brussels-CE1 and Paris-CE1 share the same configuration listings as those illustrated in Code Listing 10.1 and 10.15.

Code Listing 10.44 illustrates the BGP configuration for London-CE1. Besides establishing an EBGP session with London-PE (200.200.200.1), London-CE1 also forms an IBGP session with London-CE2 (10.10.10.2).

Code Listing 10.44:   BGP configuration for London-CE1

```
hostname London-CE1
!
interface Serial1/0
 bandwidth 64
 ip address 200.200.200.2 255.255.255.0
!
interface Serial1/3
      bandwidth 64
 ip address 10.10.10.1 255.255.255.0
!
router bgp 65021
 no synchronization
 neighbor 10.10.10.2 remote-as 65021 ! --- Declare London-CE2 as IBGP peer.
 neighbor 10.10.10.2 next-hop-self
 neighbor 200.200.200.1 remote-as 65000 ! --- Declare London-PE as EBGP peer.
 no auto-summary
```

Code Listing 10.45 illustrates the BGP configuration for London-CE2. Besides establishing an EBGP session with London-PE (202.202.202.1), London-CE2 also forms an IBGP session with London-CE1 (10.10.10.1).

Code Listing 10.45:   BGP configuration for London-CE2

```
hostname London-CE2
!
interface Ethernet0/0
 ip address 202.202.202.2 255.255.255.0
```

```
!
interface Serial0/0
 bandwidth 64
 ip address 10.10.10.2 255.255.255.0
 clockrate 64000
!
router bgp 65021
 no synchronization
 neighbor 10.10.10.1 remote-as 65021 ! --- Declare London-CE1 as IBGP peer.
 neighbor 202.202.202.1 remote-as 65000 ! --- Declare London-PE as EBGP
peer.
 no auto-summary
```

10.5.5 PE configurations

Code Listings 10.46 to 10.48 illustrate the MPLS VPN configurations for Brussels-PE, London-PE, and Paris-PE, respectively. The configurations reflect the RT and RD numbering scheme defined in Tables 10.9 to 10.11.

The VRF configuration for Brussels-PE highlighted in Code Listing 10.46 is based on the RD and RT numbering scheme defined in Table 10.9.

Code Listing 10.46:   MPLS VPN configuration for Brussels-PE

```
hostname Brussels-PE
!
ip vrf site2
 rd 65000:102
 route-target export 65000:203
 route-target import 65000:303
!
ip cef
!
interface Loopback0
 ip address 223.0.0.1 255.255.255.255
 ip router isis
!
interface Serial1/0
 encapsulation frame-relay
 frame-relay lmi-type ansi
!
interface Serial1/0.1 point-to-point
 bandwidth 64
 ip address 10.1.1.5 255.255.255.252
 ip router isis
 tag-switching ip
 frame-relay interface-dlci 102
```

```
!
interface Serial1/0.2 point-to-point
 bandwidth 64
 ip address 10.3.3.5 255.255.255.252
 ip router isis
 tag-switching ip
 frame-relay interface-dlci 103
!
interface Serial1/7
 bandwidth 64
 ip vrf forwarding site2
 ip address 130.130.130.1 255.255.255.0
 clockrate 64000
!
router isis
 net 49.0001.0000.0000.0001.00
 is-type level-1
 metric-style wide
!
router bgp 65000
 no synchronization
 neighbor 223.0.0.2 remote-as 65000
 neighbor 223.0.0.2 update-source Loopback0
 neighbor 223.0.0.3 remote-as 65000
 neighbor 223.0.0.3 update-source Loopback0
 no auto-summary
 !
 address-family ipv4 vrf site2
 redistribute connected
 neighbor 130.130.130.2 remote-as 65011
 neighbor 130.130.130.2 activate
 no auto-summary
 no synchronization
 exit-address-family
 !
 address-family vpnv4
 neighbor 223.0.0.2 activate
 neighbor 223.0.0.2 send-community extended
 neighbor 223.0.0.3 activate
 neighbor 223.0.0.3 send-community extended
 no auto-summary
 exit-address-family
```

The VRF configuration for Paris-PE highlighted in Code Listing 10.47 is based on the RD and RT numbering scheme defined in Table 10.11.

Code Listing 10.47: MPLS VPN configuration for Paris-PE

```
hostname Paris-PE
!
ip vrf site1
 rd 65000:101
 route-target export 65000:203
 route-target import 65000:303
!
ip cef
!
interface Loopback0
 ip address 223.0.0.3 255.255.255.255
 ip router isis
!
interface FastEthernet0/0
 ip vrf forwarding site1
 ip address 31.3.3.1 255.255.255.0
!
interface Serial1/0
 encapsulation frame-relay
 frame-relay lmi-type ansi
!
interface Serial1/0.1 point-to-point
 ip address 10.3.3.6 255.255.255.252
 ip router isis
 tag-switching ip
 frame-relay interface-dlci 301
!
interface Serial1/0.2 point-to-point
 ip address 10.2.2.6 255.255.255.252
 ip router isis
 tag-switching ip
 frame-relay interface-dlci 302
!
router isis
 net 49.0001.0000.0000.0003.00
 is-type level-1
 metric-style wide
!
router bgp 65000
 no synchronization
 neighbor 223.0.0.1 remote-as 65000
 neighbor 223.0.0.1 update-source Loopback0
```

```
neighbor 223.0.0.2 remote-as 65000
neighbor 223.0.0.2 update-source Loopback0
no auto-summary
!
address-family ipv4 vrf site1
redistribute connected
neighbor 31.3.3.2 remote-as 65031
neighbor 31.3.3.2 activate
no auto-summary
no synchronization
exit-address-family
!
address-family vpnv4
neighbor 223.0.0.1 activate
neighbor 223.0.0.1 send-community extended
neighbor 223.0.0.2 activate
neighbor 223.0.0.2 send-community extended
no auto-summary
exit-address-family
```

The VRF configuration for London-PE highlighted in Code Listing 10.48 is based on the RD and RT numbering scheme defined in Table 10.10. Notice that the Allowas-in BGP feature is configured for EBGP peer, London-CE2 (202.202.202.2). This is to ensure that London-PE will still accept the EBGP updates coming from London-CE2 even though these same updates have previously been sent to London-CE1 via another EBGP session and contain London-PE's ASN (65000). In this case, London-PE will accept all updates from London-CE2 as long as the updates do not contain its ASN 65000 in the AS-path more than four times.

Code Listing 10.48:   MPLS VPN configuration for London-PE

```
hostname London-PE
!
ip vrf hub
 rd 65000:88
 route-target export 65000:88
 route-target import 65000:203
!
ip vrf spoke
 rd 65000:99
 route-target export 65000:303
 route-target import 65000:88
```

```
!
ip cef
!
interface Loopback0
 ip address 223.0.0.2 255.255.255.255
 ip router isis
!
interface FastEthernet0/0
 ip vrf forwarding spoke
 ip address 202.202.202.1 255.255.255.0
!
interface Serial1/0
 encapsulation frame-relay
 frame-relay lmi-type ansi
!
interface Serial1/0.1 point-to-point
 ip address 10.2.2.5 255.255.255.252
 ip router isis
 tag-switching ip
 frame-relay interface-dlci 203
!
interface Serial1/0.2 point-to-point
 ip address 10.1.1.6 255.255.255.252
 ip router isis
 tag-switching ip
 frame-relay interface-dlci 201
!
interface Serial1/7
 bandwidth 64
 ip vrf forwarding hub
 ip address 200.200.200.1 255.255.255.0
 clockrate 64000
!
router isis
 net 49.0001.0000.0000.0002.00
 is-type level-1
 metric-style wide
!
router bgp 65000
 no synchronization
 redistribute connected
 neighbor 223.0.0.1 remote-as 65000
 neighbor 223.0.0.1 update-source Loopback0
 neighbor 223.0.0.3 remote-as 65000
 neighbor 223.0.0.3 update-source Loopback0
```

```
no auto-summary
!
address-family ipv4 vrf spoke
redistribute connected
neighbor 202.202.202.2 remote-as 65021
neighbor 202.202.202.2 activate

! --- London-PE will accept all BGP updates from London-CE2 as long as the
updates
! --- do not contain its ASN (65000) in the AS path more than four times.
neighbor 202.202.202.2 allowas-in 4
no auto-summary
no synchronization
exit-address-family
!
address-family ipv4 vrf hub
redistribute connected
neighbor 200.200.200.2 remote-as 65021
neighbor 200.200.200.2 activate
no auto-summary
no synchronization
exit-address-family
!
address-family vpnv4
neighbor 223.0.0.1 activate
neighbor 223.0.0.1 send-community extended
neighbor 223.0.0.3 activate
neighbor 223.0.0.3 send-community extended
no auto-summary
exit-address-family
```

10.5.6 Verifying VRFs configured in the PE routers

Code Listings 10.49 to 10.51 are used to verify the detailed VRF configurations for the three PE routers.

The "show ip vrf detail" command can be used to verify the VRF configuration for Brussels-PE, and the command output in Code Listing 10.49 corresponds with the RD and RT numbering scheme configured previously in Code Listing 10.46 for Brussels-PE.

Code Listing 10.49: Detailed VRF configuration for Brussels-PE

```
Brussels-PE#show ip vrf detail
VRF site2; default RD 65000:102
   Interfaces:
     Serial1/7
```

```
Connected addresses are not in global routing table
Export VPN route-target communities
   RT:65000:203
Import VPN route-target communities
   RT:65000:303
No import route-map
No export route-map
```

The "show ip vrf detail" command can be used to verify the VRF configuration for Paris-PE, and the command output in Code Listing 10.50 corresponds with the RD and RT numbering scheme configured previously in Code Listing 10.47 for Paris-PE.

Code Listing 10.50:   Detailed VRF configuration for Paris-PE

```
Paris-PE#show ip vrf detail
VRF site1; default RD 65000:101
   Interfaces:
      FastEthernet0/0
   Connected addresses are not in global routing table
   Export VPN route-target communities
      RT:65000:203
   Import VPN route-target communities
      RT:65000:303
   No import route-map
   No export route-map
```

The "show ip vrf detail" command can be used to verify the VRF configuration for London-PE, and the command output in Code Listing 10.51 corresponds with the RD and RT numbering scheme configured previously in Code Listing 10.48 for London-PE.

Code Listing 10.51:   Detailed VRF configuration for London-PE

```
London-PE#show ip vrf detail
VRF hub; default RD 65000:88
   Interfaces:
      Serial1/7
   Connected addresses are not in global routing table
   Export VPN route-target communities
      RT:65000:88
   Import VPN route-target communities
      RT:65000:203
```

```
No import route-map
No export route-map
```
VRF spoke; default RD 65000:99
```
Interfaces:
   FastEthernet0/0
Connected addresses are not in global routing table
```
Export VPN route-target communities
   RT:65000:303
Import VPN route-target communities
   RT:65000:88
```
No import route-map
No export route-map
```

10.5.7 Monitoring BGP VPNv4 tables of the PE routers

Code Listings 10.52 to 10.54 illustrate the entire content of the BGP VPNv4 table for the three PE routers.

The "show ip bgp vpnv4 all" command in Code Listing 10.52 displays the entire content of the BGP VPNv4 table for Brussels-PE. From the table, the per-VRF BGP VPNv4 information associated with VRF Site 2 includes prefixes 31.5.5.4/30 and 31.3.3.0/24 from Paris-CE1 (VRF Site 1) received via London-PE (VRF Spoke). This command output concurs with the intersite connectivity requirements defined in section 10.5.2.

Code Listing 10.52:   Brussels-PE BGP VPNv4 table

```
Brussels-PE#show ip bgp vpnv4 all
BGP table version is 35, local router ID is 223.0.0.1
Status codes: s suppressed, d damped, h history, * valid, > best, i - internal
Origin codes: i - IGP, e - EGP, ? - incomplete

   Network          Next Hop          Metric LocPrf Weight Path
<Output Omitted>
Route Distinguisher: 65000:102 (default for vrf site2)
<Output Omitted>
*>i31.3.3.0/24      223.0.0.2                100     0 65021 65000 ?
*>i31.5.5.4/30      223.0.0.2                100     0 65021 65000 65031 i
<Output Omitted>
```

The "show ip bgp vpnv4 all" command in Code Listing 10.53 displays the entire content of the BGP VPNv4 table for Paris-PE. From the table, the per-VRF BGP VPNv4 information associated with VRF Site 1 includes prefixes 8.8.8.0/24 and 130.130.130.0/24 from Brussels-CE1 (VRF Site 2) via London-PE

(VRF Spoke). This command output concurs with the intersite connectivity requirements defined in section 10.5.2.

Code Listing 10.53:   Paris-PE BGP VPNv4 table

```
Paris-PE#show ip bgp vpnv4 all
BGP table version is 173, local router ID is 223.0.0.3
Status codes: s suppressed, d damped, h history, * valid, > best, i - internal
Origin codes: i - IGP, e - EGP, ? - incomplete

   Network          Next Hop            Metric LocPrf Weight Path
<Output Omitted>
Route Distinguisher: 65000:101 (default for vrf site1)
*>i8.8.8.0/24      223.0.0.2              100       0 65021 65000 65011 i
<Output Omitted>
*>i130.130.130.0/24 223.0.0.2            100       0 65021 65000 ?
<Output Omitted>
```

The "show ip bgp vpnv4 all" command in Code Listing 10.54 displays the entire content of the BGP VPNv4 table for London-PE. From the table, the per-VRF BGP VPNv4 information associated with VRF Hub includes prefix 8.8.8.0/24 as well as 130.130.130.0/24 from Brussels-CE1 (VRF Site 2), and prefix 31.5.5.4/30 as well as 31.3.3.0/24 from Paris-CE1 (VRF Site 1). Also, the per-VRF BGP VPNv4 information associated with VRF Spoke includes prefix 8.8.8.0/24 as well as 130.130.130.0/24 from Brussels-CE1 (VRF Site 2), and prefix 31.5.5.4/30 as well as 31.3.3.0/24 from Paris-CE1 (VRF Site 1), all received via London-CE2 (202.202.202.2) instead of directly from the sites themselves. This command output concurs with the intersite connectivity requirements defined in section 10.5.2. Note that the Allowas-in BGP feature is effective as London-PE is accepting updates containing its own ASN (65000) from London-CE2.

Code Listing 10.54:   London-PE BGP VPNv4 table

```
London-PE#show ip bgp vpnv4 all
BGP table version is 30, local router ID is 223.0.0.2
Status codes: s suppressed, d damped, h history, * valid, > best, i - internal
Origin codes: i - IGP, e - EGP, ? - incomplete

   Network          Next Hop            Metric LocPrf Weight Path
Route Distinguisher: 65000:88 (default for vrf hub)
*>i8.8.8.0/24      223.0.0.1              0    100       0 65011 i
```

```
*>i31.3.3.0/24       223.0.0.3                    0    100      0 ?
*>i31.5.5.4/30       223.0.0.3                    0    100      0 65031 i
*>i130.130.130.0/24 223.0.0.1                     0    100      0 ?
<Output Omitted>
Route Distinguisher: 65000:99 (default for vrf spoke)
*>  8.8.8.0/24       202.202.202.2                0 65021 65000 65011 i
*>  31.3.3.0/24      202.202.202.2                0 65021 65000 ?
*>  31.5.5.4/30      202.202.202.2                0 65021 65000 65031 i
*>  130.130.130.0/24 202.202.202.2                0 65021 65000 ?
<Output Omitted>
```

10.5.8 Verifying the intersite connectivity

Code Listings 10.55 to 10.58 illustrate the IPv4 table for the four CE routers.

Code Listing 10.55 displays the IP routing table for Brussels-CE1 (Site 2). The BGP routes listed in Code Listing 10.52 for VRF Site 2 are installed into the IP routing table, thus fulfilling the intersite connectivity requirements defined in section 10.5.2.

Code Listing 10.55: Brussels-CE1 IPv4 forwarding table

```
Brussels-CE1#show ip route
<Output Omitted>

     8.0.0.0/24 is subnetted, 1 subnets
C       8.8.8.0 is directly connected, Loopback0
     130.130.0.0/24 is subnetted, 1 subnets
C       130.130.130.0 is directly connected, Serial1/0
     31.0.0.0/8 is variably subnetted, 2 subnets, 2 masks
B       31.5.5.4/30 [20/0] via 130.130.130.1, 00:17:05
B       31.3.3.0/24 [20/0] via 130.130.130.1, 00:17:05
```

Code Listing 10.56 displays the IP routing table for Paris-CE1 (Site 1). The BGP routes listed in Code Listing 10.53 for VRF Site 1 are installed into the IP routing table, thus fulfilling the intersite connectivity requirements defined in section 10.5.2.

Code Listing 10.56: Paris-CE1 IPv4 forwarding table

```
Paris-CE1#show ip route
<Output Omitted>

     8.0.0.0/24 is subnetted, 1 subnets
B       8.8.8.0 [20/0] via 31.3.3.1, 00:16:15
     130.130.0.0/24 is subnetted, 1 subnets
```

```
B      130.130.130.0 [20/0] via 31.3.3.1, 00:16:15
       31.0.0.0/8 is variably subnetted, 2 subnets, 2 masks
C      31.5.5.4/30 is directly connected, Loopback0
C      31.3.3.0/24 is directly connected, Ethernet0/0
```

Code Listing 10.57 displays the IP routing table for London-CE1 (Hub). The BGP routes listed in Code Listing 10.54 for VRF Hub are installed into the IP routing table, thus fulfilling the intersite connectivity requirements defined in section 10.5.2.

Code Listing 10.57:   London-CE1 IPv4 forwarding table

```
London-CE1#show ip route
<Output Omitted>


C      200.200.200.0/24 is directly connected, Serial1/0
       19.0.0.0/32 is subnetted, 1 subnets
       8.0.0.0/24 is subnetted, 1 subnets
B         8.8.8.0 [20/0] via 200.200.200.1, 00:19:12
       10.0.0.0/24 is subnetted, 1 subnets
C         10.10.10.0 is directly connected, Serial1/3
       130.130.0.0/24 is subnetted, 1 subnets
B         130.130.130.0 [20/0] via 200.200.200.1, 00:19:12
       31.0.0.0/8 is variably subnetted, 2 subnets, 2 masks
B         31.5.5.4/30 [20/0] via 200.200.200.1, 00:19:12
B         31.3.3.0/24 [20/0] via 200.200.200.1, 00:19:12
```

Code Listing 10.58 displays the IP routing table for London-CE2 (Spoke). The BGP routes listed in Code Listing 10.57 are propagated across to London-CE2 via IBGP and are installed into the IP routing table. London-CE2 then advertises the same BGP routes back to London-PE, which in turn installs the routes into VRF Spoke as illustrated in Code Listing 10.54, thus fulfilling the intersite connectivity requirements defined in section 10.5.2.

Code Listing 10.58:   London-CE2 IPv4 forwarding table

```
London-CE2#show ip route
<Output Omitted>


C      202.202.202.0/24 is directly connected, Ethernet0/0
       8.0.0.0/24 is subnetted, 1 subnets
B         8.8.8.0 [200/0] via 10.10.10.1, 00:19:39
       10.0.0.0/24 is subnetted, 1 subnets
```

```
C       10.10.10.0 is directly connected, Serial0/0
        130.130.0.0/24 is subnetted, 1 subnets
B         130.130.130.0 [200/0] via 10.10.10.1, 00:19:39
        31.0.0.0/8 is variably subnetted, 2 subnets, 2 masks
B         31.5.5.4/30 [200/0] via 10.10.10.1, 00:19:39
B         31.3.3.0/24 [200/0] via 10.10.10.1, 00:19:39
```

10.6   Case Study 10.5: Single ASN Deployment for All Customer Sites

10.6.1   Case overview

This case study is derived directly from Case Study 10.4 with only slight modifications. It provides a scenario whereby the customer uses a single ASN for all the sites in a hub-and-spoke VPN topology as illustrated in Figure 10.12. As such, a similar problem discussed in section 10.5.3 will occur, this time in the hub site: a spoke site (Paris Site 1 or Brussels Site 2) originating a route will prepend ASN 65021 and send it to the SP's MPLS VPN backbone; the SP will

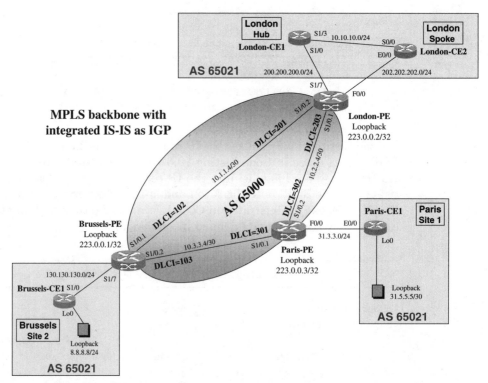

Figure 10.12: MPLS VPN design and topology for Case Study 10.5.

prepend its own ASN 65000 and send the update to the hub site (London-CE1); the hub site will then ignore and drop the update because it contains its own ASN 65021.

The solution to this problem is for the SP to overwrite the customer's ASN with its own using the BGP AS-override feature, which was illustrated in section 9.7 together with the BGP SOO extended community attribute. In this case study, we shall illustrate the AS-override feature again, this time with a hub-and-spoke VPN topology. In this scenario, the AS-override option has to be enabled on all BGP sessions between the PE and CE routers.

10.6.2 CE configurations

In this section, we illustrate only the BGP configuration for Brussels-CE1 and Paris-CE1. The configurations of London-CE1 and London-CE2 remain the same as those listed in Code Listings 10.44 and 10.45.

Code Listing 10.59 illustrates the modified EBGP configuration for Brussels-CE1. Notice that Brussels-CE1 now shares the same BGP ASN 65021 as London-CE1, London-CE2, and Paris-CE1.

Code Listing 10.59:   EBGP configuration for Brussels-CE1

```
hostname Brussels-CE1
!
interface Loopback0
 ip address 8.8.8.8 255.255.255.0
!
interface Serial1/0
 bandwidth 64
 ip address 130.130.130.2 255.255.255.0
!
router bgp 65021
 no synchronization
 network 8.8.8.0 mask 255.255.255.0
 neighbor 130.130.130.1 remote-as 65000
 no auto-summary
```

Code Listing 10.60 illustrates the modified EBGP configuration for Paris-CE1. Notice that Paris-CE1 now shares the same BGP ASN 65021 as London-CE1, London-CE2, and Brussels-CE1.

Code Listing 10.60:   EBGP configuration for Paris-CE1

```
hostname Paris-CE1
!
interface Loopback0
 ip address 31.5.5.5 255.255.255.252
!
interface Ethernet0/0
 ip address 31.3.3.2 255.255.255.0
!
router bgp 65021
 no synchronization
 network 31.5.5.4 mask 255.255.255.252
 neighbor 31.3.3.1 remote-as 65000
 no auto-summary
```

10.6.3   PE configurations

To overcome the problem discussed in section 10.6.1, in which the CE routers will ignore and drop BGP updates that contain the same ASN as their own, Code Listings 10.61 to 10.63 enable the BGP AS-override feature on the PE routers for all the EBGP sessions between the PE and CE routers.

Code Listing 10.61 shows the portion of the configuration for enabling the BGP AS-override feature on Brussels-PE. The rest of the MPLS VPN configuration for Brussels-PE is identical to Code Listing 10.46.

Code Listing 10.61:   MPLS VPN configuration for Brussels-PE

```
hostname Brussels-PE
!
<Output Omitted>
!
router bgp 65000
<Output Omitted>
!
 address-family ipv4 vrf site2
 redistribute connected

! --- Brussels-CE1 (Brussels Site2) uses the same ASN 65021
! --- as the rest of the CE routers.
 neighbor 130.130.130.2 remote-as 65021
 neighbor 130.130.130.2 activate
```

```
! --- The BGP AS-Override feature is enabled
! --- between Brussels-PE and Brussels-CE1
! --- so that Brussels-PE will overwrite the customer's
! --- ASN (65021) with its own (65000) in BGP updates
! --- that are sent to Brussels-CE1. This is to prevent
! --- Brussels-CE1 from ignoring and dropping updates
! --- that contain its own ASN (65021).
 neighbor 130.130.130.2 as-override
 no auto-summary
 no synchronization
 exit-address-family
 !

<Output Omitted>
```

Code Listing 10.62 shows the portion of the configuration for enabling the BGP AS-override feature on Paris-PE. The rest of the MPLS VPN configuration for Paris-PE is identical to Code Listing 10.47.

Code Listing 10.62: MPLS VPN configuration for Paris-PE

```
hostname Paris-PE
!
<Output Omitted>
!
router bgp 65000
<Output Omitted>
 !
 address-family ipv4 vrf site1
 redistribute connected

! --- Paris-CE1 (Paris Site1) uses the same ASN 65021
! --- as the rest of the CE routers.
 neighbor 31.3.3.2 remote-as 65021
 neighbor 31.3.3.2 activate

! --- The BGP AS-Override feature is enabled
! --- between Paris-PE and Paris-CE1
! --- so that Paris-PE will overwrite the customer's
! --- ASN (65021) with its own (65000) in BGP updates
! --- that are sent to Paris-CE1. This is to prevent
! --- Paris-CE1 from ignoring and dropping updates
! --- that contain its own ASN (65021).
```

```
neighbor 31.3.3.2 as-override
no auto-summary
no synchronization
exit-address-family
 !
<Output Omitted>
```

Code Listing 10.63 shows the portion of the configuration for enabling the BGP AS-override feature on London-PE. The rest of the MPLS VPN configuration for London-PE is identical to Code Listing 10.48.

Code Listing 10.63:   MPLS VPN configuration for London-PE

```
hostname London-PE
!
<Output Omitted>
!
router bgp 65000
<Output Omitted>
 !
 address-family ipv4 vrf spoke
 redistribute connected
 neighbor 202.202.202.2 remote-as 65021
 neighbor 202.202.202.2 activate
 neighbor 202.202.202.2 allowas-in 4
 no auto-summary
 no synchronization
 exit-address-family
 !
 address-family ipv4 vrf hub
 redistribute connected

! --- London-CE1 (London Hub) uses the same ASN 65021
! --- as the rest of the CE routers.
 neighbor 200.200.200.2 remote-as 65021
 neighbor 200.200.200.2 activate

! --- The BGP AS-Override feature is enabled
! --- between London-PE and London-CE1
! --- so that London-PE will overwrite the customer's
! --- ASN (65021) with its own (65000)
! --- in BGP updates that are sent to London-CE1.
! --- This is to prevent London-CE1 from ignoring
```

```
! --- and dropping updates that contain its own ASN (65021).
neighbor 200.200.200.2 as-override
no auto-summary
no synchronization
exit-address-family
!
<Output Omitted>
```

10.6.4 Monitoring BGP VPNv4 tables of the PE routers

Code Listings 10.64 to 10.66 illustrate the entire content of the BGP VPNv4 table for the three PE routers.

In Code Listing 10.64, from the per-VRF BGP VPNv4 information associated with VRF Site 2, we can see that the AS-path for prefix 31.5.5.4/30 is now "65021 65000 65000".

Code Listing 10.64:   Brussels-PE BGP VPNv4 table

```
Brussels-PE#show ip bgp vpnv4 all
BGP table version is 184, local router ID is 223.0.0.1
Status codes: s suppressed, d damped, h history, * valid, > best, i - internal
Origin codes: i - IGP, e - EGP, ? - incomplete

   Network          Next Hop            Metric LocPrf Weight Path
<Output Omitted>
Route Distinguisher: 65000:102 (default for vrf site2)
<Output Omitted>
*>i31.3.3.0/24      223.0.0.2               100    0 65021 65000 ?
*>i31.5.5.4/30      223.0.0.2               100    0 65021 65000 65000 i
<Output Omitted>
```

In Code Listing 10.65, from the per-VRF BGP VPNv4 information associated with VRF Site 1, we can see that the AS-path for prefix 8.8.8.0/24 is now "65021 65000 65000."

Code Listing 10.65:   Paris-PE BGP VPNv4 table

```
Paris-PE#show ip bgp vpnv4 all
BGP table version is 15, local router ID is 223.0.0.3
Status codes: s suppressed, d damped, h history, * valid, > best, i - internal
Origin codes: i - IGP, e - EGP, ? - incomplete
```

```
    Network              Next Hop            Metric LocPrf Weight Path
<Output Omitted>
Route Distinguisher: 65000:101 (default for vrf site1)
*>i8.8.8.0/24          223.0.0.2               100      0 65021 65000 65000 i
<Output Omitted>
*>i130.130.130.0/24 223.0.0.2                  100       0 65021 65000 ?
<Output Omitted>
```

Code Listing 10.66 shows that before London-PE send prefixes 8.8.8.0/24 and 31.5.5.4/30 to London-CE1, the AS-path for both routes is still reflected as "65021" (see VRF Hub), which is the same ASN that London-CE1 is using. However, after the same prefixes are sent to London-CE1 and propagated back from London-CE2 to London-PE, we can see the AS path for prefixes 8.8.8.0/24 and 31.5.5.4/30 are now both reflected as "65021 65000 65000" (see VRF Spoke).

Code Listing 10.66:   London-PE BGP VPNv4 table

```
London-PE#show ip bgp vpnv4 all
BGP table version is 68, local router ID is 223.0.0.2
Status codes: s suppressed, d damped, h history, * valid, > best, i - internal
Origin codes: i - IGP, e - EGP, ? - incomplete

    Network             Next Hop            Metric LocPrf Weight Path
Route Distinguisher: 65000:88 (default for vrf hub)
*>i8.8.8.0/24          223.0.0.1              0   100      0 65021 i
*>i31.3.3.0/24         223.0.0.3              0   100      0 ?
*>i31.5.5.4/30         223.0.0.3              0   100      0 65021 i
*>i130.130.130.0/24 223.0.0.1                 0   100      0 ?
<Output Omitted>
Route Distinguisher: 65000:99 (default for vrf spoke)
*> 8.8.8.0/24          202.202.202.2                       0 65021 65000 65000 i
*> 31.3.3.0/24         202.202.202.2                       0 65021 65000 ?
*> 31.5.5.4/30         202.202.202.2                       0 65021 65000 65000 i
*> 130.130.130.0/24 202.202.202.2                          0 65021 65000 ?
<Output Omitted>
```

10.6.5   Monitoring BGP tables of the CE routers

Code Listings 10.67 to 10.70 illustrates the BGP table for Brussels-CE1, Paris-CE1, and London-CE1.

Code Listing 10.67 displays the BGP table for Brussels-CE1 (Site 2). The AS-path of the BGP route 31.5.5.4/30 listed in Code Listing 10.64 for VRF Site 2 has been modified from "65021 65000 65000" to "65000 65000 65000 65000," indicating that the ASN 65021 has been overwritten by ASN 65000, and the entire AS-path is prepended by another "65000" when the route is traversing from Brussels-PE to Brussels-CE1. The same applies to BGP route 31.3.3.0/24. This verifies that the AS-override feature enabled between Brussels-PE and Brussels-CE1 is effective.

Code Listing 10.67:   Brussels-CE1 BGP table

```
Brussels-CE1#show ip bgp
<Output Omitted>


   Network           Next Hop        Metric LocPrf Weight Path
<Output Omitted>
*> 31.3.3.0/24       130.130.130.1         0 65000 65000 65000 ?
*> 31.5.5.4/30       130.130.130.1         0 65000 65000 65000 65000 i
<Output Omitted>
```

Code Listing 10.68 displays the BGP table for Paris-CE1 (Site 1). The AS-path of the BGP route 8.8.8.0/24 listed in Code Listing 10.65 for VRF Site 1 has been modified from "65021 65000 65000" to "65000 65000 65000 65000," indicating that the ASN 65021 has been overwritten by ASN 65000, and the entire AS path is prepended by another "65000" when the route is traversing from Paris-PE to Paris-CE1. The same applies to BGP route 130.130.130.0/24. This verifies that the AS-override feature enabled between Paris-PE and Paris-CE1 is effective.

Code Listing 10.68:   Paris-CE1 BGP table

```
Paris-CE1#show ip bgp
<Output Omitted>


   Network           Next Hop        Metric LocPrf Weight Path
*> 8.8.8.0/24        31.3.3.1              0 65000 65000 65000 65000 i
<Output Omitted>
*> 130.130.130.0/24 31.3.3.1               0 65000 65000 65000 ?
<Output Omitted>
```

Code Listing 10.69 displays the BGP table for London-CE1 (Hub). The AS-path of the BGP route 8.8.8.0/24 listed in Code Listing 10.66 for VRF Hub has been modified from "65021" to "65000 65000," indicating that the ASN 65021 has been overwritten by ASN 65000, and the entire AS-path is prepended by another "65000" when the route is traversing from London-PE to London-CE1. The same applies to BGP route 31.5.5.4/30. This verifies that the AS-override feature enabled between London-PE and London-CE1 is effective.

Code Listing 10.69:   London-CE1 BGP table

```
London-CE1#show ip bgp
<Output Omitted>

    Network            Next Hop            Metric LocPrf Weight Path
*> 8.8.8.0/24         200.200.200.1                        0 65000 65000 i
*> 31.3.3.0/24        200.200.200.1                        0 65000 ?
*> 31.5.5.4/30        200.200.200.1                        0 65000 65000 i
*> 130.130.130.0/24 200.200.200.1                          0 65000 ?
<Output Omitted>
```

10.7   Internet Access and MPLS VPN

In the previous sections, we discussed the various types of MPLS VPN topologies that can be deployed in an SP environment. There is one more concern: VPN sites may require Internet access. Typically, there are two design models that you can use to provide Internet access in an MPLS VPN:

1. Internet access can be implemented via a separate VPN.

2. Internet access can be implemented via global routing in the PE routers.

In the following sections, we briefly discuss the advantages and disadvantages of these two models and the implications of their usage. The actual deployment of the two models is beyond the scope of this book.

10.7.1   Internet access with a separate VPN

The main advantage of implementing Internet access as a separate VPN is the isolation between the Internet and the SP's backbone, which reinforces security. In this design model, the Internet gateways will appear as CE routers to the MPLS VPN backbone, and Internet access is facilitated by incorporating the Internet VPN with the customer VPN in the customer VRFs with the help of an overlapping VPN topology.

The major disadvantage of this design is the scalability of this solution. The Internet VPN cannot carry full Internet routing (approximately 100,000 routes). Therefore, the Internet gateways (CE routers) should announce only a default route toward the PE routers. In addition, local Internet routes should be injected into the Internet VPN to optimize local routing.

10.7.2 Internet access with global routing

Internet access with global routing is implemented in the same manner as a traditional Internet service provider (ISP) backbone. In this case, the global routing table on the PE routers is used to forward traffic toward the Internet destinations. The VPN customer can have access to the global routing table in two ways:

1. Use a separate logical link for Internet access: This option is equivalent to the traditional way of providing Internet access to the VPN customers. The main advantage of this method is its simplicity. The major disadvantage is the requirement to have either separate physical links or a single physical link with WAN encapsulation (for example, frame relay) that supports logical links (subinterfaces).

2. Packet leaking between the VRF and the global routing table: This option allows packets originating in a VPN to leak out into the global address space via a global next-hop using the "ip route vrf *name prefix mask next-hop* global" command, and packets originating in the global address space to be forwarded toward a CE router via a VRF interface in a VPN using the "ip route *prefix mask interface*" command. The main advantage of this method is that it can be implemented over any LAN or WAN media, providing the all the necessary flexibility. The major disadvantage is that the VPN and Internet traffic are mixed over the same link, thus providing weaker security as compared to traditional Internet access methods.

10.8 MPLS VPN Performance Optimization

Another important aspect to consider when deploying MPLS VPN is its performance optimization. In the following sections, we examine some of the performance optimization mechanisms and features that are commonly used to fine-tune the performance of MPLS VPN.

10.8.1 BGP route advertisement

In an MPLS VPN environment, each PE router is required to advertise the best BGP routes within each VRF to all its VRF neighbors, and this occurs at both ingress and egress of the MPLS VPN network. With EBGP CE neighbors, advertisement of these routes occurs every 30 seconds; with MP-IBGP PE neighbors, the route advertisement occurs every 5 seconds. Moreover, during this period, any network changes are not announced to other BGP speakers immediately; instead they are batched together and are sent only when the advertisement interval expires. Therefore, we might need to fine-tune the BGP advertisement interval to improve the convergence time between sites. The advertisement interval can be adjusted with the "neighbor advertisement-interval ⟨0–600⟩" command and has to be changed for both PE and CE routers (if BGP is used as the PE-CE routing protocol).

10.8.2 BGP scanner process

The BGP scanner process will also have an affect on the site-to-site convergence time. It is used to validate BGP updates by verifying their respective next-hop reachability. By default, the BGP scanner process is invoked every 60 seconds for IPv4 and VPNv4 prefixes and can be adjusted with the "bgp scan-time ⟨5–60⟩" command. Take note of the BGP table size when adjusting this timer as large BGP table and small scan time can place a very heavy burden on the CPU.

10.8.3 BGP import process

The BGP import process uses a separate invocation of the regular BGP scanner process, so we also have to take this timer into consideration. By default, the BGP import process is performed once every 60 seconds and can be adjusted using the "bgp scan-time [import] ⟨5–60⟩" command under address-family vpnv4.

10.8.4 VRF route limit

SPs providing MPLS VPN services have to take into consideration that any customer can generate any number of routes to the PE routers intentionally or unintentionally, thus depleting valuable resources in the PE routers. To prevent the total depletion of the resources from happening, the resources used by individual customers have to be limited. This can be achieved with the VRF route limit, which limits the overall number of routes in a VRF regardless of where the routes are originated. The route limit is configured for each individual VRF using the "maximum routes ⟨limit⟩" command.

10.9 Summary

In this chapter, we explored the various MPLS VPN topologies that do not share the same connectivity requirements. We first examined the overlapping VPN topology in Case Study 10.1, which connects part of the networks of two separate VPNs by using a third VPN that is created within the MPLS VPN containing sites from the two different VPNs.

The central services VPN topology and hybrid VPN topology were discussed in Case Study 10.2 and 10.3, respectively. These topologies are used when a set of different client-VPNs are required to share a common set of servers. These servers reside in the central services VPN and are accessible to other client-VPNs. However, the client-VPNs are not able to communicate with each other. In other words, they can see the servers but not one another.

The most interesting topology is the unorthodox hub-and-spoke VPN topology, which was illustrated in Case Study 10.4. The hub-and-spoke VPN topology is deployed when the customer wants to preserve the centralized control inherent with overlay VPN hub-and-spoke topology, in which all traffic is exchanged via a central site (or sites). To force traffic to go through the central site, two links are required for the central site—one (the hub link) is importing routes from other CE sites, and the other (the spoke link) is exporting them to other CE sites. Case Study 10.4 also demonstrated how the BGP Allowas-in feature can be used to prevent returning traffic from being dropped by the PE router.

If a single ASN is deployed for all customer sites, the BGP AS-override feature has to be enabled on all BGP sessions between the PE and CE routers as illustrated in Case Study 10.5 to prevent routing updates from being ignored by the CE routers. On top of all these, two other important aspects to consider when deploying MPLS VPN—Internet access and performance optimization—are also covered in the concluding portion of the chapter.

Scaling MPLS VPN with Route Reflectors

11.1 MPLS VPN Scaling Overview

As discussed in Chapter 1, MPLS VPN happens to reap the best of both worlds by retaining the advantages of peer-to-peer and overlay models. MPLS VPN does not require complex filters or dedicated routers; it supports optimal routing between sites, and each PE router needs only the routing information for the directly connected VPNs (or attached VRFs) it supports.

Internet (IPv4) and VPNv4 routing information are exchanged between PE routers via MP-BGP (IBGP). This engages the full IBGP mesh prerequisite between the PE routers that require the same VPN information. The IBGP full mesh among PE routers results in the flooding of all VPN routes to all PE routers and thus creates a significant scaling issue when a large number of routes are propagated.

Existing BGP techniques such as route reflectors can be deployed to scale MP-BGP (IBGP) route distribution. The deployment of route reflectors (RRs) breaks up the IBGP mesh. Each PE router needs to establish only one BGP session to each RR. Despite this, no single RR (PE router) can hold all Internet and VPN routing information. Hence, to improve the scalability of MPLS VPN and to reduce excessive processing burden on the RR, we can further segment additional routing information using partitioned RRs. This chapter focuses on how to deploy partitioned VPN RRs and consolidate VPN routes based on BGP standard communities.

11.2. MPLS VPN RR Partitioning

The MPLS VPN backbone (or core) carries both Internet and VPN routes. Dedicated VPN RRs can be deployed solely for VPN routes. This can be achieved by removing Internet routes (or deactivating IPv4 sessions) from the PE routers that are doubling as RRs. However, for large-scale MPLS VPN backbone implementation, a single dedicated VPN RR cannot hold all the VPN routes. Partitioned VPN RRs can be deployed to mitigate this scaling issue. In this case, VPN routing information is partitioned and consolidated based on route targets (see Chapter 8 for more details on route targets) or other BGP attributes. We shall further discuss dedicated VPN RRs in section 11.2.1 and partitioned VPN RRs in section 11.2.2.

11.2.1. Dedicated VPN Route Reflectors

Traditional RRs supporting Internet routes can also be used for reflecting VPN routes, which gives service providers the competitive advantage to quickly deploy pilot services. However, this standard implementation is not scalable when the number of VPN customer increases. To improve scalability, dedicated VPN RRs can be deployed instead. In this case, PE routers still carry Internet routes together with a subset of VPN routes. This is achieved by selectively activating IPv4 and VPNv4 sessions (address-family IPv4 and address-family VPNv4) on the PE routers. Note that RRs for each address family must be redundant to avoid a single point of failure.

Figure 11.1 illustrates a simple deployment of dedicated Internet and VPN RRs in which:

- VPN-101 is supported by Bern-PE, Frankfurt-PE, Vienna-PE, and Munich-PE.

- VPN-102 is supported by Zurich-PE, Frankfurt-PE, Vienna-PE, and Munich-PE.

- Internet is supported by Bern-PE, Zurich-PE, and Geneva-PE.

In this scenario, Geneva-PE is the dedicated Internet RR. Bern-PE and Zurich-PE are the nonreflecting PE routers, or rather the RR-clients to Geneva-PE. Likewise, Frankfurt-PE is the dedicated VPN RR reflecting VPN-101 and

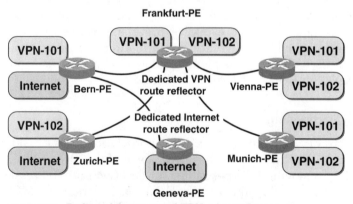

Figure 11.1: Dedicated Internet and VPN route reflectors.

VPN-102 routes, which are advertised from Bern-PE, Zurich-PE, Vienna-PE, and Munich-PE. In this instance, Bern-PE, Zurich-PE, Vienna-PE, and Munich-PE become the RR-clients to Frankfurt-PE.

11.2.2   Partitioned VPN route reflectors

With the further expansion of VPN customers, dedicated VPN RRs (PE routers) will no longer be able to carry full Internet routing together with VPN routes. This is because excessive processing overheads are incurred on dedicated VPN RRs with the additional growth. As the number of VPN routes continues to increase, the solution to alleviate this scaling problem is to deploy partitioned VPN RRs. Partitioned VPN RRs selectively segments VPN routes based on route targets (RT) or BGP communities. Each RR only stores the routes for a set of VPNs. Therefore, no single VPN RR needs to store the routing information on all the VPNs. In this case, the nonreflecting PE routers will peer to RRs according to the VPNs they support.

Figure 11.2 illustrates a simple deployment of partitioned VPN RRs in which:

- VPN-101 is supported by Bern-PE, Frankfurt-PE, and Vienna-PE

- VPN-102 is supported by Zurich-PE, Frankfurt-PE, and Vienna-PE

- VPN-201 is supported by Bern-PE, Zurich-PE, Geneva-PE, and Munich-PE

Geneva-PE is the partitioned VPN RR that supports only VPN-201. Bern-PE, Zurich-PE, and Munich-PE are the RR-clients to Geneva-PE that propagate VPN-201 routes. Likewise, Frankfurt-PE is the partitioned VPN RR reflecting VPN-101 and VPN-102 routes, which are advertised from Bern-PE, Zurich-PE, and Vienna-PE. In this instance, Bern-PE, Zurich-PE, and Vienna-PE become the RR-clients to Frankfurt-PE.

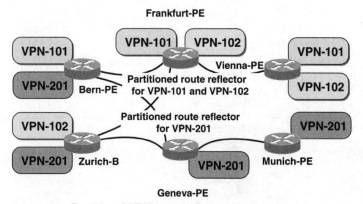

Figure 11.2: Partitioned VPN route reflectors.

11.3   Case Study 11.1: Partitioned VPN Route Reflectors with Standard BGP Communities

11.3.1   Case Overview and Network Topology

This case study illustrates an MPLS VPN over frame relay environment in which partitioned VPN RRs are deployed for scaling the MP-BGP route distribution in the MPLS VPN backbone. The MPLS VPN deployment uses Cisco routers and spans five different customer sites: Brussels-CE1, Brussels-CE2, London-CE1, London-CE2, and Paris-CE1. EBGP is present on the customer side as the PE-CE routing protocol and Integrated IS-IS is configured as the IGP between the PE routers. In addition, the PE routers use IBGP or rather MP-BGP to distribute the VPN routing information with the help of BGP extended communities. The MPLS VPN topology layout is shown in Figure 11.3, and the BGP Autonomous Systems (ASs) layout is illustrated in Figure 11.4.

The MPLS VPN implementation allows the five different locations to transparently interconnect through a service provider's network. In this specific instance, the service provider network supports two different IP VPNs: VPN 101 and VPN 102. Each of these appears to its users as a private network,

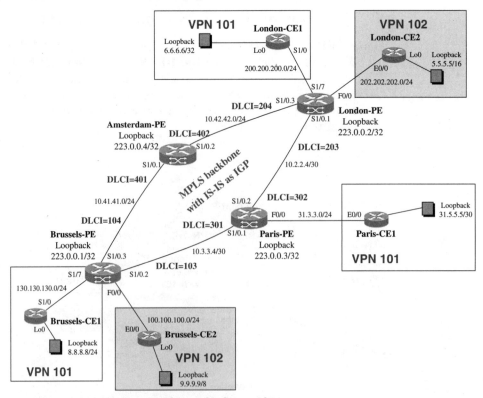

Figure 11.3: MPLS VPN topology layout for Case study 11.1.

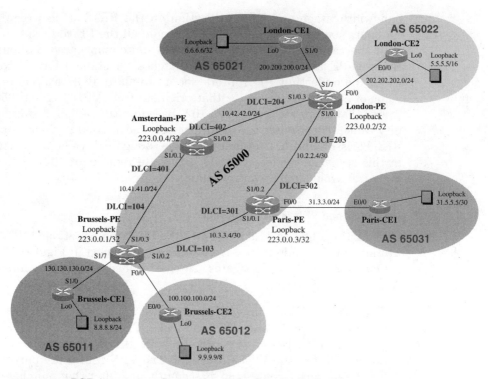

Figure 11.4: BGP Autonomous Systems layout for Case Study 11.1.

separate from all other networks. Within a VPN, each site can send IP packets to any other site belonging to the same VPN. Each VPN is associated with one or more VPN routing or forwarding instances (VRFs). A VRF consists of an IP routing table, a derived Cisco express forwarding (CEF) table, and a set of interfaces that use this forwarding table.

A partitioned RR design requires an indicative mechanism to identify routes that originate from different VPNs and also a filtering mechanism to partition these routes according to the assigned VPN RRs. These can be achieved by setting a standard BGP community attribute value based on the VRF's membership of the route and the implementation of BGP community filters. The filters can be implemented either at the outbound direction on (nonreflecting) PE routers or at the inbound direction on RRs.

11.3.2 Inbound versus Outbound Filters

Implementing inbound filters on RRs helps to reduce maintenance costs but at the same time also increases CPU utilization on the RRs. Conversely implementing outbound filters on (nonreflecting) PE routers helps to reduce bandwidth usage and CPU utilization on the RRs but also requires manual configuration and constant maintenance on all the PE routers.

In the case study, we apply a simple route-map based outbound filter at Brussels-PE and London-PE that matches on standard BGP communities. These nonreflecting PE routers set a standard BGP community attribute value of 65000:101 for routes that originate from VPN 101 and 65000:102 for routes that originate from VPN 102. Based on the assigned standard community value, these routes are partitioned and advertised to their respective VPN RRs, Paris-PE and Amsterdam-PE. Even though RR resiliency is not illustrated in this scenario, readers are advised to use redundant RRs in their network design to avoid single point of failure.

11.3.3 BGP and Partitioned VPN RR Configurations

Code Listings 11.1 to 11.9 illustrates the BGP and partitioned VPN RR configurations for the nine respective routers as portrayed in Figures 11.3 and 11.4.

Code Listing 11.1 illustrates the partitioned VPN RR configuration for Amsterdam-PE router:

- Configure MP-BGP for Amsterdam-PE under the regular (global) BGP process. Declare Brussels-PE and London-PE as MP-BGP (IBGP) peers. Here, Amsterdam-PE is the VPN RR for VPN 102 routes, and Brussels-PE and London-PE are the VPN RR-clients. The VPN routes are partitioned at the RR-clients with a route-map–based outbound filter that matches on standard BGP communities.

- Use the "no bgp default ipv4-unicast" command to disable the default behavior of propagating IPv4 routes. With this, neighbors that need to receive IPv4 routes will have to be activated for IPv4 route exchange individually. No neighbors are activated here for IPv4 route propagation because Amsterdam-PE is allocated for VPNv4 route exchange only.

- Enter the "address-family vpnv4" mode, and activate Brussels-PE and London-PE for VPNv4 route exchange.

- Declare Brussels-PE and London-PE as VPN RR-clients.

- Propagate both standard and extended communities to Brussels-PE and London-PE.

Code Listing 11.1: Partitioned VPN RR configuration for Amsterdam-PE

```
hostname Amsterdam-PE
!
ip cef ! Make sure that ip cef is enabled on Amsterdam-PE once MPLS is set up
!
interface Loopback0
```

```
 ip address 223.0.0.4 255.255.255.255
 ip router isis ! Enable IS-IS on Loopback0
!
interface Serial1/0
 encapsulation frame-relay
 frame-relay lmi-type ansi
!
interface Serial1/0.1 point-to-point
 ip address 10.41.41.2 255.255.255.0
 ip router isis ! Enable IS-IS on Serial1/0.1
 tag-switching ip ! Enable MPLS on subinterface Serial1/0.1
 frame-relay interface-dlci 401
!
interface Serial1/0.2 point-to-point
ip address 10.42.42.2 255.255.255.0
ip router isis ! Enable IS-IS on Serial1/0.2
tag-switching ip ! Enable MPLS on subinterface Serial1/0.2
frame-relay interface-dlci 402
!
! Configure integrated IS-IS as the IGP
router isis
 net 49.0001.0000.0000.0004.00 ! Configure Amsterdam-PE in Area 0x490001
 is-type level-1 ! Configure IS-IS Level-1 routing globally on Amsterdam-PE

 ! Use new style of TLVs to carry wider metric
 ! 24 bits for interface metric and 32 bits for total path metric
metric-style wide
!
! Configure MP-BGP for AS 65000
router bgp 65000
 no synchronization
 no bgp default ipv4-unicast ! Disable default exchange of IPv4 routes
 neighbor 223.0.0.1 remote-as 65000 ! Declare Brussels-PE as IBGP neighbor
 neighbor 223.0.0.1 update-source Loopback0
 neighbor 223.0.0.2 remote-as 65000 ! Declare London-PE as IBGP neighbor
 neighbor 223.0.0.2 update-source Loopback0
 no auto-summary
 !
 address-family vpnv4 ! Enter the address-family VPNv4 mode
! Activate Brussels-PE for the advertisement of VPNv4 NLRI
 neighbor 223.0.0.1 activate
! Declare Brussels-PE as VPNv4 RR-client
 neighbor 223.0.0.1 route-reflector-client

 ! Propagate both standard and extended BGP communities to Brussels-PE
neighbor 223.0.0.1 send-community both
```

```
! Activate London-PE for the advertisement of VPNv4 NLRI
 neighbor 223.0.0.2 activate
! Declare London-PE as VPNv4 RR-client
 neighbor 223.0.0.2 route-reflector-client

! Propagate both standard and extended BGP communities to London-PE
 neighbor 223.0.0.2 send-community both
 no auto-summary
 exit-address-family
!
! Specify 32-bit community values in AS:NN format
ip bgp-community new-format
```

Code Listing 11.2 illustrates the BGP configuration for Brussels-CE1 router. In this example, the PE-CE routing protocol is EBGP. Brussels-CE1 in AS 65011 peers to its EBGP neighbor Brussels-PE in AS 65000, and subnet 8.8.8.0/24 is advertised.

Code Listing 11.2:   BGP configuration for Brussels-CE1

```
hostname Brussels-CE1
!
interface Loopback0
 ip address 8.8.8.8 255.255.255.0
!
interface Serial1/0
 bandwidth 64
 ip address 130.130.130.2 255.255.255.0
!
! Regular BGP Configuration
!
router bgp 65011 ! Configure Brussels-CE1 in AS 65011
 no synchronization
 network 8.8.8.0 mask 255.255.255.0 ! Advertise subnet 8.8.8.0/24
! Peer to EBGP neighbor Brussels-PE in AS 65000
 neighbor 130.130.130.1 remote-as 65000
 no auto-summary
```

Code Listing 11.3 illustrates the BGP configuration for Brussels-CE2 router. In this example, the PE-CE routing protocol is EBGP. Brussels-CE2 in AS 65012 peers to its EBGP neighbor Brussels-PE in AS 65000, and network 9.0.0.0/8 is advertised.

Code Listing 11.3:   BGP configuration for Brussels-CE2

```
hostname Brussels-CE2
!
interface Loopback0
 ip address 9.9.9.9 255.0.0.0
!
interface Ethernet0/0
 ip address 100.100.100.2 255.255.255.0
!
! Regular BGP Configuration
!
router bgp 65012 ! Configure Brussels-CE2 in AS 65012
 no synchronization
 network 9.0.0.0 ! Advertise network 9.0.0.0/8
! Peer to EBGP neighbor Brussels-PE in AS 65000
 neighbor 100.100.100.1 remote-as 65000
 no auto-summary
```

Code Listing 11.4 illustrates the partitioned VPN RR configuration for Brussels-PE router.

Code Listing 11.4:   Partitioned VPN RR configuration for Brussels-PE

```
hostname Brussels-PE
!
ip vrf vpn101 ! Create VRF for VPN 101
 rd 65000:101 ! Specify the route distinguisher, 65000:101, used for VPN 101

! Set up the import and export properties for the BGP extended communities
 route-target export 65000:101
 route-target import 65000:101
!
ip vrf vpn102 ! Create VRF for VPN 102
 rd 65000:102 ! Specify the route distinguisher, 65000:102, used for VPN 102

! Set up the import and export properties for the BGP extended communities
 route-target export 65000:102
 route-target import 65000:102
!
ip cef ! Make sure that ip cef is enabled on Brussels-PE once MPLS is set
up
!
```

```
interface Loopback0
 ip address 223.0.0.1 255.255.255.255
 ip router isis ! Enable IS-IS on Loopback0
!
interface FastEthernet0/0
! Configure the forwarding details for FastEthernet0/0
 ip vrf forwarding vpn102
! Remember to set up the IP address after doing this
ip address 100.100.100.1 255.255.255.0
!
interface Serial1/0
 encapsulation frame-relay
 frame-relay lmi-type ansi
!
interface Serial1/0.2 point-to-point
 bandwidth 64
 ip address 10.3.3.5 255.255.255.252
 ip router isis ! Enable IS-IS on Serial1/0.2
 tag-switching ip ! Enable MPLS on subinterface Serial1/0.2
 frame-relay interface-dlci 103
!
interface Serial1/0.3 point-to-point
 bandwidth 64
 ip address 10.41.41.1 255.255.255.0
 ip router isis ! Enable IS-IS on Serial1/0.3
 tag-switching ip ! Enable MPLS on subinterface Serial1/0.3
 frame-relay interface-dlci 104
!
interface Serial1/7
 bandwidth 64
 ip vrf forwarding vpn101 ! Configure the forwarding details for Serial1/7
! Remember to set up the IP address after doing this
 ip address 130.130.130.1 255.255.255.0
 clockrate 64000
!
! Configure integrated IS-IS as the IGP
router isis
 net 49.0001.0000.0000.0001.00 ! Configure Brussels-PE in Area 0x490001
 is-type level-1 ! Configure IS-IS level 1 routing globally on Brussels-PE

 ! Use new style of TLVs to carry wider metric
 ! 24 bits for interface metric and 32 bits for total path metric
 metric-style wide
!
! Configure MP-BGP for AS 65000
router bgp 65000
```

```
 no synchronization
 no bgp default ipv4-unicast ! Disable default exchange of IPv4 routes
! Advertise Internet host route 223.0.0.1/32
 network 223.0.0.1 mask 255.255.255.255
 neighbor 223.0.0.3 remote-as 65000 ! Declare Paris-PE as IBGP neighbor
 neighbor 223.0.0.3 update-source Loopback0
! Activate Paris-PE for the advertisement of IPv4 NLRI
 neighbor 223.0.0.3 activate
 neighbor 223.0.0.4 remote-as 65000 ! Declare Amsterdam-PE as IBGP neighbor
 neighbor 223.0.0.4 update-source Loopback0
 no auto-summary
 !
! Enter the address-family IPv4 mode for VRF vpn102
 address-family ipv4 vrf vpn102
 redistribute connected ! Redistribute FastEthernet0/0 into MP-BGP

! Declare Brussels-CE2 as CE EBGP neighbor in AS 65012
 neighbor 100.100.100.2 remote-as 65012
 neighbor 100.100.100.2 activate ! Activate Brussels-CE2

! Apply inbound route-map setcomm102 to tag routes from Brussels-CE2
 neighbor 100.100.100.2 route-map setcomm102 in
 no auto-summary
 no synchronization
 exit-address-family
 !
! Enter the address-family IPv4 mode for VRF vpn101
 address-family ipv4 vrf vpn101
 redistribute connected ! Redistribute Serial1/7 into MP-BGP

 ! Declare Brussels-CE1 as CE EBGP neighbor in AS 65011
 neighbor 130.130.130.2 remote-as 65011
 neighbor 130.130.130.2 activate ! Activate Brussels-CE1

 ! Apply inbound route-map setcomm101 to tag routes from Brussels-CE1
 neighbor 130.130.130.2 route-map setcomm101 in
 no auto-summary
 no synchronization
 exit-address-family
 !
 address-family vpnv4 ! Enter the address-family VPNv4 mode
! Activate Paris-PE for the advertisement of VPNv4 NLRI
 neighbor 223.0.0.3 activate

! Propagate both standard and extended BGP communities to Paris-PE
 neighbor 223.0.0.3 send-community both
```

```
! Apply outbound route-map rr-vpn101
! For partitioning VPN 101 routes from Brussels-CE1 to Paris-PE
neighbor 223.0.0.3 route-map rr-vpn101 out
! Activate Amsterdam-PE for the advertisement of VPNv4 NLRI
neighbor 223.0.0.4 activate

! Propagate both standard and extended BGP communities to Amsterdam-PE
neighbor 223.0.0.4 send-community both

! Apply outbound route-map rr-vpn102
! For partitioning VPN 102 routes from Brussels-CE2 to Amsterdam-PE
neighbor 223.0.0.4 route-map rr-vpn102 out
no auto-summary
exit-address-family
!
! Specify 32-bit community values in AS:NN format
ip bgp-community new-format
!
! Standard community-list 1 matches only routes
! that have a community value of 65000:101
ip community-list 1 permit 65000:101

! Standard community-list 2 matches only routes
! that have a community value of 65000:102
ip community-list 2 permit 65000:102
!
! Route-map setcomm102 sets the standard community value 65000:102
route-map setcomm102 permit 10
 set community 65000:102
!
! Route-map setcomm101 sets the standard community value 65000:101
route-map setcomm101 permit 10
 set community 65000:101
!
! Route-map rr-vpn101 matches routes that have
! a standard community value of 65000:101
route-map rr-vpn101 permit 10
 match community 1
!
! Route-map rr-vpn102 matches routes that have
! a standard community value of 65000:102
route-map rr-vpn102 permit 10
 match community 2
```

The following briefly explains the configuration of Brussels-PE shown in Code Listing 11.4.

- Integrated IS-IS is the IGP running between the four PE routers over frame relay.

- The PE routers use MP-BGP to distribute VPN 101 and VPN 102 routing information across the MPLS VPN backbone (service provider's network) with the help of BGP extended communities.

- For MPLS to be operating properly, Cisco Express Forwarding has to be enabled first using the "ip cef" command.

- MPLS is then enabled on Serial1/0.2 and Serial1/0.3 using the "mpls ip" or "tag-switching ip" command.

- Once MPLS is set up, create VRF for VPN 101 connected on Serial1/7, and create VRF for VPN 102 connected on FastEthernet0/0, using the "ip vrf <VPN routing/forwarding instance name>" command.

- When doing this, specify route distinguisher 65000:101 for VPN 101 and route distinguisher 65000:102 for VPN 102 using the "rd <VPN route distinguisher>" command. The route distinguisher is used to extend the IP address so that you can identify which VPN it belongs to.

- Set up the import and export properties for the BGP extended communities using the "route-target [export|import|both] <target VPN extended community>" command. These are used for filtering the import and export process.

- Configure the forwarding details for the Serial1/7 and FastEthernet0/0 using the "ip vrf forwarding <table name>" command. Remember to set up the IP address after doing this.

- Configure MP-BGP for Brussels-PE under the regular (global) BGP process. Declare Paris-PE and Amsterdam-PE as MP-BGP (IBGP) peers. Here, Amsterdam-PE is the partitioned RR for VPN 102 routes, and Paris-PE is the partitioned RR for VPN 101 as well as Internet (IPv4) routes.

- Use the "no bgp default ipv4-unicast" command to disable the default behavior of propagating IPv4 routes. With this, neighbors that need to receive IPv4 routes will have to be activated for IPv4 route exchange individually. In this instance, only Paris-PE is activated for IPv4 route propagation since it is the allocated RR for IPv4 routes.

- Advertise Internet host route 223.0.0.1/32.

- Now configure EBGP as the PE-CE routing protocol for Brussels-CE1 and Brussels-CE2 by selecting the "address-family ipv4 vrf <VPN routing/forwarding instance name>" command. This is applicable only to the forwarding interfaces for the previously defined VRFs, vpn101 and vpn102.

- As there can be only one BGP process per router, CE EBGP neighbors have to be specified within the per-VRF BGP context, and not under the global BGP process.

- Activate the EBGP neighbors.

- Non-BGP per-VRF routes, such as directly connected interfaces Serial1/7 and FastEthernet0/0, have to be redistributed into their respective per-VRF BGP context to be propagated by MP-BGP to other PE routers.
- Apply inbound route-map setcomm101 to tag VPN 101 routes from Brussels-CE1 with a standard community value of 65000:101.
- Apply inbound route-map setcomm102 to tag VPN 102 routes from Brussels-CE2 with a standard community value of 65000:102.
- Enter the "address-family vpnv4" mode, and activate Paris-PE and Amsterdam-PE for VPNv4 route exchange.
- Apply outbound route-map rr-vpn101 for partitioning VPN 101 routes from Brussels-CE1 to Paris-PE. The route-map matches on standard BGP community value 65000:101.
- Apply outbound route-map rr-vpn102 for partitioning VPN 102 routes from Brussels-CE2 to Amsterdam-PE. The route-map matches on standard BGP community value 65000:102.
- Propagate both standard and extended communities to Paris-PE and Amsterdam-PE.

Code Listing 11.5 illustrates the BGP configuration for London-CE1 router. In this example, the PE-CE routing protocol is EBGP. London-CE1 in AS 65021 peers to its EBGP neighbor London-PE in AS 65000, and host route 6.6.6.6/32 is advertised.

Code Listing 11.5:   BGP configuration for London-CE1

```
hostname London-CE1
!
interface Loopback0
 ip address 6.6.6.6 255.255.255.255
!
interface Serial1/0
 bandwidth 64
 ip address 200.200.200.2 255.255.255.0
!
! Regular BGP Configuration
!
router bgp 65021 ! Configure London-CE1 in AS 65021
 no synchronization
 network 6.6.6.6 mask 255.255.255.255 ! Advertise host route 6.6.6.6/32
! Peer to EBGP neighbor London-PE in AS 65000
 neighbor 200.200.200.1 remote-as 65000
 no auto-summary
```

Code Listing 11.6 illustrates the BGP configuration for London-CE2 router. In this example, the PE-CE routing protocol is EBGP. London-CE2 in AS 65022 peers to its EBGP neighbor London-PE in AS 65000, and subnet 5.5.0.0/16 is advertised.

Code Listing 11.6:   BGP configuration for London-CE2

```
hostname London-CE2
!
interface Loopback0
 ip address 5.5.5.5 255.255.0.0
!
interface Ethernet0/0
 ip address 202.202.202.2 255.255.255.0
!
! Regular BGP Configuration
!
router bgp 65022 ! Configure London-CE2 in AS 65022
 no synchronization
 network 5.5.0.0 mask 255.255.0.0 ! Advertise subnet 5.5.0.0/16
! Peer to EBGP neighbor London-PE in AS 65000
 neighbor 202.202.202.1 remote-as 65000
 no auto-summary
```

Code Listing 11.7 illustrates the partitioned VPN RR configuration for London-PE router.

Code Listing 11.7:   Partitioned VPN RR configuration for London-PE

```
hostname London-PE
!
ip vrf vpn101 ! Create VRF for VPN 101
 rd 65000:101 ! Specify the route distinguisher, 65000:101, used for VPN 101

! Set up the import and export properties for the BGP extended communities
 route-target export 65000:101
 route-target import 65000:101
!
ip vrf vpn102 ! Create VRF for VPN 102
 rd 65000:102 ! Specify the route distinguisher, 65000:102, used for VPN 102

! Set up the import and export properties for the BGP extended communities
 route-target export 65000:102
```

```
   route-target import 65000:102
!
ip cef ! Make sure that ip cef is enabled on London-PE once MPLS is set up
!
interface Loopback0
 ip address 223.0.0.2 255.255.255.255
 ip router isis ! Enable IS-IS on Loopback0
!
interface FastEthernet0/0
! Configure the forwarding details for FastEthernet0/0
 ip vrf forwarding vpn102
! Remember to set up the IP address after doing this
 ip address 202.202.202.1 255.255.255.0
!
interface Serial1/0
 encapsulation frame-relay
 frame-relay lmi-type ansi
!
interface Serial1/0.1 point-to-point
 ip address 10.2.2.5 255.255.255.252
 ip router isis ! Enable IS-IS on Serial1/0.1
 tag-switching ip ! Enable MPLS on subinterface Serial1/0.1
 frame-relay interface-dlci 203
!
interface Serial1/0.3 point-to-point
 bandwidth 64
 ip address 10.42.42.1 255.255.255.0
 ip router isis ! Enable IS-IS on Serial1/0.3
 tag-switching ip ! Enable MPLS on subinterface Serial1/0.3
 frame-relay interface-dlci 204
!
interface Serial1/7
 bandwidth 64
 ip vrf forwarding vpn101 ! Configure the forwarding details for Serial1/7
 ! Remember to set up the IP address after doing this
 ip address 200.200.200.1 255.255.255.0
 clockrate 64000
!
! Configure Integrated IS-IS as the IGP
router isis
 net 49.0001.0000.0000.0002.00 ! Configure London-PE in Area 0x490001
 is-type level-1 ! Configure IS-IS Level-1 routing globally on London-PE

 ! Use new style of TLVs to carry wider metric
 ! 24 bits for interface metric and 32 bits for total path metric
```

```
metric-style wide
!
! Configure MP-BGP for AS 65000
router bgp 65000
 no synchronization
 no bgp default ipv4-unicast ! Disable default exchange of IPv4 routes
 neighbor 223.0.0.3 remote-as 65000 ! Declare Paris-PE as IBGP neighbor
 neighbor 223.0.0.3 update-source Loopback0
! Activate Paris-PE for the advertisement of IPv4 NLRI
 neighbor 223.0.0.3 activate
 neighbor 223.0.0.4 remote-as 65000 ! Declare Amsterdam-PE as IBGP neighbor
 neighbor 223.0.0.4 update-source Loopback0
 no auto-summary
!
! Enter the address-family IPv4 mode for VRF vpn102
 address-family ipv4 vrf vpn102
 redistribute connected ! Redistribute FastEthernet0/0 into MP-BGP

 ! Declare London-CE2 as CE EBGP neighbor in AS 65022
 neighbor 202.202.202.2 remote-as 65022
 neighbor 202.202.202.2 activate ! Activate London-CE2

 ! Apply inbound route-map setcomm102 to tag routes from London-CE2
 neighbor 202.202.202.2 route-map setcomm102 in
 no auto-summary
 no synchronization
 exit-address-family
!
! Enter the address-family IPv4 mode for VRF vpn101
 address-family ipv4 vrf vpn101
 redistribute connected ! Redistribute Serial1/7 into MP-BGP

 ! Declare London-CE1 as CE EBGP neighbor in AS 65021
 neighbor 200.200.200.2 remote-as 65021
 neighbor 200.200.200.2 activate ! Activate London-CE1

 ! Apply inbound route-map setcomm101 to tag routes from London-CE1
 neighbor 200.200.200.2 route-map setcomm101 in
 no auto-summary
 no synchronization
 exit-address-family
!
 address-family vpnv4 ! Enter the address-family VPNv4 mode
! Activate Paris-PE for the advertisement of VPNv4 NLRI
 neighbor 223.0.0.3 activate
```

```
! Propagate both standard and extended BGP communities to Paris-PE
 neighbor 223.0.0.3 send-community both

 ! Apply outbound route-map rr-vpn101
 ! For partitioning VPN 101 routes from London-CE1 to Paris-PE
 neighbor 223.0.0.3 route-map rr-vpn101 out
! Activate Amsterdam-PE for the advertisement of VPNv4 NLRI
 neighbor 223.0.0.4 activate

 ! Propagate both standard and extended BGP communities to Amsterdam-PE
 neighbor 223.0.0.4 send-community both

 ! Apply outbound route-map rr-vpn102
 ! For partitioning VPN 102 routes from London-CE2 to Amsterdam-PE
 neighbor 223.0.0.4 route-map rr-vpn102 out
 no auto-summary
 exit-address-family
!
! Specify 32-bit community values in AS:NN format
ip bgp-community new-format
!
! Standard community-list 1 matches only routes
! that have a community value of 65000:101
ip community-list 1 permit 65000:101

! Standard community-list 2 matches only routes
! that have a community value of 65000:102
ip community-list 2 permit 65000:102
!
! Route-map setcomm102 sets the standard community value 65000:102
route-map setcomm102 permit 10
 set community 65000:102
!
! Route-map setcomm101 sets the standard community value 65000:101
route-map setcomm101 permit 10
 set community 65000:101
!
! Route-map rr-vpn101 matches routes that have
! a standard community value of 65000:101
route-map rr-vpn101 permit 10
 match community 1
!
! Route-map rr-vpn102 matches routes that have
! a standard community value of 65000:102
route-map rr-vpn102 permit 10
 match community 2
```

The following briefly explains the configuration of London-PE shown in Code Listing 11.7.

- Configure MP-BGP for London-PE under the regular (global) BGP process. Declare Paris-PE and Amsterdam-PE as MP-BGP (IBGP) peers. Here, Amsterdam-PE is the partitioned RR for VPN 102 routes, and Paris-PE is the partitioned RR for VPN 101 as well as Internet (IPv4) routes.

- Use the "no bgp default ipv4-unicast" command to disable the default behavior of propagating IPv4 routes. With this, neighbors that need to receive IPv4 routes will have to be activated for IPv4 route exchange individually. In this instance, only Paris-PE is activated for IPv4 route propagation since it is the allocated RR for IPv4 routes.

- Now configure EBGP as the PE-CE routing protocol for London-CE1 and London-CE2 by selecting the "address-family ipv4 vrf <VPN routing/forwarding instance name>" command. This is applicable only to the forwarding interfaces for the previously defined VRFs, VPN101 and VPN102.

- As there can be only one BGP process per router, CE EBGP neighbors have to be specified within the per-VRF BGP context, and not under the global BGP process.

- Activate the EBGP neighbors.

- Non-BGP per-VRF routes such as directly connected interfaces Serial1/7 and FastEthernet0/0 have to be redistributed into their respective per-VRF BGP context to be propagated by MP-BGP to other PE routers.

- Apply inbound route-map setcomm101 to tag VPN 101 routes from London-CE1 with a standard community value of 65000:101.

- Apply inbound route-map setcomm102 to tag VPN 102 routes from London-CE2 with a standard community value of 65000:102.

- Enter the "address-family vpnv4" mode, and activate Paris-PE and Amsterdam-PE for VPNv4 route exchange.

- Apply outbound route-map rr-vpn101 for partitioning VPN 101 routes from London-CE1 to Paris-PE. The route-map matches on standard BGP community value 65000:101.

- Apply outbound route-map rr-vpn102 for partitioning VPN 102 routes from London-CE2 to Amsterdam-PE. The route-map matches on standard BGP community value 65000:102.

- Propagate both standard and extended communities to Paris-PE and Amsterdam-PE.

Code Listing 11.8 illustrates the BGP configuration for Paris-CE1 router. In this example, the PE-CE routing protocol is EBGP. Paris-CE1 in AS 65031 peers to its EBGP neighbor Paris-PE in AS 65000, and subnet 31.5.5.4/30 is advertised.

Code Listing 11.8:   BGP configuration for Paris-CE1

```
hostname Paris-CE1
!
interface Loopback0
 ip address 31.5.5.5 255.255.255.252
!
interface Ethernet0/0
 ip address 31.3.3.2 255.255.255.0
!
! Regular BGP Configuration
!
router bgp 65031 ! Configure Paris-CE1 in AS 65031
 no synchronization
 network 31.5.5.4 mask 255.255.255.252 ! Advertise subnet 31.5.5.4/30
! Peer to EBGP neighbor Paris-PE in AS 65000
 neighbor 31.3.3.1 remote-as 65000
 no auto-summary
```

Code Listing 11.9 illustrates the partitioned VPN RR configuration for Paris-PE router.

Code Listing 11.9:   Partitioned VPN RR Configuration for Paris-PE

```
hostname Paris-PE
!
ip vrf vpn101 ! Create VRF for VPN 101
 rd 65000:101 ! Specify the route distinguisher, 65000:101, used for VPN 101

! Set up the import and export properties for the BGP extended communities
 route-target export 65000:101
 route-target import 65000:101
!
ip cef ! Make sure that ip cef is enabled on Paris-PE once MPLS is set up
!
interface Loopback0
 ip address 223.0.0.3 255.255.255.255
 ip router isis ! Enable IS-IS on Loopback0
!
interface FastEthernet0/0
! Configure the forwarding details for FastEthernet0/0
 ip vrf forwarding vpn101
! Remember to set up the IP address after doing this
```

```
  ip address 31.3.3.1 255.255.255.0
!
interface Serial1/0
 encapsulation frame-relay
 frame-relay lmi-type ansi
!
interface Serial1/0.1 point-to-point
 ip address 10.3.3.6 255.255.255.252
 ip router isis ! Enable IS-IS on Serial1/0.1
 tag-switching ip ! Enable MPLS on subinterface Serial1/0.1
 frame-relay interface-dlci 301
!
 interface Serial1/0.2 point-to-point
 ip address 10.2.2.6 255.255.255.252
 ip router isis ! Enable IS-IS on Serial1/0.2
 tag-switching ip ! Enable MPLS on subinterface Serial1/0.2
 frame-relay interface-dlci 302
!
! Configure Integrated IS-IS as the IGP
router isis
 net 49.0001.0000.0000.0003.00 ! Configure Paris-PE in Area 0x490001
 is-type level-1 ! Configure IS-IS level 1 routing globally on Paris-PE


 ! Use new style of TLVs to carry wider metric
 ! 24 bits for interface metric and 32 bits for total path metric
 metric-style wide
!
! Configure MP-BGP for AS 65000
router bgp 65000
 no synchronization
 no bgp default ipv4-unicast ! Disable default exchange of IPv4 routes
 neighbor 223.0.0.1 remote-as 65000 ! Declare Brussels-PE as IBGP neighbor
 neighbor 223.0.0.1 update-source Loopback0
! Activate Brussels-PE for the advertisement of IPv4 NLRI
 neighbor 223.0.0.1 activate
! Declare Brussels-PE as IPv4 RR-client
 neighbor 223.0.0.1 route-reflector-client
 neighbor 223.0.0.2 remote-as 65000 ! Declare London-PE as IBGP neighbor
 neighbor 223.0.0.2 update-source Loopback0
! Activate London-PE for the advertisement of IPv4 NLRI
 neighbor 223.0.0.2 activate
! Declare London-PE as IPv4 RR-client
 neighbor 223.0.0.2 route-reflector-client
 no auto-summary
 !
```

```
! Enter the address-family IPv4 mode for VRF vpn101
address-family ipv4 vrf vpn101
redistribute connected ! Redistribute FastEthernet0/0 into MP-BGP

! Declare Paris-CE1 as CE EBGP neighbor in AS 65031
neighbor 31.3.3.2 remote-as 65031
neighbor 31.3.3.2 activate ! Activate Paris-CE1
no auto-summary
no synchronization
exit-address-family
!
address-family vpnv4 ! Enter the address-family VPNv4 mode
! Activate Brussels-PE for the advertisement of VPNv4 NLRI
neighbor 223.0.0.1 activate
! Declare Brussels-PE as VPNv4 RR-client
neighbor 223.0.0.1 route-reflector-client

! Propagate both standard and extended BGP communities to Brussels-PE
neighbor 223.0.0.1 send-community both
! Activate London-PE for the advertisement of VPNv4 NLRI
neighbor 223.0.0.2 activate
! Declare London-PE as VPNv4 RR-client
neighbor 223.0.0.2 route-reflector-client

! Propagate both standard and extended BGP communities to London-PE
neighbor 223.0.0.2 send-community both
no auto-summary
exit-address-family
!
! Specify 32-bit community values in AS:NN format
ip bgp-community new-format
```

The following briefly explains the configuration of Paris-PE shown in Code Listing 11.9.

- Configure MP-BGP for Paris-PE under the regular (global) BGP process. Declare Brussels-PE and London-PE as MP-BGP (IBGP) peers. Here, Paris-PE is the VPN and Internet (IPv4) RR for VPN 101 as well as Internet routes, and Brussels-PE and London-PE are the VPN as well as Internet RR clients. The VPN routes are partitioned at the RR clients with a route-map–based outbound filter that matches on standard BGP communities.

- Use the "no bgp default ipv4-unicast" command to disable the default behavior of propagating IPv4 routes. With this, neighbors that need to receive IPv4 routes will have to be activated for IPv4 route exchange individually. In this instance, both Brussels-PE and London-PE are activated for IPv4 route propagation since they are also IPv4 RR-clients.

- Now configure EBGP as the PE-CE routing protocol for Paris-CE1 by selecting the "address-family ipv4 vrf <VPN routing/forwarding instance name>" command. This is applicable only to the forwarding interface for the previously defined VRF, VPN101.

- As there is only one BGP process per router, the CE EBGP neighbor has to be specified within its per-VRF BGP context, and not under the global BGP process.

- Activate the EBGP neighbor.

- Non-BGP per-VRF route such as directly connected interface FastEthernet0/0 has to be redistributed into its per-VRF BGP context to be propagated by MP-BGP to other PE routers.

- Enter the "address-family vpnv4" mode, and activate Brussels-PE and London-PE for VPNv4 route exchange.

- Propagate both standard and extended communities to Brussels-PE and London-PE.

11.3.4. Monitoring IGP Routes between the PE Routers

Integrated IS-IS (refer to Chapter 9 for a brief overview on Integrated IS-IS) is configured as the IGP between Amsterdam-PE, Brussels-PE, London-PE, and Paris-PE. When a RR reflects a route, it does not change the BGP next-hop attribute. Therefore, it is essential that the IGP is functioning properly so that the next-hop addresses are reachable to all the PE routers. Only Level-1 IS-IS routing (intra-area routing) is implemented in the case study. Code Listings 11.10 to 11.13 illustrate the IPv4 routing tables for the four PE routers.

Code Listing 11.10 indicates that Paris-PE (223.0.0.3 with load balancing on two equal-cost paths), London-PE (223.0.0.2), and Brussels-PE (223.0.0.1) are reachable from Amsterdam-PE.

Code Listing 11.10:   IPv4 forwarding table for Amsterdam-PE

```
Amsterdam-PE#show ip route
<Output Omitted>

     223.0.0.0/32 is subnetted, 4 subnets
C      223.0.0.4 is directly connected, Loopback0
i L1   223.0.0.3 [115/30] via 10.41.41.1, Serial1/0.1
              [115/30] via 10.42.42.1, Serial1/0.2
i L1   223.0.0.2 [115/20] via 10.42.42.1, Serial1/0.2
i L1   223.0.0.1 [115/20] via 10.41.41.1, Serial1/0.1
```

```
      10.0.0.0/8 is variably subnetted, 4 subnets, 2 masks
C       10.42.42.0/24 is directly connected, Serial1/0.2
C       10.41.41.0/24 is directly connected, Serial1/0.1
i L1  10.2.2.4/30 [115/20] via 10.42.42.1, Serial1/0.2
i L1  10.3.3.4/30 [115/20] via 10.41.41.1, Serial1/0.1
```

Code Listing 11.11 indicates that Amsterdam-PE (223.0.0.4), Paris-PE (223.0.0.3), and London-PE (223.0.0.2 with load balancing on two equal-cost paths) are reachable from Brussels-PE.

Code Listing 11.11:   IPv4 forwarding table for Brussels-PE

```
Brussels-PE#show ip route
<Output Omitted>


      223.0.0.0/32 is subnetted, 4 subnets
i L1  223.0.0.4 [115/20] via 10.41.41.2, Serial1/0.3
i L1  223.0.0.3 [115/20] via 10.3.3.6, Serial1/0.2
i L1  223.0.0.2 [115/30] via 10.3.3.6, Serial1/0.2
                 [115/30] via 10.41.41.2, Serial1/0.3
C       223.0.0.1 is directly connected, Loopback0
      10.0.0.0/8 is variably subnetted, 4 subnets, 2 masks
i L1  10.42.42.0/24 [115/20] via 10.41.41.2, Serial1/0.3
C       10.41.41.0/24 is directly connected, Serial1/0.3
i L1  10.2.2.4/30 [115/20] via 10.3.3.6, Serial1/0.2
C       10.3.3.4/30 is directly connected, Serial1/0.2
```

Code Listing 11.12 indicates that Amsterdam-PE (223.0.0.4), Paris-PE (223.0.0.3), and Brussels-PE (223.0.0.1 with load balancing on two equal-cost paths) are reachable from London-PE.

Code Listing 11.12:   IPv4 forwarding table for London-PE

```
London-PE#show ip route
<Output Omitted>


      223.0.0.0/32 is subnetted, 4 subnets
i L1  223.0.0.4 [115/20] via 10.42.42.2, Serial1/0.3
i L1  223.0.0.3 [115/20] via 10.2.2.6, Serial1/0.1
C       223.0.0.2 is directly connected, Loopback0
i L1  223.0.0.1 [115/30] via 10.2.2.6, Serial1/0.1
                 [115/30] via 10.42.42.2, Serial1/0.3
```

```
       10.0.0.0/8 is variably subnetted, 4 subnets, 2 masks
i L1   10.41.41.0/24 [115/20] via 10.42.42.2, Serial1/0.3
C      10.42.42.0/24 is directly connected, Serial1/0.3
i L1   10.3.3.4/30 [115/20] via 10.2.2.6, Serial1/0.1
C      10.2.2.4/30 is directly connected, Serial1/0.1
```

Code Listing 11.13: indicates that Amsterdam-PE (223.0.0.4 with load balancing on two equal-cost paths), London-PE (223.0.0.2), and Brussels-PE (223.0.0.1) are reachable from Paris-PE.

Code Listing 11.13:   IPv4 forwarding table for Paris-PE

```
Paris-PE#show ip route
<Output Omitted>


       223.0.0.0/32 is subnetted, 4 subnets
i L1   223.0.0.4 [115/30] via 10.3.3.5, Serial1/0.1
                  [115/30] via 10.2.2.5, Serial1/0.2
C      223.0.0.3 is directly connected, Loopback0
i L1   223.0.0.2 [115/20] via 10.2.2.5, Serial1/0.2
i L1   223.0.0.1 [115/20] via 10.3.3.5, Serial1/0.1
       10.0.0.0/8 is variably subnetted, 4 subnets, 2 masks
i L1   10.42.42.0/24 [115/20] via 10.2.2.5, Serial1/0.2
i L1   10.41.41.0/24 [115/20] via 10.3.3.5, Serial1/0.1
C      10.2.2.4/30 is directly connected, Serial1/0.2
C      10.3.3.4/30 is directly connected, Serial1/0.1
```

11.3.5   Monitoring and verifying partitioned VPN RR operations

11.3.5.1   Propagating network 9.0.0.0/8 from VPN 102 at Brussels-CE2 to London-CE2.
First, let us trace and examine the traffic flow when network 9.0.0.0/8 is advertised from VPN 102 at Brussels-CE2 to London-CE2.

Code Listing 11.14 illustrates that Brussels-PE has received network 9.0.0.0/8 with a route distinguisher (RD) value of 65000:102 (for VPN 102) from Brussels-CE2 (via next-hop 100.100.100.2). The route originates from AS 65012 and is advertised to Brussels-PE via EBGP (PE-CE routing protocol).

Code Listing 11.14:   Brussels-PE BGP VPNv4 table

```
Brussels-PE#show ip bgp vpnv4 all
BGP table version is 17, local router ID is 223.0.0.1
Status codes: s suppressed, d damped, h history, * valid, > best, i - internal
Origin codes: i - IGP, e - EGP, ? - incomplete
```

```
     Network              Next Hop              Metric LocPrf Weight Path
<Output Omitted>
Route Distinguisher: 65000:102 (default for vrf vpn102)
<Output Omitted>
*> 9.0.0.0              100.100.100.2                0             0 65012 i
<Output Omitted>
```

Code Listing 11.15 further illustrates that network 9.0.0.0/8 is set with a standard BGP community value of 65000:102 when it is received by Brussels-PE. Brussels-PE then advertises the route to Amsterdam-PE (223.0.0.4).

Code Listing 11.15:   Brussels-PE BGP VPNv4 table for VPN 102 with subnet 9.0.0.0/8

```
Brussels-PE#show ip bgp vpnv4 all 9.9.9.9
BGP routing table entry for 65000:102:9.0.0.0/8, version 16
Paths: (1 available, best #1, table vpn102)
  Advertised to non peer-group peers:
  223.0.0.4
  65012
    100.100.100.2 from 100.100.100.2 (9.9.9.9)
      Origin IGP, metric 0, localpref 100, valid, external, best
      Community: 65000:102
      Extended Community: RT:65000:102
```

Code Listing 11.16 illustrates that Amsterdam-PE has received network 9.0.0.0/8 from Brussels-PE (via next-hop 223.0.0.1). Since Amsterdam-PE is the partitioned VPN RR for VPN 102, the BGP VPNv4 table comprises only routes that originate from VPN 102 (route distinguisher = 65000:102).

Code Listing 11.16:   Amsterdam-PE BGP VPNv4 table

```
Amsterdam-PE#show ip bgp vpnv4 all
BGP table version is 7, local router ID is 223.0.0.4
Status codes: s suppressed, d damped, h history, * valid, > best, i - internal
Origin codes: i - IGP, e - EGP, ? - incomplete

     Network              Next Hop              Metric LocPrf Weight Path
Route Distinguisher: 65000:102
<Output Omitted>
*>i9.0.0.0              223.0.0.1                0     100       0 65012 i
```

Code Listing 11.17 further illustrates that network 9.0.0.0/8 has been bypassed (or permitted) by the outbound standard BGP community list implemented at Brussels-PE. The BGP VPNv4 table indicates that the route is received from an RR client (Brussels-PE), and the standard BGP community value for the route remains as 65000:102. Amsterdam-PE then reflects the route to the other RR-client, London-PE (223.0.0.2).

Code Listing 11.17: Amsterdam-PE BGP VPNv4 table for VPN 102 with subnet 9.0.0.0/8

```
Amsterdam-PE#show ip bgp vpnv4 all 9.9.9.9
BGP routing table entry for 65000:102:9.0.0.0/8, version 7
Paths: (1 available, best #1, table NULL)
  Advertised to non peer-group peers:
  223.0.0.2
  65012, (Received from a RR-client)
    223.0.0.1 (metric 20) from 223.0.0.1 (223.0.0.1)
      Origin IGP, metric 0, localpref 100, valid, internal, best
      Community: 65000:102
      Extended Community: RT:65000:102
```

Code Listing 11.18 illustrates the neighbor information on Amsterdam-PE's IBGP peer Brussels-PE (223.0.0.1). From the information, we can gather that Brussels-PE is the (VPNv4) RR client of Amsterdam-PE and has the capability to advertise as well as receive VPNv4 routes. Meanwhile, BGP community attributes are also sent to this peer for the VPNv4 address family.

Code Listing 11.18: Amsterdam-PE BGP VPNv4 neighbor Brussels-PE

```
Amsterdam-PE#show ip bgp vpnv4 all neighbor 223.0.0.1
BGP neighbor is 223.0.0.1,  remote AS 65000, internal link
  BGP version 4, remote router ID 223.0.0.1
  BGP state = Established, up for 00:17:13
  <Output Omitted>
  Neighbor capabilities:
    <Output Omitted>
    Address family VPNv4 Unicast: advertised and received
  <Output Omitted>

 For address family: VPNv4 Unicast
  BGP table version 7, neighbor version 7
  <Output Omitted>
  Route-Reflector Client
  Community attribute sent to this neighbor
  <Output Omitted>
```

Code Listing 11.19 illustrates the neighbor information on Amsterdam-PE's IBGP peer London-PE (223.0.0.2). From this data, we can gather that London-PE is the (VPNv4) RR client of Amsterdam-PE and has the capability to advertise as well as receive VPNv4 routes. Meanwhile, BGP community attributes are also sent to this peer for the VPNv4 address family.

Code Listing 11.19: Amsterdam-PE BGP VPNv4 neighbor London-PE

```
Amsterdam-PE#show ip bgp vpnv4 all neighbor 223.0.0.2
BGP neighbor is 223.0.0.2,   remote AS 65000, internal link
  BGP version 4, remote router ID 223.0.0.2
  BGP state = Established, up for 00:33:53
  <Output Omitted>
  Neighbor capabilities:
    <Output Omitted>
    Address family VPNv4 Unicast: advertised and received
  <Output Omitted>

  For address family: VPNv4 Unicast
  BGP table version 7, neighbor version 7
  Index 2, Offset 0, Mask 0x4
  Route-Reflector Client
  Community attribute sent to this neighbor
  <Output Omitted>
```

Code Listing 11.20 illustrates that London-PE has received network 9.0.0.0/8 from Amsterdam-PE with Brussels-PE (223.0.0.1) as the next-hop. When an RR reflects a route, it does not change the BGP next-hop attribute. Therefore, when Amsterdam-PE reflects 9.0.0.0/8 to London-PE, it preserves the next-hop attribute (223.0.0.1) that belongs to Brussels-PE.

Code Listing 11.20: London-PE BGP VPNv4 table

```
London-PE#show ip bgp vpnv4 all
BGP table version is 53, local router ID is 223.0.0.2
Status codes: s suppressed, d damped, h history, * valid, > best, i - internal
Origin codes: i - IGP, e - EGP, ? - incomplete

   Network          Next Hop            Metric LocPrf Weight Path
<Output Omitted>
Route Distinguisher: 65000:102 (default for vrf vpn102)
<Output Omitted>
*>i9.0.0.0          223.0.0.1                    0    100      0 65012 i
<Output Omitted>
```

Code Listing 11.21 further illustrates that network 9.0.0.0/8 has a standard BGP community value of 65000:102 and is reflected to London-PE from Amsterdam-PE (223.0.0.4). London-PE then advertises the route to London-CE2 (202.202.202.2).

In this instance, the cluster list is identified by the Router-ID (223.0.0.4) of the RR (Amsterdam-PE) since there is only a single RR serving VPN 102. The Router ID (223.0.0.1) of the originating IBGP router (Brussels-PE) is stored in the Originator ID. A router receiving an IBGP route with Originator ID set to its own Router ID ignores that route.

Together, the cluster list and Originator ID BGP attributes are used to avoid routing information loops during route reflection when more than one route reflector is deployed to avoid single point of failure. Even though RR resiliency is not illustrated in this scenario, readers are strongly encouraged to incorporate redundant RRs in their network design.

Code Listing 11.21: London-PE BGP VPNv4 table for VPN 102 with subnet 9.0.0.0/8

```
London-PE#show ip bgp vpnv4 all 9.9.9.9
BGP routing table entry for 65000:102:9.0.0.0/8, version 53
Paths: (1 available, best #1, table vpn102)
  Advertised to non peer-group peers:
  202.202.202.2
  65012
    223.0.0.1 (metric 30) from 223.0.0.4 (223.0.0.1)
      Origin IGP, metric 0, localpref 100, valid, internal, best
      Community: 65000:102
      Extended Community: RT:65000:102
      Originator: 223.0.0.1, Cluster list: 223.0.0.4
```

Code Listing 11.22 illustrates that network 9.0.0.0/8 from VPN 102 at Brussels-CE2 (AS 65012) has successfully propagated across the service provider's network (AS 65000) via Brussels-PE, Amsterdam-PE (RR), and London-PE (202.202.202.1) to London-CE2.

Code Listing 11.22: London-CE2 BGP table

```
London-CE2#show ip bgp
BGP table version is 40, local router ID is 5.5.5.5
Status codes: s suppressed, d damped, h history, * valid, > best, i - internal
Origin codes: i - IGP, e - EGP, ? - incomplete

   Network          Next Hop            Metric LocPrf Weight Path
<Output Omitted>
*> 9.0.0.0          202.202.202.1                         0 65000 65012 i
<Output Omitted>
```

Code Listing 11.23 further illustrates that network 9.0.0.0/8 is a valid route and is installed into the IP routing table of London-CE2 as an EBGP route.

Code Listing 11.23:   IP forwarding table for London-CE2

```
London-CE2#show ip route
<Output Omitted>
C     202.202.202.0/24 is directly connected, Ethernet0/0
      5.0.0.0/16 is subnetted, 1 subnets
C         5.5.0.0 is directly connected, Loopback0
B     9.0.0.0/8 [20/0] via 202.202.202.1, 00:31:02
```

11.3.5.2   Propagating subnet 5.5.0.0/16 from VPN 102 at London-CE2 to Brussels-CE2.
Next, let us trace and examine the traffic flow in the reverse direction when subnet 5.5.0.0/16 is advertised from VPN 102 at London-CE2 to Brussels-CE2.

Code Listing 11.24 illustrates that London-PE has received subnet 5.5.0.0/16 with a route distinguisher (RD) value of 65000:102 (for VPN 102) from London-CE2 (via next-hop 202.202.202.2). The route originates from AS 65022 and is advertised to London-PE via EBGP (PE-CE routing protocol).

Code Listing 11.24:   London-PE BGP VPNv4 table

```
London-PE#show ip bgp vpnv4 all
BGP table version is 53, local router ID is 223.0.0.2
Status codes: s suppressed, d damped, h history, * valid, > best, i - internal
Origin codes: i - IGP, e - EGP, ? - incomplete

   Network          Next Hop            Metric LocPrf Weight Path
<Output Omitted>
Route Distinguisher: 65000:102 (default for vrf vpn102)
*> 5.5.0.0/16       202.202.202.2            0             0 65022 i
<Output Omitted>
```

Code Listing 11.25 further illustrates that subnet 5.5.0.0/16 is set with a standard BGP community value of 65000:102 when it is received by London-PE. London-PE then advertises the route to Amsterdam-PE (223.0.0.4).

Code Listing 11.25:   London-PE BGP VPNv4 table for VPN 102 with subnet 5.5.0.0/16

```
London-PE#show ip bgp vpnv4 all 5.5.5.5
BGP routing table entry for 65000:102:5.5.0.0/16, version 17
Paths: (1 available, best #1, table vpn102)
```

```
Advertised to non peer-group peers:
223.0.0.4
65022
  202.202.202.2 from 202.202.202.2 (5.5.5.5)
    Origin IGP, metric 0, localpref 100, valid, external, best
    Community: 65000:102
    Extended Community: RT:65000:102
```

Code Listing 11.26 illustrates that Amsterdam-PE has received subnet 5.5.0.0/16 from London-PE (via next-hop 223.0.0.2). Since Amsterdam-PE is the partitioned VPN RR for VPN 102, the BGP VPNv4 table comprises only routes that originate from VPN 102 (route distinguisher = 65000:102).

Code Listing 11.26:   Amsterdam-PE BGP VPNv4 table

```
Amsterdam-PE#show ip bgp vpnv4 all
BGP table version is 7, local router ID is 223.0.0.4
Status codes: s suppressed, d damped, h history, * valid, > best, i - internal
Origin codes: i - IGP, e - EGP, ? - incomplete

   Network            Next Hop            Metric LocPrf Weight Path
Route Distinguisher: 65000:102
*>i5.5.0.0/16         223.0.0.2                0    100      0 65022 i
<Output Omitted>
```

Code Listing 11.27 further illustrates that subnet 5.5.0.0/16 has been bypassed (or permitted) by the outbound standard BGP community list implemented at London-PE. The BGP VPNv4 table indicates that the route is received from an RR client (London-PE) and the standard BGP community value for the route remains as 65000:102. Amsterdam-PE then reflects the route to the other RR client, Brussels-PE (223.0.0.1).

Code Listing 11.27:   Amsterdam-PE BGP VPNv4 table for VPN 102 with subnet 5.5.0.0/16

```
Amsterdam-PE#show ip bgp vpnv4 all 5.5.5.5
BGP routing table entry for 65000:102:5.5.0.0/16, version 2
Paths: (1 available, best #1, table NULL)
  Advertised to non peer-group peers:
  223.0.0.1
  65022, (Received from a RR-client)
    223.0.0.2 (metric 20) from 223.0.0.2 (223.0.0.2)
      Origin IGP, metric 0, localpref 100, valid, internal, best
      Community: 65000:102
      Extended Community: RT:65000:102
```

Code Listing 11.28 illustrates that Brussels-PE has received subnet 5.5.0.0/16 from Amsterdam-PE with London-PE (223.0.0.2) as the next-hop. When an RR reflects a route, it does not change the BGP next-hop attribute. Therefore, when Amsterdam-PE reflects 5.5.0.0/16 to Brussels-PE, it preserves the next-hop attribute (223.0.0.2) that belongs to London-PE.

Code Listing 11.28:   Brussels-PE BGP VPNv4 table

```
Brussels-PE#show ip bgp vpnv4 all
BGP table version is 17, local router ID is 223.0.0.1
Status codes: s suppressed, d damped, h history, * valid, > best, i - internal
Origin codes: i - IGP, e - EGP, ? - incomplete

   Network          Next Hop            Metric LocPrf Weight Path
<Output Omitted>
Route Distinguisher: 65000:102 (default for vrf vpn102)
*>i5.5.0.0/16        223.0.0.2              0    100        0 65022 i
<Output Omitted>
```

Code Listing 11.29 further illustrates that subnet 5.5.0.0/16 has a standard BGP community value of 65000:102 and is reflected to Brussels-PE from Amsterdam-PE (223.0.0.4). Brussels-PE then advertises the route to Brussels-CE2 (100.100.100.2). In this instance, the cluster list is identified by the Router-ID (223.0.0.4) of the RR (Amsterdam-PE) since there is only a single RR serving VPN 102. The Router-ID (223.0.0.2) of the originating IBGP router (London-PE) is stored in the Originator-ID.

Code Listing 11.29:   Brussels-PE BGP VPNv4 table for VPN 102 with subnet 5.5.0.0/16

```
Brussels-PE#show ip bgp vpnv4 all 5.5.5.5
BGP routing table entry for 65000:102:5.5.0.0/16, version 9
Paths: (1 available, best #1, table vpn102)
  Advertised to non peer-group peers:
  100.100.100.2
  65022
    223.0.0.2 (metric 30) from 223.0.0.4 (223.0.0.4)
      Origin IGP, metric 0, localpref 100, valid, internal, best
      Community: 65000:102
      Extended Community: RT:65000:102
      Originator: 223.0.0.2, Cluster list: 223.0.0.4
```

Code Listing 11.30 illustrates that subnet 5.5.0.0/16 from VPN 102 at London-CE2 (AS 65022) has successfully propagated across the service provider's network (AS 65000) via London-PE, Amsterdam-PE (RR), and Brussels-PE (100.100.100.1) to Brussels-CE2.

Code Listing 11.30:   Brussels-CE2 BGP table

```
Brussels-CE2#show ip bgp
BGP table version is 46, local router ID is 9.9.9.9
Status codes: s suppressed, d damped, h history, * valid, > best, i - internal
Origin codes: i - IGP, e - EGP, ? - incomplete

   Network          Next Hop         Metric LocPrf Weight Path
*> 5.5.0.0/16       100.100.100.1                      0 65000 65022 i
<Output Omitted>
```

Code Listing 11.31 further illustrates that subnet 5.5.0.0/16 is a valid route and is installed into the IP routing table of Brussels-CE2 as an EBGP route.

Code Listing 11.31:   IP forwarding table for Brussels-CE2

```
Brussels-CE2#show ip route
<Output Omitted>

     100.0.0.0/24 is subnetted, 1 subnets
C       100.100.100.0 is directly connected, Ethernet0/0
     5.0.0.0/16 is subnetted, 1 subnets
B       5.5.0.0 [20/0] via 100.100.100.1, 00:15:27
C    9.0.0.0/8 is directly connected, Loopback0
```

11.3.5.3 Propagating subnet 8.8.8.0/24 from VPN 101 at Brussels-CE1 to London-CE1.

Now, let us trace and examine the traffic flow when subnet 8.8.8.0/24 is advertised from VPN 101 at Brussels-CE1 to London-CE1.

Code Listing 11.32 illustrates that Brussels-PE has received subnet 8.8.8.0/24 with a route distinguisher (RD) value of 65000:101 (for VPN 101) from Brussels-CE1 (via next-hop 130.130.130.2). The route originates from AS 65011 and is advertised to Brussels-PE via EBGP (PE-CE routing protocol).

Code Listing 11.32:   Brussels-PE BGP VPNv4 table

```
Brussels-PE#show ip bgp vpnv4 all
BGP table version is 17, local router ID is 223.0.0.1
Status codes: s suppressed, d damped, h history, * valid, > best, i - internal
Origin codes: i - IGP, e - EGP, ? - incomplete
```

```
       Network              Next Hop              Metric LocPrf Weight Path
Route Distinguisher: 65000:101 (default for vrf vpn101)
<Output Omitted>
*> 8.8.8.0/24           130.130.130.2              0            0 65011 i
<Output Omitted>
```

Code Listing 11.33 further illustrates that subnet 8.8.8.0/24 is set with a standard BGP community value of 65000:101 when it is received by Brussels-PE. Brussels-PE then advertises the route to Paris-PE (223.0.0.3).

Code Listing 11.33: Brussels-PE BGP VPNv4 table for VPN 101 with subnet 8.8.8.0/24

```
Brussels-PE#show ip bgp vpnv4 all 8.8.8.8
BGP routing table entry for 65000:101:8.8.8.0/24, version 14
Paths: (1 available, best #1, table vpn101)
  Advertised to non peer-group peers:
  223.0.0.3
  65011
    130.130.130.2 from 130.130.130.2 (8.8.8.8)
      Origin IGP, metric 0, localpref 100, valid, external, best
      Community: 65000:101
      Extended Community: RT:65000:101
```

Code Listing 11.34 illustrates that Paris-PE has received subnet 8.8.8.0/24 from Brussels-PE (via next-hop 223.0.0.1). Since Paris-PE is the partitioned VPN RR for VPN 101, the BGP VPNv4 table comprises only routes that originate from VPN 101 (route distinguisher = 65000:101). Notice that subnets 31.3.3.0/24 (originating from Paris-PE itself in AS 65000) and 31.5.5.4/30 (from Paris-CE1 in AS 65031) are also listed.

Code Listing 11.34: Paris-PE BGP VPNv4 table

```
Paris-PE#show ip bgp vpnv4 all
BGP table version is 21, local router ID is 223.0.0.3
Status codes: s suppressed, d damped, h history, * valid, > best, i - internal
Origin codes: i - IGP, e - EGP, ? - incomplete

       Network              Next Hop              Metric LocPrf Weight Path
Route Distinguisher: 65000:101 (default for vrf vpn101)
<Output Omitted>
*>i8.8.8.0/24           223.0.0.1              0     100      0 65011 i
*> 31.3.3.0/24          0.0.0.0               0          32768 ?
*> 31.5.5.4/30          31.3.3.2             0              0 65031 i
<Output Omitted>
```

Code Listing 11.35 further illustrates subnet 8.8.8.0/24 has been bypassed (or permitted) by the outbound standard BGP community list implemented at Brussels-PE. The BGP VPNv4 table indicates that the route is received from an RR client (Brussels-PE) and the standard BGP community value for the route remains as 65000:101. Paris-PE then reflects the route to the other RR client, London-PE (223.0.0.2), and to nonclient, Paris-CE1 (31.3.3.2).

Code Listing 11.35:   Paris-PE BGP VPNv4 table for VPN 101 with subnet 8.8.8.0/24

```
Paris-PE#show ip bgp vpnv4 all 8.8.8.8
BGP routing table entry for 65000:101:8.8.8.0/24, version 21
Paths: (1 available, best #1, table vpn101)
  Advertised to non peer-group peers:
  31.3.3.2 223.0.0.2
  65011, (Received from a RR-client)
    223.0.0.1 (metric 20) from 223.0.0.1 (223.0.0.1)
      Origin IGP, metric 0, localpref 100, valid, internal, best
      Community: 65000:101
      Extended Community: RT:65000:101
```

Code Listing 11.36 illustrates the neighbor information on Paris-PE's IBGP peer Brussels-PE (223.0.0.1). From this data, we can gather that Brussels-PE is the (IPv4 and VPNv4) RR client of Paris-PE and has the capability to advertise as well as receive both IPv4 and VPNv4 routes. Meanwhile, BGP community attributes are sent only to this peer for the VPNv4 address family.

Code Listing 11.36:   Paris-PE BGP VPNv4 neighbor Brussels-PE

```
Paris-PE#show ip bgp vpnv4 all neighbor 223.0.0.1
BGP neighbor is 223.0.0.1,  remote AS 65000, internal link
  BGP version 4, remote router ID 223.0.0.1
  BGP state = Established, up for 00:09:48
  <Output Omitted>
  Neighbor capabilities:
    <Output Omitted>
    Address family IPv4 Unicast: advertised and received
    Address family VPNv4 Unicast: advertised and received
    <Output Omitted>

  For address family: IPv4 Unicast
  BGP table version 22, neighbor version 22
  Index 1, Offset 0, Mask 0x2
  Route-Reflector Client
  <Output Omitted>
```

```
For address family: VPNv4 Unicast
 BGP table version 21, neighbor version 21
 Index 2, Offset 0, Mask 0x4
 Route-Reflector Client
 Community attribute sent to this neighbor
 <Output Omitted>
```

Code Listing 11.37 illustrates the neighbor information on Paris-PE's IBGP peer London-PE (223.0.0.2). From the information, we can gather that London-PE is the (IPv4 and VPNv4) RR client of Paris-PE and has the capability to advertise as well as receive both IPv4 and VPNv4 routes. Meanwhile, BGP community attributes are sent only to this peer for the VPNv4 address family. In other words, besides being a partitioned VPN RR for VPN 101, Paris-PE also serves as a dedicated Internet RR in this case study.

Code Listing 11.37:   Paris-PE BGP VPNv4 neighbor London-PE

```
Paris-PE#show ip bgp vpnv4 all neighbor 223.0.0.2
BGP neighbor is 223.0.0.2,   remote AS 65000, internal link
  BGP version 4, remote router ID 223.0.0.2
  BGP state = Established, up for 01:29:11
  <Output Omitted>
  Neighbor capabilities:
    <Output Omitted>
    Address family IPv4 Unicast: advertised and received
    Address family VPNv4 Unicast: advertised and received
  <Output Omitted>

 For address family: IPv4 Unicast
  BGP table version 22, neighbor version 22
  Index 2, Offset 0, Mask 0x4
  Route-Reflector Client
  <Output Omitted>

 For address family: VPNv4 Unicast
  BGP table version 21, neighbor version 21
  Index 3, Offset 0, Mask 0x8
  Route-Reflector Client
  Community attribute sent to this neighbor
  <Output Omitted>
```

Code Listing 11.38 illustrates that London-PE has received subnet 8.8.8.0/24 from Paris-PE with Brussels-PE (223.0.0.1) as the next-hop. When an RR reflects a route, it does not change the BGP next-hop attribute. Therefore, when Paris-PE reflects 8.8.8.0/24 to London-PE, it preserves the next-hop attribute (223.0.0.1) that belongs to Brussels-PE. Notice that subnets 31.3.3.0/24 (from Paris-PE, which is in the same AS 65000 as London-PE) and 31.5.5.4/30 (from Paris-CE1 in AS 65031) are also received by London-PE.

Code Listing 11.38: London-PE BGP VPNv4 table

```
London-PE#show ip bgp vpnv4 all
BGP table version is 53, local router ID is 223.0.0.2
Status codes: s suppressed, d damped, h history, * valid, > best, i - internal
Origin codes: i - IGP, e - EGP, ? - incomplete

    Network          Next Hop          Metric LocPrf Weight Path
Route Distinguisher: 65000:101 (default for vrf vpn101)
<Output Omitted>
*>i8.8.8.0/24        223.0.0.1              0    100      0 65011 i
*>i31.3.3.0/24       223.0.0.3              0    100      0 ?
*>i31.5.5.4/30       223.0.0.3              0    100      0 65031 i
<Output Omitted>
```

Code Listing 11.39 further illustrates that subnet 8.8.8.0/24 has a standard BGP community value of 65000:101 and is reflected to London-PE from Paris-PE (223.0.0.3). London-PE then advertises the route to London-CE1 (200.200.200.2). In this instance, the cluster list is identified by the Router-ID (223.0.0.3) of the RR (Paris-PE) since there is only a single RR serving VPN 101. The Router-ID (223.0.0.1) of the originating IBGP router (Brussels-PE) is stored in the Originator-ID.

Code Listing 11.39: London-PE BGP VPNv4 table for VPN 101 with subnet 8.8.8.0/24

```
London-PE#show ip bgp vpnv4 all 8.8.8.8
BGP routing table entry for 65000:101:8.8.8.0/24, version 52
Paths: (1 available, best #1, table vpn101)
  Advertised to non peer-group peers:
  200.200.200.2
  65011
    223.0.0.1 (metric 30) from 223.0.0.3 (223.0.0.1)
      Origin IGP, metric 0, localpref 100, valid, internal, best
      Community: 65000:101
      Extended Community: RT:65000:101
      Originator: 223.0.0.1, Cluster list: 223.0.0.3
```

Code Listing 11.40 illustrates that subnet 8.8.8.0/24 from VPN 101 at Brussels-CE1 (AS 65011) has successfully propagated across the service provider's network (AS 65000) via Brussels-PE, Paris-PE (RR), and London-PE (200.200.200.1) to London-CE1. Notice that subnets 31.3.3.0/24 (from Paris-PE in AS 65000) and 31.5.5.4/30 (from Paris-CE1 in AS 65031) are also propagated to London-CE1 via London-PE (200.200.200.1).

Code Listing 11.40:   London-CE1 BGP table

```
London-CE1#show ip bgp
BGP table version is 70, local router ID is 6.6.6.6
Status codes: s suppressed, d damped, h history, * valid, > best, i - internal
Origin codes: i - IGP, e - EGP, ? - incomplete

   Network              Next Hop          Metric LocPrf Weight Path
<Output Omitted>
*> 8.8.8.0/24           200.200.200.1                       0 65000 65011 i
*> 31.3.3.0/24          200.200.200.1                       0 65000 ?
*> 31.5.5.4/30          200.200.200.1                       0 65000 65031 i
<Output Omitted>
```

Code Listing 11.41 further illustrates that subnet 8.8.8.0/24 is a valid route and is installed into the IP routing table of London-CE1 as an EBGP route. Notice that subnets 31.5.5.4/30 (from Paris-CE1) and 31.3.3.0/24 (from Paris-PE) are also installed in the IP routing table as EBGP routes.

Code Listing 11.41:   IP forwarding table for London-CE1

```
London-CE1#show ip route
<Output Omitted>

C    200.200.200.0/24 is directly connected, Serial1/0
     6.0.0.0/32 is subnetted, 1 subnets
C       6.6.6.6 is directly connected, Loopback0
     8.0.0.0/24 is subnetted, 1 subnets
B       8.8.8.0 [20/0] via 200.200.200.1, 00:29:26
     31.0.0.0/8 is variably subnetted, 2 subnets, 2 masks
B       31.5.5.4/30 [20/0] via 200.200.200.1, 01:48:38
B       31.3.3.0/24 [20/0] via 200.200.200.1, 01:48:38
```

11.3.5.4 Propagating host route 6.6.6.6/32 from VPN 101 at London-CE1 to Brussels-CE1. Next, let us trace and examine the traffic flow in the reverse direction when host route 6.6.6.6/32 is advertised from VPN 101 at London-CE1 to Brussels-CE1.

Code Listing 11.42 illustrates that London-PE has received host route 6.6.6.6/32 with a route distinguisher (RD) value of 65000:101 (for VPN 101) from London-CE1 (via next-hop 200.200.200.2). The route originates from AS 65021 and is advertised to London-PE via EBGP (PE-CE routing protocol).

Code Listing 11.42:   London-PE BGP VPNv4 table

```
London-PE#show ip bgp vpnv4 all
BGP table version is 53, local router ID is 223.0.0.2
Status codes: s suppressed, d damped, h history, * valid, > best, i - internal
Origin codes: i - IGP, e - EGP, ? - incomplete

   Network          Next Hop            Metric LocPrf Weight Path
Route Distinguisher: 65000:101 (default for vrf vpn101)
*> 6.6.6.6/32       200.200.200.2            0             0 65021 i
<Output Omitted>
```

Code Listing 11.43 further illustrates that host route 6.6.6.6/32 is set with a standard BGP community value of 65000:101 when it is received by London-PE. London-PE then advertises the route to Paris-PE (223.0.0.3).

Code Listing 11.43:   London-PE BGP VPNv4 table for VPN 101 with host route 6.6.6.6/32

```
London-PE#show ip bgp vpnv4 all 6.6.6.6
BGP routing table entry for 65000:101:6.6.6.6/32, version 16
Paths: (1 available, best #1, table vpn101)
  Advertised to non peer-group peers:
  223.0.0.3
  65021
    200.200.200.2 from 200.200.200.2 (6.6.6.6)
      Origin IGP, metric 0, localpref 100, valid, external, best
      Community: 65000:101
      Extended Community: RT:65000:101
```

Code Listing 11.44 illustrates that Paris-PE has received host route 6.6.6.6/32 from London-PE (via next-hop 223.0.0.2). Since Paris-PE is the partitioned VPN RR for VPN 101, the BGP VPNv4 table comprises only routes that originate from VPN 101 (route distinguisher = 65000:101). Notice that subnets 31.3.3.0/24 (originating from Paris-PE itself in AS 65000) and 31.5.5.4/30 (from Paris-CE1 in AS 65031) are also listed.

Code Listing 11.44: Paris-PE BGP VPNv4 table

```
Paris-PE#show ip bgp vpnv4 all
BGP table version is 21, local router ID is 223.0.0.3
Status codes: s suppressed, d damped, h history, * valid, > best, i - internal
Origin codes: i - IGP, e - EGP, ? - incomplete

   Network           Next Hop           Metric LocPrf Weight Path
Route Distinguisher: 65000:101 (default for vrf vpn101)
*>i6.6.6.6/32        223.0.0.2               0    100        0 65021 i
<Output Omitted>
*> 31.3.3.0/24       0.0.0.0                 0           32768 ?
*> 31.5.5.4/30       31.3.3.2                0               0 65031 i
<Output Omitted>
```

Code Listing 11.45 further illustrates that host route 6.6.6.6/32 has been bypassed (or permitted) by the outbound standard BGP community list implemented at London-PE. The BGP VPNv4 table indicates that the route is received from an RR client (London-PE) and the standard BGP community value for the route remains as 65000:101. Paris-PE then reflects the route to the other RR client, Brussels-PE (223.0.0.1), and to nonclient, Paris-CE1 (31.3.3.2).

Code Listing 11.45: Paris-PE BGP VPNv4 table for VPN 101 with host route 6.6.6.6/32

```
Paris-PE#show ip bgp vpnv4 all 6.6.6.6
BGP routing table entry for 65000:101:6.6.6.6/32, version 7
Paths: (1 available, best #1, table vpn101)
  Advertised to non peer-group peers:
  31.3.3.2 223.0.0.1
  65021, (Received from a RR-client)
    223.0.0.2 (metric 20) from 223.0.0.2 (223.0.0.2)
      Origin IGP, metric 0, localpref 100, valid, internal, best
      Community: 65000:101
      Extended Community: RT:65000:101
```

Code Listing 11.46 illustrates that Brussels-PE has received host route 6.6.6.6/32 from Paris-PE with London-PE (223.0.0.2) as the next-hop. When an RR reflects a route, it does not change the BGP next-hop attribute. Therefore, when Paris-PE reflects 6.6.6.6/32 to Brussels-PE, it preserves the next-hop attribute (223.0.0.2), which belongs to London-PE. Notice that subnets 31.3.3.0/24 (from Paris-PE, which is in the same AS 65000 as Brussels-

PE) and 31.5.5.4/30 (from Paris-CE1 in AS 65031) are also received by Brussels-PE.

Code Listing 11.46:   Brussels-PE BGP VPNv4 table

```
Brussels-PE#show ip bgp vpnv4 all
BGP table version is 17, local router ID is 223.0.0.1
Status codes: s suppressed, d damped, h history, * valid, > best, i - internal
Origin codes: i - IGP, e - EGP, ? - incomplete

   Network              Next Hop            Metric LocPrf Weight Path
Route Distinguisher: 65000:101 (default for vrf vpn101)
*>i6.6.6.6/32          223.0.0.2                0    100        0 65021 i
<Output Omitted>
*>i31.3.3.0/24         223.0.0.3                0    100        0 ?
*>i31.5.5.4/30         223.0.0.3                0    100        0 65031 i
<Output Omitted>
```

Code Listing 11.47 further illustrates that host route 6.6.6.6/32 has a standard BGP community value of 65000:101 and is reflected to Brussels-PE from Paris-PE (223.0.0.3). Brussels-PE then advertises the route to Brussels-CE1 (130.130.130.2). In this instance, the cluster list is identified by the Router-ID (223.0.0.3) of the RR (Paris-PE) since there is only a single RR serving VPN 101. The Router-ID (223.0.0.2) of the originating IBGP router (London-PE) is stored in the Originator-ID.

Code Listing 11.47:   Brussels-PE BGP VPNv4 table for VPN 101 with host route 6.6.6.6/32

```
Brussels-PE#show ip bgp vpnv4 all 6.6.6.6
BGP routing table entry for 65000:101:6.6.6.6/32, version 6
Paths: (1 available, best #1, table vpn101)
  Advertised to non peer-group peers:
  130.130.130.2
  65021
    223.0.0.2 (metric 30) from 223.0.0.3 (223.0.0.3)
      Origin IGP, metric 0, localpref 100, valid, internal, best
      Community: 65000:101
      Extended Community: RT:65000:101
      Originator: 223.0.0.2, Cluster list: 223.0.0.3
```

Code Listing 11.48 illustrates that host route 6.6.6.6/32 from VPN 101 at London-CE1 (AS 65021) has successfully propagated across the service provider's network (AS 65000) via London-PE, Paris-PE (RR), and Brussels-PE (130.130.130.1) to Brussels-CE1. Notice that subnets 31.3.3.0/24 (from Paris-PE in AS 65000) and 31.5.5.4/30 (from Paris-CE1 in AS 65031) are also propagated to Brussels-CE1 via Brussels-PE (130.130.130.1).

Code Listing 11.48:   Brussels-CE1 BGP table

```
Brussels-CE1#show ip bgp
BGP table version is 92, local router ID is 8.8.8.8
Status codes: s suppressed, d damped, h history, * valid, > best, i - internal
Origin codes: i - IGP, e - EGP, ? - incomplete

   Network              Next Hop          Metric LocPrf Weight Path
*> 6.6.6.6/32           130.130.130.1                      0 65000 65021 i
<Output Omitted>
*> 31.3.3.0/24          130.130.130.1                      0 65000 ?
*> 31.5.5.4/30          130.130.130.1                      0 65000 65031 i
<Output Omitted>
```

Code Listing 11.49 further illustrates that host route 6.6.6.6/32 is a valid route and is installed into the IP routing table of Brussels-CE1 as an EBGP route. Notice that subnets 31.5.5.4/30 (from Paris-CE1) and 31.3.3.0/24 (from Paris-PE) are also installed in the IP routing table as EBGP routes.

Code Listing 11.49:   IP forwarding table for Brussels-CE1

```
Brussels-CE1#show ip route
<Output Omitted>


     6.0.0.0/32 is subnetted, 1 subnets
B        6.6.6.6 [20/0] via 130.130.130.1, 00:13:11
     8.0.0.0/24 is subnetted, 1 subnets
C        8.8.8.0 is directly connected, Loopback0
     130.130.0.0/24 is subnetted, 1 subnets
C        130.130.130.0 is directly connected, Serial1/0
     31.0.0.0/8 is variably subnetted, 2 subnets, 2 masks
B        31.5.5.4/30 [20/0] via 130.130.130.1, 00:13:11
B        31.3.3.0/24 [20/0] via 130.130.130.1, 00:13:11
```

11.3.5.5 Propagating Internet host route 223.0.0.1/32 from Brussels-PE to London-PE. For illustrative purposes, Paris-PE is deployed as a hybrid RR (see Code Listings 11.36 and 11.37) in this case study. It can reflect both VPN 101 and

Internet (IPv4) routes. At this point, let us trace and examine the traffic flow when Internet host route 223.0.0.1/32 is advertised from Brussels-PE to London-PE.

Code Listing 11.50 illustrates that Brussels-PE is the originator (with next-hop 0.0.0.0) of Internet host route 223.0.0.1/32.

Code Listing 11.50: Brussels-PE BGP IPv4 table

```
Brussels-PE#show ip bgp
BGP table version is 2, local router ID is 223.0.0.1
Status codes: s suppressed, d damped, h history, * valid, > best, i - internal
Origin codes: i - IGP, e - EGP, ? - incomplete

   Network          Next Hop            Metric LocPrf Weight Path
*> 223.0.0.1/32     0.0.0.0                  0           32768 i
```

Code Listing 11.51 illustrates that Paris-PE has received Internet host route 223.0.0.1/32 from Brussels-PE (through next-hop 223.0.0.1) via IBGP.

Code Listing 11.51: Paris-PE BGP IPv4 table

```
Paris-PE#show ip bgp
BGP table version is 22, local router ID is 223.0.0.3
Status codes: s suppressed, d damped, h history, * valid, > best, i - internal
Origin codes: i - IGP, e - EGP, ? - incomplete

   Network          Next Hop            Metric LocPrf Weight Path
*>i223.0.0.1/32     223.0.0.1                0    100      0 i
```

Code Listing 11.52 illustrates that London-PE has successfully received Internet host route 223.0.0.1/32 from Paris-PE with Brussels-PE (223.0.0.1) as the next-hop. When an RR reflects a route, it does not change the BGP next-hop attribute. Therefore, when Paris-PE reflects 223.0.0.1/32 to London-PE, it preserves the next-hop attribute (223.0.0.1), which belongs to Brussels-PE.

Code Listing 11.52: London-PE BGP IPv4 table

```
London-PE#show ip bgp
BGP table version is 18, local router ID is 223.0.0.2
Status codes: s suppressed, d damped, h history, * valid, > best, i - internal
Origin codes: i - IGP, e - EGP, ? - incomplete
```

```
           Network              Next Hop                  Metric  LocPrf  Weight  Path
           *>i223.0.0.1/32      223.0.0.1                      0     100        0  i
```

11.4 Route Reflector Performance Tuning and Optimization

When designing and implementing a sizable network, performance optimization is always an important factor to consider. The RR plays a crucial role in scaling the route distribution in the MPLS VPN backbone. It has become a critical component that must be fine-tuned for optimized performance. Performance optimization begins with good network design:

- For better stability and faster convergence, deploy separate RRs to reflect IPv4 and VPNv4 routes. This also simplifies the fulfillment of service level agreements (SLAs).

- Implement the route reflector server model in which a specific routing device is dedicated for the RR function and is not placed in a forwarding path. This helps to conserve CPU and memory for faster convergence.

- Utilize a high-end routing device that has supreme CPU power and huge DRAM memory pool (512 MB).

Other performance-enhancement features, and parameters, such as BGP peer groups, input queue size, TCP window size, and TCP maximum segment size (MSS), are discussed in the subsequent sections.

11.4.1 BGP peer groups

Besides simplifying BGP configuration, BGP peer groups can also be used to improve scalability. In an RR core design, we can reduce the resource (CPU and I/O memory) consumption of RRs by implementing BGP peer groups for RR clients and redundant RRs.

Without peer groups, BGP must walk the BGP table for every peer (with the BGP scanner process), filter prefixes through outbound policies, and generate updates that are sent to every single peer. The BGP scanner process is used to check reachable next-hop and to process any "network" commands with the BGP process. It is invoked every 60 seconds (scan time) by default for IPv4 and VPNv4 prefixes. Beware that the scanner process can be CPU intensive, involving large BGP table and small scan time.

In a BGP peer group, all peer group members must have a common outbound policy. This makes it possible to sent the same update packets to each group member, reducing the number of CPU cycles and the amount of I/O memory that BGP requires when it advertises routes to every peer individually.

11.4.2 Input queue size

When BGP peers receive thousands of update packets and send TCP acknowledgments back to the advertising BGP speaker accordingly, the BGP speaker will receive a flood of TCP ACKs in a short period. If the ACKs arrive at a rate that is too high for the route processor, the packets are stored temporarily in inbound interface queues. By default, Cisco router interfaces use an input queue size of 75 packets. In addition, special control packets such as BGP updates use a special queue with selective packet discard (SPD). This special queue holds 100 packets. TCP ACKs can devour these 175 slots of input buffer swiftly during BGP convergence, and packets that arrive subsequently will be dropped. Therefore, increasing the interface input queue size helps to reduce the number of dropped packets (TCP ACKs), which in turn lessens the processing burden and decreases BGP convergence times.

Based on the formula: {*Hold-Queue Size = Window Size / (MSS) · Number of BGP Peers*}, we can determine the optimum input queue size that is required on the RR to prevent input drops. The TCP window size can be modified using the "ip tcp window-size ⟨bytes⟩" command and the input hold-queue size can be set using the "hold-queue ⟨length⟩ in" command. A typical input queue size of 1000 can resolve the problems created by input queue drops.

11.4.3 Path MTU

By default, the maximum segment size (MSS) is 536 bytes. It is the maximum number of bytes that can be transported by TCP in a single packet. TCP fragments packets in a transmit queue into 536 byte chunks before passing them down to the IP layer. The command "show ip bgp neighbors | include max data" can be used to display the MSS of BGP peers. With a 536-byte MSS, packets are not likely to be fragmented at an IP device along the path to the destination because most links use an MTU of at least 1500 bytes. However, smaller packets also increase the amount of bandwidth used to transport overheads.

Since BGP establishes a TCP session to all peers, a 536-byte MSS will affect BGP convergence times significantly. We can alleviate this problem by enabling the path MTU (PMTU) feature, using the "ip tcp path-mtu-discovery" command. PMTU dynamically determines the largest MSS value without generating packets that need to be fragmented. It allows TCP to determine the smallest MTU size among all links in a TCP session and then uses this MTU value, less the number of bytes taken up by the IP and TCP headers, as the MSS for the session. PMTU is described in RFC 1191.

For traversing Ethernet segments, the MSS is 1460 bytes, and for traversing Packet over SONET (POS) segments the MSS is 4430 bytes. The augmentation in MSS from 536 (default) to 1460 or 4430 bytes reduces TCP/IP overhead and increases throughput, which helps BGP to converge more quickly.

11.5 Summary

Chapter 11 discussed how route reflectors can be used to scale MP-BGP route distribution in the MPLS VPN backbone. Since no single BGP router can hold all Internet and VPN routing information, partitioned route reflectors can be deployed to further enhance MPLS VPN scalability. A partitioned RR design requires an indicative mechanism to identify routes that originate from different VPNs and also a filtering mechanism to partition these routes according to the assigned VPN RRs. This can be achieved by setting a standard BGP community attribute value according to the VRF's membership of the route and the implementation of BGP community filters.

The case study in this chapter gives some valuable insight into how to partition VPN routes based on BGP standard communities. In the case study, we apply a simple route-map–based outbound filter at the RR clients that matches on standard BGP communities. These nonreflecting PE routers set a standard BGP community attribute value for routes that originate from different VPNs. According to the assigned standard community value, these routes are then partitioned and advertised to their respective VPN RRs with the help of BGP community lists. The concluding section of the chapter explained briefly how to optimize and fine-tune the RR for better performance.

Acronyms and Abbreviations

3DES	Triple DES
ABR	Area border router
ACK	Acknowledgment
ACL	Access control list
AD	Administrative distance
AH	Authentication header
ARP	Address resolution protocol
ASBR	Autonomous system border router
ASCII	American Standard Code for Information Interchange
AS	Autonomous system
ASN	Autonomous system number
ASN.l	Abstract syntax notation
ATM	Asynchronous transfer mode
ATT	Attached
BGP	Border gateway protocol
C	Customer
CA	Certificate authority
CBC	Cipher block chaining
CEF	Cisco express forwarding
CFB	Cipher feedback
CE	Customer edge
CIDR	Classless inter-domain routing
CLNS	Connectionless network service
CONFED	Confederation
CoS	Class of service

CPE	Customer premises equipment
CPU	Central processing unit
CR-LDP	Constraint-based label distribution protocol
CRL	Certificate revocation list
CUG	Closed user group
DES	Data encryption standard
D-H	Diffie-Hellman
DER	Distinguished encoding rules
DF	Don't fragment
DN	Distinguished name
DoS	Denial of service
DS	Digital signal
DSL	Digital subscriber line
EBGP	External BGP
ECB	Electronic code book
EGP	Exterior gateway protocol
EIGRP	Enhanced interior gateway routing protocol
ES	End system
ESP	Encapsulating security payload
FEC	Forwarding equivalence class
FIB	Forwarding information base
FQDN	Fully qualified domain name
GRE	Generic routing encapsulation
HDLC	High-level data link control
HMAC	Hashed message authentication code
HSRP	Hot standby router protocol
HTTP	Hypertext transfer protocol
IANA	Internet assigned numbers authority
IBGP	Internal BGP
ICMP	Internet control message protocol
ID	Identifier
IEEE	Institute of Electrical and Electronics Engineers
IETF	Internet Engineering Task Force
IG	Inside global
IGP	Interior gateway protocol
IKE	Internet key exchange
IL	Inside local
IOS	Internetwork Operating System
IP	Internet protocol

IPv4	Internet protocol version 4
IPv6	Internet protocol version 6
IPSec	IP security
IPX	Internetwork packet exchange
IS	Intermediate system
IS-IS	Intermediate system to intermediate system
ISAKMP	Internet security association and key management protocol
ISDN	Integrated services digital network
ISO	International Organization for Standardization
ISP	Internet service provider
IV	Initialization vector
KE	Key exchange
L1	Level 1
L2	Level 2
LAN	Local area network
LDAP	Lightweight directory access protocol
LDP	Label distribution protocol
LFIB	Label forwarding information base
LLC	Logical link control
LSA	Link-state advertisement
LSDB	Link-state database
LSP	Label switched path
LSR	Label switch router
MAC	Message authentication code
MD5	Message digest 5
MIM	Man-in-middle
MM	Main mode
MOD	Modulus
MP-BGP	Multiprotocol BGP
MP-IBGP	Multiprotocol internal BGP
MPLS	Multiprotocol label switching
MSS	Maximum segment size
MTU	Maximum transfer unit
MED	Multi-exit discriminator
NAT	Network address translation
NBMA	Nonbroadcast multiaccess
NET	Network entity title
NH	Next hop
NLRI	Network layer reachability information

NSAP	Network service access point
NSSA	Not-so-stubby area
NTP	Network time protocol
NVRAM	Nonvolatile random access memory
OFB	Output feedback
OG	Outside global
OL	Outside local
OSI	Open system interconnect
OSPF	Open shortest path first
P	Provider
PAT	Port address translation
PC	Portable computer
PE	Provider edge
PEM	Privacy enhanced mail
PFS	Perfect forward secrecy
PHP	Penultimate hop popping
PKCS	Public-key cryptography standard
PKI	Public key infrastructure
PMTU	Path maximum transfer unit
PMTUD	Path maximum transfer unit discovery
PoS	Packet over sonet
PPP	Point-to-point protocol
PRC	Partial route calculation
PREF	Preference
QM	Quick mode
RA	Registration authority
RD	Route distinguisher
RFC	Request for comment
RIP	Routing information protocol
RIPv2	Routing information protocol version 2
RR	Route reflector
RSA	Rivest-Shamir-Adleman
RSVP	Resource reservation protocol
RT	Route-target
SA	Security association
SADB	SA database
SCEP	Simple certificate enrollment protocol
SDH	Synchronous digital hierarchy

SHA	Secure hash algorithm
SLA	Service-level agreement
SONET	Synchronous optical network
SOO	Site-of-origin
SP	Service provider
SPD	Selective packet discard
SPI	Security parameter index
SPF	Shortest path first
TCP	Transmission control protocol
TDM	Time division multiplexing
TE	Traffic engineering
TED	Tunnel endpoint discovery
TLV	Type, length, and value
ToS	Type of service
TTL	Time-to-live
UDP	User datagram protocol
VC	Virtual circuit
VPN	Virtual private network
VPNv4	VPN-IPv4
VPDN	Virtual private dialup network
VRF	Virtual routing and forwarding
WAN	Wide area network
XOR	Exclusively OR

Bibliography

Adams, C., and Lloyd, S., *Understanding Public-Key Infrastructure: Concepts, Standards, and Deployment Considerations*, Macmillian Technical Publishing, Indianapolis, IN, 1999.

Andersson, L. et. al., "LDP Specification," RFC 3036, January 2001.

Ash, J. et. al., "Applicability Statement for CR-LDF," RFC 3213, January 2002.

Ash, J. et. al., "LSP Modification Using CR-LDF," RFC 3214, January 2002.

Awduche, D. et. al., "Requirements for Traffic Engineering Over MPLS," RFC 2702, September 1999.

Awduche, D. et. al., "RSVP-TE: Extensions to RSVP for LSP Tunnels," RFC 3209, December 2001.

Awduche, D. et. al., "Applicability Statement for Extensions to RSVP for LSP-Tunnels," RFC 3210, December 2001.

Bates, T., Chandra, R., and Chen, E., "BGP Route Reflection—An Alternative to Full Mesh IBGF," RFC 2796, April 2000.

Bates, T., Rekhter, Y., Chandra, R., and Katz, D., "Multiprotocol Extensions for BGP-4," RFC 2858, June 2000.

Chandra, R., Traina, P., and Li, T., "BGP Communities Attribute," RFC 1997, August 1996.

Chandra, R. and Scudder, J., "Capabilities Advertisement with BGP-4," RFC 2842, May 2000.

Chen, E. and Bates, T., "An Application of the BGP Community Attribute in Multi-home Routing," RFC 1998, August 1996.

Chen, E., "Route Refresh Capability for BGP-4," RFC 2918, September 2000.

Christian, P., "Generic Routing Encapsulation over CLNS Networks," RFC 3147, July 2001.

Cisco Systems, Inc., "Cisco IOS IP Configuration Guide, Release 12.2," 2001.

Cisco Systems, Inc., "Cisco IOS Security Configuration Guide, Release 12.2," 2001.

Cisco Systems, Inc., "Cisco IOS Switching Services Configuration Guide, Release 12.2," 2001.

Cisco Systems, Inc., "Cisco IOS Enterprise VPN Configuration Guide," Part Number 78-6342-04, 2001.

Cisco Systems, Inc., "Cisco Managed MPLS VPN Solution Overview Guide," Part Number OL-1225-01, 2001.

Cisco Systems, Inc., "A Comparison Between IPSec and MPLS VPNs," White Paper, 2000.

Cisco Systems, Inc., "Building MPLS-Based VPNs and Services for Service Provider Core Networks," White Paper, 2001.

Cisco Systems, Inc., "Designing MPLS Extensions for Customer Edge Routers," Product Bulletin, No. 1575, 2001.

Cisco Systems, Inc., "OSPF Sham-Link Support for MPLS VPN," Cisco IOS Release 12.2(8)T, 2002.

Cisco Systems, Inc., "Configuring BGP on Cisco Routers, Revision 1.0: Student Guide," 2000.

Cisco Systems, Inc., "Cisco Secure Virtual Private Network, Revision 2.0: Student Guide," 2001.

Cisco Systems, Inc., "Implementing Cisco MPLS, Revision 1.0: Student Guide," 2001.

Comer, Douglas E., *Internetworking with TCP/IP, Volume 1: Principles, Protocols, and Architecture*, Prentice Hall, Englewood Cliffs, NJ, 1991.

Coltun, R., and Fuller, V., "The OSPF NSSA Optior," RFC 1587, March 1994.

Coltun, R., "The OSPF Opaque LSA Optior," RFC 2370, July 1998.

Conta, A., and Deering, S., "Generic Packet Tunneling in IPv6 Specification," RFC 2473, December 1998.

deSouza, O., and Rodrigues, M., "Guidelines for Running OSPF Over Frame Relay Networks," RFC 1586, March 1994.

Dommety, G., "Key and Sequence Number Extensions to GRE," RFC 1326, September 2000.

Doraswamy, N., and Harkins, D., *IPSec The New Security Standard for the Internets, Intranets, and Virtual Private Networks*, Prentice Hall, Upper Saddle River, NJ, 1999.

Egevang, K., and Francis, P., "The IP Network Address Translator (NAT)," RFC 1631, May 1994.

Farinacci, D., Li, T., Hanks, S., Meyer, D. and Traina, P. "Generic Routing Encapsulation (GRE)," RFC 2890, March 2000.

Glenn, R., and Kent, S., "The NULL Encryption Algorithm and Its Use With IPSec," RFC 2410, November 1998.

Hanks, S., Li, T., Farinacci, D. and Traina, P. "Generic Routing Encapsulation (GRE)," RFC 1701, October 1994.

Hanks, S., Li, T., Farinacci, D., and Traina, P., "Generic Routing Encapsulation over IPv4 Networks," RFC 1702, October 1994.

Harkins, D., and Carrel, D., "The Internet Key Exchange (IKE)," RFC 2409, November 1998.

Housley, R. et. al., "Internet X.509 Public Key Infrastructure Certificate and CRL Profile," RFC 2459, January 1999.

Jamoussi, B. et. al., "Constraint-Based LSP Setup using LDP," RFC 3212, January 2002.

Kaliski, B., "PKCS #7: Cryptographic Message Syntax Version 1.5," RFC 2315, March 1998.

Kent, S., and Atkinson, R., "Security Architecture for the Internet Protocol," RFC 2401, November 1998.

Kent, S., and Atkinson, R., "IP Authentication Header," RFC 2402, November 1998.

Kent, S., and Atkinson, R., "IP Encapsulating Security Payload (ESP)," RFC 2406, November 1998.

Krawczyk, H., Bellare, M., and Canetti R., "HMAC: Keyed-Hashing for Message Authentication," RFC 2104, February 1997.

Li, T., Cole, B., Morton, P., and Li, D., "Cisco Hot Standby Router Protocol (HSRP)," RFC 2281, March 1998.

Li, T., and Smit, H., "IS-IS extensions for Traffic Engineering," Internet Draft draft-ietf-isis-traffic-04.txt, August 2001.

Liu, X., Madson, C., McGrew, D., and Nourse, A., "Cisco Systems' Simple Certificate Enrollment Protocol (SCEP)," Internet Draft draft-nourse-scep-06.txt, May 2002.

Madson, C., and Glenn, R., "The Use of HMAC-MD5-96 within ESP and AH," RFC 2403, November 1998.

Madson, C., and Glenn, R., "The Use of HMAC-SHA-1-96 within ESP and AH," RFC 2404, November 1998.

Madson, C., and Doraswamy, N., "The ESP DES-CBC Cipher Algorithm With Explicit IV," RFC 2405, November 1998.

Maughan, D., Schertler, M., Schneider, M., and Turner, J., "Internet Security Association and Key Management Protocol (ISAKMP)," RFC 2408, November 1998.

Mogul, J., and Deering, S., "Path MTU Discovery," RFC 1191, November 1990.

Moy, J., "Extending OSPF to Support Demand Circuits," RFC 1793, April 1995.

Moy, J., "OSPF Version 2," RFC 2328, April 1998.

Muthukrishnan, K., and Malis, A., "A Core MPLS IP VPN Architecture," RFC 2917, September 2000.

Nystrom, M., and Kaliski, B., "PKCS #10: Certification Request Syntax Specification Version 1.7," RFC 2986, November 2000.

Orman, H., "The OAKLEY Key Determination Protocol," RFC 2412, November 1998.

Piper, D., "The Internet IP Security Domain of Interpretation for ISAKMP," RFC 2407, November 1998.

Rekhter, Y., and Li, T., "A Border Gateway Protocol 4 (BGP-4)," RFC 1771, March 1995.

Rekhter, Y. et. al., "Address Allocation for Private Internets," RFC 1918, February 1996.

Rekhter, Y., and Rosen, E., "Carrying Label Information in BGP-4," RFC 3107, May 2001.

Rosen, E., and Rekhter, Y., "BGP/MPLS VPNs," RFC 2547, March 1999.

Rosen, E., Viswanathan, A., and Callon, R., "Multiprotocol Label Switching Architecture," RFC 3031, January 2001.

Rosen, E. et. al., "MPLS Label Stack Encoding," RFC 3032, January 2001.

Rosen, E. et. al., "OSPF as the PE/CE Protocol in BGP/MPLS VPNs," Internet Draft draft-rosen-vpns-ospf-bgp-mpls-05.txt, July 2002.

Sangli, R., Tappan, D., and Rekhter, Y., "BGP Extended Communities Attribute," Internet Draft draft-ietf-idr-bgp-ext-communities-05.txt, May 2002.

Schneier, Bruce, *Applied Cryptography: Protocols, Algorithms, and Source Code in C*, Second editor, John Wiley & Sons, New York, 1996.

Smith, Richard E., *Internet Cryptography*, Addison Wesley, Reading, MA, 1997.

Srisuresh, P., and Holdrege, M., "IP Network Address Translator (NAT) Terminology and Considerations," RFC 2663, August 1999.

Stallings, W., "Multiprotocol Label Switching (MPLS)," Internet Protocol Journal 4 (3): 2–14, September 2001.

Thayer, R., Doraswamy, N., and Glenn, R., "IP Security Document Roadmap," RFC 2411, November 1998.

Traina, P., McPherson, D., and Scudder, J., "Autonomous System Confederations for BGP," RFC 3065, February 2001.

Tsuchiya, P., "Mutual Encapsulation Considered Dangerous," RFC 1326, May 1992.

Varadhan, K., "BGP OSPF Interaction," RFC 1403, January 1993.

Woodburn, R., and Mills, D., "A Scheme for an Internet Encapsulation Protocol: Version 1," RFC 1241, July 1991.

Index